PROBABILITY ALGEBRAS AND STOCHASTIC SPACES

Probability and Mathematical Statistics

A Series of Monographs and Textbooks

---

1. Thomas Ferguson. Mathematical Statistics: A Decision Theoretic Approach. 1967
2. Howard Tucker. A Graduate Course in Probability. 1967
3. K. R. Parthasarathy. Probability Measures on Metric Spaces. 1967
4. P. Révész. The Laws of Large Numbers. 1968
5. H. P. McKean, Jr. Stochastic Integrals. 1969
6. B. V. Gnedenko, Yu. K. Belyayev, and A. D. Solovyev. Mathematical Methods of Reliability Theory. 1969
7. Demetrios A. Kappos. Probability Algebras and Stochastic Spaces. 1970

PROBABILITY ALGEBRAS AND STOCHASTIC SPACES

Demetrios A. Kappos

UNIVERSITY OF ATHENS
ATHENS, GREECE

1969

ACADEMIC PRESS New York and London

ACADEMIC PRESS INC.
111 Fifth Avenue, New York, New York 10003

United Kingdom Edition published by
ACADEMIC PRESS INC. (LONDON) LTD.
Berkeley Square House, Berkeley Square, London, W1X 6BA.

Standard Book Number: 12-397650-2

Library of Congress Catalog Card Number: 70-84234

AMS 1968 Subject Classifications 2865, 4606, 4610, 6005

PRINTED IN THE UNITED STATES OF AMERICA

PREFACE

During the last twenty years, many excellent textbooks have enriched the mathematical literature on probability theory. With the help of classical set-theoretic measure and integral theory, these books introduce the fundamental concepts of probability theory, then formulate and study old and recent problems germane to the theory. There is however an alternate way to introduce the main notions of probability theory, a way which is more naturally adapted to the empirical origins of the subject. The present volume is an exploration of this alternate development.

There are three fundamental notions of probability theory: Event, probability of an event, random variable. A given set of events forms a Boolean algebra with respect to the logical connectives (considered as operations) " or ", " and ", and " not ". Probability is a normed measure on a Boolean algebra of events. It is natural to consider probability as finitely additive and strictly positive, i.e., equal to zero only for the impossible event. Therefore, a Boolean algebra $\mathfrak{A}$ endowed with a finitely additive and strictly positive probability p can be considered as a probability algebra $(\mathfrak{A}, p)$. In all empirical cases a Boolean algebra of events can be endowed with an additive and strictly positive probability. Moreover, the σ-(countable) additivity of probability, which has important mathematical consequences in the theory, can always be obtained by a metric extension of a probability algebra $(\mathfrak{A}, p)$ to a probability σ-algebra $(\tilde{\mathfrak{A}}, \tilde{p})$, in which $\tilde{\mathfrak{A}}$ is a Boolean σ-algebra and $\tilde{p}$ a countably additive and strictly positive probability.

It is well-known that a Boolean algebra $\mathfrak{A}$ is always isomorphic to a field (Boolean algebra) of subsets of a set (= space) Ω. Thus the investigation of events and their probabilities can be reduced to a study of normed measures on fields of sets. Moreover, it is always possible to represent the Boolean algebra $\mathfrak{A}$ of events by a field $\mathbf{A}$ of subsets of a set, so that the normed measure (the probability) P on $\mathbf{A}$ is set-theoretically countably additive. Let $(\Omega, \mathbf{A}, P)$ be a so-called probability space, which represents, set-theoretically, a probability algebra $(\mathfrak{A}, p)$; then $(\Omega, \mathbf{A}, P)$ can always be extended to a complete probability σ-space $(\Omega, \tilde{\mathbf{A}}, \tilde{P})$, in which $\tilde{\mathbf{A}}$ is a σ-field containing $\mathbf{A}$ and P a complete, normed, countably additive measure on $\tilde{\mathbf{A}}$. The elements of $\tilde{\mathbf{A}}$ can be considered as events; in the case, however, in which the cardinality of Ω is $> \aleph_0$ it may happen that there exist non-empty sets $E \in \tilde{\mathbf{A}}$ (i.e., events different from the impossible event) of measure (= probability) zero, which have no probabilistic interpretation. We can overcome this difficulty by considering directly the probability σ-algebra $(\tilde{\mathfrak{A}}, \tilde{p})$ instead of the probability σ-space $(\Omega, \tilde{\mathbf{A}}, \tilde{P})$.

Our aim is to develop the fundamental notions of probability theory in this " point-free " way. This, however, requires knowledge of lattice theory. We find that lattice theory also provides simplicity and generality, since it deals with classes of random variables which are the elements of the so-called stochastic spaces. The space of all elementary random variables defined over a probability algebra in a " point-free " way is a base for the stochastic space of all random variables, which can be obtained from it by lattice-theoretic extension processes. There are, however, problems in which one wants to consider individual samples and cannot work without points; then one can always assign a suitable probability σ-space to the probability algebra under consideration. Conversely, one can assign to every probability σ-space $(\Omega, \mathbf{A}, P)$ a probability σ-algebra $(\mathfrak{A}, p)$ in which $\mathfrak{A}$ is the quotient Boolean σ-algebra $\mathbf{A}/\mathbf{N}$, where $\mathbf{N}$ is the σ-ideal of all sets of probability zero. Thus the two theories are equivalent.

In the lattice-theoretic treatment of probability theory, a structural classification of all possible probability algebras (therefore an analogous classification of all possible stochastic spaces) can easily be obtained, which provides us with a representation of every probability σ-algebra by a probability σ-space $(\Omega, \mathbf{A}, P)$ in which Ω is the cartesian product of factors equal to the interval $[0, 1]$ of the real line, $\mathbf{A}$ a σ-subfield of the σ-field of all Lebesgue product measurable subsets of Ω, and P the product measure of $\mathbf{A}$. A corresponding representation of random variables by Lebesgue measurable functions defined on Ω can also be obtained.

The origin of this book is in a set of lectures which I gave in the academic year 1963–64 at the Catholic University of America, Washington, D.C. The Statistical Laboratory there issued a mimeographed version of my notes under the title " Lattices and their Applications to Probability ". The present text is a revised and expanded version of these notes, maintaining the central mathematical ideas of the lectures, namely probability algebras and stochastic spaces (i.e., spaces of random variables). The part of the notes devoted to pure lattice theory has been shortened and the most important concepts and theorems of this theory have been stated in two appendices, mostly without proofs.

In addition to the material in the mimeographed edition, the present volume contains a general way to introduce the concept of random variables taking values in spaces endowed with any algebraic or topological structure. In particular, we study the cases in which the space of the values is a lattice group, or vector lattice. When the space of the values is a Banach space, a theory of expectation and moments is stated. A theory of expectation can be easily stated in more general cases of spaces of values: for example, locally convex vector spaces or topological vector spaces, for which an integration theory is known.

We have restricted ourselves on the introduction and study only of the fundamental mathematical notions. We mention only a few facts about the concepts of independence and conditional probabilities and expectations. Certainly, it would be interesting to state the theory of stochastic processes and, especially, the theory of martingales. But this would go beyond the scope of the present monograph, or it would have to be published in a second volume.

The author wishes to express his gratitude to Dr. Eugene Lukacs of the Catholic University, who made it possible for him to give lectures and publish them. It is a pleasure to offer thanks to F. Papangelou and G. Anderson who read critically the manuscript of the lectures and made valuable suggestions during the mimeographed edition of them.

He expresses his best thanks to Miss Susan Papadopoulou who read all the manuscript of the present edition carefully and made valuable suggestions. In many cases she has helped him to give simpler formulations and proofs.

Thanks are finally due to Mr. Constantine Halatsis for his efficient typing of the manuscript and the composition of the several indices.

Athens, Greece. D. A. Kappos.

April 1969.

CONTENTS

PROBABILITY ALGEBRAS AND STOCHASTIC SPACES

I

PROBABILITY ALGEBRAS

1. DEFINITIONS AND PROPERTIES

1.1. A *probability algebra* (*pr algebra*) $(\mathfrak{A}, p)$ consists of a nonempty set $\mathfrak{A}$ of elements denoted by lower-case latin letters: $a, b, c, \ldots, x, y, \ldots$, called *events* and a real-valued function p on $\mathfrak{A}$, called a *probability* (pr). In the set $\mathfrak{A}$ two binary operations $a \vee b$ (*a or b*) and $a \wedge b$ (*a and b*) and one unitary operation a^c (*not a*) are defined, which introduce in $\mathfrak{A}$ the algebraic structure of a Boolean algebra.†

The probability p satisfies the following conditions:

1.1.1. p is *strictly positive*, i.e., $p(x) \geqslant 0$, for every $x \in \mathfrak{A}$ and $p(x) = 0$ if and only if $x = \emptyset$, where $\emptyset$ is the zero of $\mathfrak{A}$.

1.1.2. p is *normed*, i.e., $p(e) = 1$, where e is the unit of $\mathfrak{A}$.

1.1.3. p is *additive* i.e., $p(a \vee b) = p(a) + p(b)$ if a and b are disjoint.‡

We shall call the unit e the *sure event* and the zero $\emptyset$ the *impossible event* of the event algebra $\mathfrak{A}$. Every $x \in \mathfrak{A}$ with $x \neq \emptyset$ and $x \neq e$ is called a *possible event* of the algebra $\mathfrak{A}$.

---

† We consider the theory of Boolean algebras as known; *cf.* also Appendix 1 of this book.

‡ We say that the event a and the event b are disjoint (exclude each other, or are mutually exclusive, or are incompatible) if $a \wedge b = \emptyset$.

1.2. Properties of the pr p.

1.2.1. If $x_1, x_2, \ldots, x_n$ are pairwise disjoint events of $\mathfrak{A}$, then

$$p(x_1 \vee x_2 \vee \ldots \vee x_n) = p(x_1)+p(x_2)+\ldots+p(x_n).$$

Proof by induction.

1.2.2. If $x \leqslant y$, then $p(x) \leqslant p(y)$ and, moreover, $p(x) < p(y)$ if $x \neq y$ (i.e., if $x < y$, then $p(x) < p(y)$).

Proof. If $x \leqslant y$, then $z = y - x$ is such that $y = x \vee z$ and x and z are disjoint. According to 1.1.3, we have $p(y) = p(x)+p(z)$, i.e., $p(x) \leqslant p(y)$. If $x \neq y$, then $z \neq \emptyset$ and $p(z) > 0$, i.e., $p(x) < p(y)$.

1.2.3. If $x \leqslant y$, then $p(y-x) = p(y)-p(x)$. In fact, according to the proof of 1.2.2: $p(y) = p(x)+p(z)$, i.e., $p(z) = p(y-x) = p(y)-p(x)$.

1.2.4. For every pair $x \in \mathfrak{A}$ and $y \in \mathfrak{A}$, we have

$$p(x)+p(y) = p(x \vee y)+p(x \wedge y).$$

In fact, we have

$$x \vee y = (x - x \wedge y) \vee (y - x \wedge y) \vee (x \wedge y).$$

The terms of the join on the right are pairwise disjoint, thus

$$\begin{aligned} p(x \vee y) &= p(x - x \wedge y)+p(y - x \wedge y)+p(x \wedge y) \\ &= p(x)-p(x \wedge y)+p(y)-p(x \wedge y)+p(x \wedge y) \\ &= p(x)+p(y)-p(x \wedge y); \end{aligned}$$

i.e.,

$$p(x)+p(y) = p(x \vee y)+p(x \wedge y).$$

1.2.5. For every $x \in \mathfrak{A}$ we have $p(x^c) = 1-p(x)$. Proof obvious.

1.2.6. We have $p(x) = 1$ if and only if $x = e$. Proof obvious.

By induction, one can prove the following generalization of property 1.2.4:

1.2.7. For three, or more generally for $n \geqslant 2$ events $x_\nu \in \mathfrak{A}$, $\nu = 1, 2, \ldots, n$ we have:

$$\sum_{\nu=1}^{n} p\,(x) = \sum_{\nu=1}^{n} p\left(\bigvee_{\rho \in S^n,\, \nu} \bigwedge_{i \leqslant \nu} x_{\rho_i} \right), \qquad n \geqslant 2$$

where $S^{n,\nu}$ is the set of all finite sequences $\rho = \{\rho_1, \rho_2, \dots, \rho_\nu\}$ with $1 \leqslant \rho_1 < \rho_2 < \dots < \rho_\nu \leqslant n$.

2. PROBABILITY SUBALGEBRAS

2.1. Let $(\mathfrak{A}, p)$ be a pr algebra and $\mathfrak{B}$ a Boolean subalgebra of $\mathfrak{A}$; then the restriction of the function p to $\mathfrak{B}$ is a probability on $\mathfrak{B}$. The pr algebra $(\mathfrak{B}, p)$ is then called a *probability subalgebra* of $(\mathfrak{A}, p)$. Hence to every Boolean subalgebra $\mathfrak{B}$ of $\mathfrak{A}$ there corresponds a pr subalgebra $(\mathfrak{B}, p)$ of the pr algebra $(\mathfrak{A}, p)$.

2.2. Let $\mathfrak{S}$ be a non-empty subset of $\mathfrak{A}$; then there exists a smallest Boolean subalgebra, $b(\mathfrak{S})$, of $\mathfrak{A}$ containing $\mathfrak{S}$; the pr subalgebra $(b(\mathfrak{S}), p)$ is then called the *pr subalgebra generated* by $\mathfrak{S}$ in $(\mathfrak{A}, p)$.

2.3. There is a pr subalgebra of $(\mathfrak{A}, p)$, namely, the pr algebra $(\mathfrak{U}, p)$ where $\mathfrak{U} = \{\emptyset, e\}$ and $p(\emptyset) = 0$, $p(e) = 1$. We call $(\mathfrak{U}, p)$ the *improper* pr algebra.

2.4. *Remark.* Throughout this book every pr algebra and every pr subalgebra will be supposed as *not improper.*

2.5. Every possible event $x \in \mathfrak{A}$, i.e., $x \neq \emptyset$, $x \neq e$, generates a pr subalgebra $(\mathfrak{X}, p)$ in $(\mathfrak{A}, p)$, where $\mathfrak{X} = \{\emptyset, x, x^c, e\}$.

3. ISOMETRIC PROBABILITY ALGEBRAS

3.1. Let $(\mathfrak{A}_1, p_1)$ and $(\mathfrak{A}_2, p_2)$ be two pr algebras. We say that the pr algebra $(\mathfrak{A}_1, p_1)$ is *isometric* to the pr algebra $(\mathfrak{A}_2, p_2)$ if and only if there exists an isomorphic map ϕ of the Boolean algebra $\mathfrak{A}_1$ onto the Boolean algebra $\mathfrak{A}_2$ such that $p_1(x) = p_2(\phi(x))$ for every $x \in \mathfrak{A}_1$.

4. EXAMPLES

4.1. Let $E = \{a_1, a_2, \dots, a_n\}$ be a finite set of points $a_1, a_2, \dots, a_n$, $n \geqslant 2$. We take the class $\mathfrak{P}(E)$ of all the subsets of E to be the Boolean algebra $\mathfrak{A}_n = \mathfrak{P}(E)$. Let $p_1, p_2, \dots, p_n$ be positive real numbers with $0 < p_\nu < 1$, $\nu = 1, 2, \dots, n$, and $p_1 + p_2 + \dots + p_n = 1$. For every subset

$$\{a_{i_1}, a_{i_2}, \dots, a_{i_k}\}$$

of E, we define the probability as follows:

$$p(\{a_{i_1}, a_{i_2}, \dots, a_{i_k}\}) = p_{i_1} + p_{i_2} + \dots + p_{i_k},$$

and we define, for the empty set, $p(\emptyset) = 0$. Then $(\mathfrak{A}_n, p)$ is a pr algebra with a finite number (2^n) of events.

4.2. Let $E = \{a_1, a_2, \dots,\}$ be a denumerable set of points a_ν, $\nu = 1, 2, \dots$. Let, moreover, p_ν, $\nu = 1, 2, \dots$, be positive real numbers with $0 < p_\nu < 1$, $\nu = 1, 2, \dots$, and $\sum_{\nu=1}^{\infty} p_\nu = 1$.

We take the class $\mathfrak{P}(E)$ of all subsets of E to be the Boolean algebra $\mathfrak{A}_{\aleph_0}$ and assign to every finite subset $\{a_{i_1}, a_{i_2}, \dots, a_{i_k}\} \subseteq E$ or infinite subset $\{a_{j_1}, a_{j_2}, \dots,\} \subseteq E$ the probability defined by

$$p(\{a_{i_1}, a_{i_2}, \dots, a_{i_k}\}) = p_{i_1} + p_{i_2} + \dots + p_{i_k}$$

or

$$p(\{a_{j_1}, a_{j_2}, \dots\}) = \sum_{\nu=1}^{\infty} p_{j_\nu}$$

respectively. We define, for the empty set, $p(\emptyset) = 0$.

Then $(A_{\aleph_0}, p)$ is a pr algebra with an uncountable number of events because, as it is well known, the cardinality of $A_{\aleph_0} = \mathfrak{P}(E)$, with E a denumerable set, is equal to the cardinality of the continuum, i.e., $> \aleph_0$.

4.3. Let $X = \{x_i\}$, $i \in I$, be a class of symbols x_i, $i \in I$, where I is a set of indices with a cardinality $|I| \equiv \aleph \geqslant \aleph_0$; then there exists a Boolean algebra $\mathfrak{B}_\aleph$, the so-called *free Boolean algebra* $\mathfrak{B}_\aleph$ with $\aleph$ generators, which is characterized by the following properties:

(α) The symbols $\{x_i\}$, $i \in I$, are elements of $\mathfrak{B}_\aleph$ and the class $X = \{x_i\}$, $i \in I$, generates the Boolean algebra $\mathfrak{B}_\aleph$.

(β) Every map ϕ of the class X into an arbitrary Boolean algebra $\mathfrak{B}$ can always be extended to a homomorphism of $\mathfrak{B}_\aleph$ into $\mathfrak{B}$.

We shall not prove this statement, but we shall mention some facts about the algebraic structure of $\mathfrak{B}_\aleph$ that are important for the definition of a probability on $\mathfrak{B}_\aleph$.†

Let $\{i_1, i_2, \dots, i_n\}$, $n \geqslant 1$, be a finite subset of I; then we call the expression $y = y_{i_1} \wedge y_{i_2} \wedge \dots \wedge y_{i_n}$, where $y_{i_\nu} = x_{i_\nu}$ or $x_{i_\nu}^c$, $\nu = 1, 2, \dots, n$, a

† Details about free Boolean algebras one can find in Sikorski [5], §14.

monomial and $\{i_1, i_2, \ldots, i_n\}$ the *length* of y. A monomial is always $\neq \emptyset$ and $\neq e$. Two monomials are equal if and only if they have the same length with corresponding factors equal. Two monomials are disjoint if and only if their lengths have a non-empty intersection and at least for a common index i the corresponding factors are complementary elements. Every element $b \in \mathfrak{B}_{\aleph}$, $b \neq \emptyset$, $b \neq e$, can be represented as a finite join of pairwise disjoint monomials of the same length; there always exists exactly a smallest set $\{i_1, i_2, \ldots, i_n\} \subseteq I$ such that b can be represented as a finite join of pairwise disjoint monomials of the length $\{i_1, i_2, \ldots, i_n\}$. The element $\emptyset$ does not have a representation by means of monomials. The element e can be represented by $e = x_i \vee x_i^c$ for every $i \in I$.

Let now p_i, q_i be real numbers with $0 < p_i < 1$, $0 < q_i < 1$ and $p_i + q_i = 1$ for every $i \in I$. We define a probability p on $\mathfrak{B}_{\aleph}$ as follows:

(1) $p(x_i) = p_i$, $p(x_i^c) = q_i$, for every $i \in I$.

(2) $p(y_{i_1} \wedge y_{i_2} \wedge \ldots \wedge y_{i_n}) = p(y_{i_1})\, p(y_{i_2}) \ldots p(y_{i_n})$, for every monomial $y = y_{i_1} \wedge y_{i_2} \wedge \ldots \wedge y_{i_n}$.

Let now $b \in \mathfrak{B}_{\aleph}$ with $b \neq \emptyset$; then there exists a uniquely determined representation $b = y_1 \vee y_2 \vee \ldots \vee y_k$ where $y_1, y_2, \ldots, y_k$ are pairwise disjoint monomials of the same smallest length. Hence, we can define:

$$p(b) = p(y_1) + p(y_2) + \ldots + p(y_k).$$

It is now easy to prove that the so defined function p is a probability on $\mathfrak{B}_{\aleph}$, i.e., $(\mathfrak{B}_{\aleph}, p)$ is a pr algebra.

4.4. We say that a Boolean algebra $\mathfrak{B}$ is generated by a chain if and only if there exists a subset $\mathfrak{S}$ of $\mathfrak{B}$ which is a chain relative to the order relation $\leqslant$ and generates the Boolean algebra $\mathfrak{B}$. There is no loss of generality in assuming that $\emptyset \in \mathfrak{S}$ and $e \in \mathfrak{S}$, because this may always be achieved by adjoining the elements $\emptyset$ and e to $\mathfrak{S}$. Every element x of a Boolean algebra $\mathfrak{B}$ generated by a chain $\mathfrak{S}$ can be represented uniquely by an expression of the form:

$$x = \bigvee_{i=0}^{n-1} (s_{2i+1} \wedge s_{2i}^c), \tag{1}$$

where $s_0, s_1, \ldots, s_{2n-1}$ are elements of $\mathfrak{S}$ with $s_j < s_{j+1}$, $s_j^c = e + s_j$, $j = 0, 1, 2, \ldots, 2n-2$.

Let us now suppose that the chain $\mathfrak{S}$ can be mapped order-isomorphically into the chain $[0, 1] \equiv \{\xi \in R\colon 0 \leqslant \xi \leqslant 1\}$ so that 0 is the image of $\emptyset$ and 1 the image of e. We denote by ϕ this order isomorphic mapping of $\mathfrak{S}$ into $[0, 1]$. Then we can define a real valued function p on $\mathfrak{B}$ using for every $x \in \mathfrak{B}$ its expression in the form (1), as follows:

$$p(x) = \sum_{i=0}^{n-1} (\phi(s_{2i+1}) - \phi(s_{2i})).$$

Obviously we have $p(e) = 1$ and $p(\emptyset) = 0$, and the so defined function p is a probability on $\mathfrak{B}$. The following theorem holds:

Theorem 4.1.

A Boolean algebra $\mathfrak{B}$ generated by a chain $\mathfrak{S}$ can be endowed with a probability p if and only if $\mathfrak{S}$ is order-isomorphic to a subchain of the chain $[0, 1]$ *of all real numbers ξ with* $0 \leqslant \xi \leqslant 1$.

4.5. Let $L_\beta = [0, \beta) \equiv \{\xi \in R\colon 0 \leqslant \xi < \beta\}$ for every $\beta \in R$ with $0 < \beta \leqslant 1$ and $L_0 = \emptyset$; then the class $\mathfrak{D}$ of all half-open intervals L_β, $0 \leqslant \beta \leqslant 1$ is a class of elements of the Boolean algebra $\mathfrak{P}(E)$ of all subsets of the set $L_1 = E \equiv \{\xi \in R\colon 0 \leqslant \xi < 1\}$. $\mathfrak{D}$ is a chain relative to the inclusion relation $\subseteq$ and generates a Boolean subalgebra $b(\mathfrak{D}) = \mathfrak{A}$ of $\mathfrak{P}(E)$. According to section 4.4, every element $a \in \mathfrak{A}$ can be represented by an expression of the form:

$$a = \bigcup_{i=0}^{n-1} (L_{\beta_{2i+1}} \cap L^c_{\beta_{2i}}). \tag{I}$$

We call the Boolean algebra $\mathfrak{A}$, the *interval algebra* $\mathfrak{A}$. A probability m can be defined on $\mathfrak{A}$ according to section 4.4 as follows:

$$m(a) = \sum_{i=0}^{n-1} (\beta_{2i+1} - \beta_{2i}).$$

We call the so defined probability algebra the *probability interval algebra* $(\mathfrak{A}, m)$.

The following theorem can be easily proved:

Theorem 4.2.

Every pr algebra $(\mathfrak{B}, p)$, in which the Boolean algebra $\mathfrak{B}$ is generated by a chain $\mathfrak{S}$, is isometric to a probability subalgebra of the probability interval algebra $(\mathfrak{A}, m)$.

5. SEPARABILITY RELATIVE TO A PROBABILITY

5.1. Let $(\mathfrak{B}, p)$ be a pr algebra. We say that a class $\mathfrak{K}$ of elements of $\mathfrak{B}$ is *p-dense* in $\mathfrak{B}$ if and only if:

(s) For every $x \in \mathfrak{B}$ and for every positive real number $\varepsilon > 0$ there exists an element $a = a(x, \varepsilon) \in \mathfrak{K}$ such that $p(x+a) < \varepsilon$.

A pr algebra $(\mathfrak{B}, p)$ is called p-separable if and only if there exists a countable class $\mathfrak{K}$ of elements of $\mathfrak{B}$ which is p-dense in $\mathfrak{B}$. *Every pr subalgebra of a p-separable pr algebra is also p-separable.* The following theorem holds:

Theorem 5.1.

The pr interval algebra $(\mathfrak{A}, m)$ is m-separable.

In fact, let $\mathfrak{K}$ be the Boolean subalgebra of $\mathfrak{A}$ generated by the class of all the intervals L_β for every rational number β with $0 \leqslant \beta \leqslant 1$. $\mathfrak{K}$ is a countable class and it is m-dense in $\mathfrak{A}$.

Theorems 4.2 and 5.1 imply:

Theorem 5.2.

Every pr algebra $(\mathfrak{B}, p)$, in which the Boolean algebra $\mathfrak{B}$ is generated by a chain $\mathfrak{S}$, is p-separable.

6. COUNTABLY ADDITIVE [σ-ADDITIVE] PROBABILITIES

6.1. Let $(\mathfrak{B}, p)$ be a pr algebra. The probability p is said to be *countably additive* or *σ-additive* on $\mathfrak{B}$, if and only if the following property holds:

(I) For every countable sequence a_ν, $\nu = 1, 2, \ldots$, of pairwise disjoint events in $\mathfrak{B}$ such that $(\mathfrak{B}) \bigvee_{\nu=1}^{\infty} a_\nu = a$ exists, we have $p(a) = \sum_{\nu=1}^{\infty} p(a_\nu)$.

The following theorem holds:

Theorem 6.1.

Let $(\mathfrak{B}, p)$ be a pr algebra; then the property (I) *of the σ-additivity of p is equivalent to the so-called continuity of p on $\mathfrak{B}$, i.e., to the property;*

(II) *For every monotonically decreasing sequence* $b_1 \geqslant b_2 \geqslant \ldots$ *of events* $b_\nu \in \mathfrak{B}$, $\nu = 1, 2, \ldots$, *with* $(\mathfrak{B}) \bigwedge_{\nu=1}^{\infty} b_\nu = \emptyset$, *we have:* $\lim_{\nu \to \infty} p(b_\nu) = 0$.

Proof: (A) Let (I) be true and let $b_\nu \downarrow$ with $(\mathfrak{B}) \bigwedge_{\nu=1}^{\infty} b_\nu = \emptyset$;

Then we have:

$$b_1 = (b_1 + b_2) \vee (b_2 + b_3) \vee \ldots. \tag{1}$$

In fact, $b_1 \geqslant b_\nu + b_{\nu+1}$, for every $\nu = 1, 2, \ldots$; hence it suffices to prove that

$$x \geqslant b_\nu + b_{\nu+1} \text{ for every } \nu = 1, 2, \ldots, \text{ implies } x \geqslant b_1,$$

or equivalently

$$x \wedge (b_\nu + b_{\nu+1}) = b_\nu + b_{\nu+1}, \quad \text{for every } \nu = 1, 2, \ldots, \quad \text{implies } x \wedge b_1 = b_1.$$

We have

$$x \wedge (b_1 + b_2) = x \wedge b_1 + x \wedge b_2 = b_1 + b_2$$

$$x \wedge (b_2 + b_3) = x \wedge b_2 + x \wedge b_3 = b_2 + b_3;$$

hence

$$x \wedge b_1 + x \wedge b_3 = b_1 + b_3$$

and in general by induction,

$$x \wedge b_1 + x \wedge b_\nu = b_1 + b_\nu.$$

Hence,

$$o\text{-}\lim_{\nu \to \infty} (x \wedge b_1 + x \wedge b_\nu) = o\text{-}\lim_{\nu \to \infty} (b_1 + b_\nu)$$

or

$$o\text{-}\lim_{\nu \to \infty} x \wedge b_1 + o\text{-}\lim_{\nu \to \infty} x \wedge b_\nu = o\text{-}\lim_{\nu \to \infty} b_1 + o\text{-}\lim_{\nu \to \infty} b_\nu$$

or

$$x \wedge b_1 + (\mathfrak{B}) \bigwedge_{\nu=1}^{\infty} (x \wedge b_\nu) = b_1 + (\mathfrak{B}) \bigwedge_{\nu=1}^{\infty} b_\nu$$

or

$$x \wedge b_1 = b_1.$$

In (1), the members $b_\nu + b_{\nu+1}$, $\nu = 1, 2, \ldots$, are pairwise disjoint; i.e., according to (I),

$$p(b_1) = \sum_{\nu=1}^{\infty} p(b_\nu + b_{\nu+1}) = \sum_{\nu=1}^{\infty} (p(b_\nu) - p(b_{\nu+1})),$$

hence

$$p(b_1) = p(b_1) - \lim_{\nu \to \infty} p(b_\nu) \quad \text{or} \quad \lim_{\nu \to \infty} p(b_\nu) = 0.$$

(B) Let (II) be true and let

$$(\mathfrak{B}) \bigvee_{\nu=1}^{\infty} a_\nu = a \in \mathfrak{B}$$

with the terms $a_1, a_2, \ldots,$ pairwise disjoint. We write

$$r_\nu = a + \bigvee_{n=1}^{\nu} a_n;$$

then $r_\nu \downarrow$; i.e., r_ν is a monotonically decreasing sequence. We shall prove that

$$(\mathfrak{B}) \bigwedge_{\nu=1}^{\infty} r_\nu = \emptyset.$$

It suffices to prove that

$$x \leqslant r, \quad \text{i.e.,} \quad x \wedge r_\nu = x \quad \text{for every} \quad \nu = 1, 2, \ldots, \quad \text{implies} \quad x = \emptyset.$$

Let $x \neq \emptyset$; then x is disjoint to $\bigvee_{n=1}^{\nu} a_n$, for every $\nu = 1, 2, \ldots,$ hence x is disjoint to a_ν for every $\nu = 1, 2, \ldots$; i.e., $a_\nu \wedge x = \emptyset$, $\nu = 1, 2, \ldots$. Hence,

$$x \wedge a = x \wedge \bigvee_{\nu=1}^{\infty} a_\nu = \bigvee_{\nu=1}^{\infty} (x \wedge a_\nu) = \emptyset;$$

therefore $x \leqslant a^c$, and $x \wedge r_\nu = \emptyset$ for every $\nu = 1, 2, \ldots$. But we have $x \wedge r_\nu = x$ for every $\nu = 1, 2, \ldots$; hence, x must be equal to $\emptyset$ (contradiction). Hence,

$$\bigwedge_{\nu=1}^{\infty} r_\nu = \emptyset.$$

According to (II) we have

$$\lim_{\nu \to \infty} p(r_\nu) = 0;$$

i.e.,

$$\lim p\left(a + \bigvee_{n=1}^{\nu} a_n\right) = p(a) - \lim_{\nu \to \infty} p\left(\bigvee_{n=1}^{\nu} a_n\right) = 0;$$

thus

$$p(a) = \lim_{\nu \to \infty} p\left(\bigvee_{n=1}^{\nu} a_n\right) = \lim_{\nu \to \infty} \sum_{n=1}^{\nu} p(a_n) = \sum_{\nu=1}^{\infty} p(a_\nu).$$

Hence, in another formulation, we have proved the:

Theorem 6.2.

The continuity of a pr p on a Boolean algebra $\mathfrak{B}$ is equivalent to the σ-additivity of p on $\mathfrak{B}$.

Exercise. Prove: if p is continuous on $\mathfrak{B}$ and

$$(\mathfrak{B}) \bigwedge_{\nu=1}^{\infty} a_\nu = a \in \mathfrak{B}$$

with $a_\nu \downarrow$ then

$$\lim_{\nu \to \infty} p(a_\nu) = p(a).$$

7. PROBABILITY σ-ALGEBRAS

7.1. A pr algebra $(\mathfrak{B}, p)$ is said to be a *pr σ-algebra* if and only if the Boolean algebra $\mathfrak{B}$ is a Boolean σ-algebra and the probability p is σ-additive (equivalently, continuous) on $\mathfrak{B}$.

Let $(\mathfrak{B}, p)$ be a pr algebra and $(\mathfrak{A}, p)$ a pr subalgebra of $(\mathfrak{B}, p)$; then the σ-additivity (equivalently the continuity) of p on $\mathfrak{B}$, in general, does not imply the σ-additivity (equivalently the continuity) of p on the Boolean subalgebra $\mathfrak{A}$. The following theorem holds in this respect.

Theorem 7.1.

Let $(\mathfrak{A}, p)$ be a pr subalgebra of the pr algebra $(\mathfrak{B}, p)$. Let the probability p be continuous on $\mathfrak{B}$. Then the following two statements are equivalent:

(1) *The probability p is continuous on $\mathfrak{A}$.*

(2) *The Boolean algebra $\mathfrak{A}$ is a σ-regular Boolean subalgebra of $\mathfrak{B}$; i.e., if*

$$(\mathfrak{A}) \bigwedge_{\nu=1}^{\infty} a_\nu = a \in \mathfrak{A}$$

with $a_\nu \in \mathfrak{A}$, $\nu = 1, 2, \ldots$, then $(\mathfrak{B}) \bigwedge_{\nu=1}^{\infty} a_\nu$ exists and is also equal to $a \in \mathfrak{A}$:

$$(\mathfrak{A}) \bigwedge_{\nu=1}^{\infty} a_\nu = (\mathfrak{B}) \bigwedge_{\nu=1}^{\infty} a_\nu.$$

Proof. If (2) holds, then obviously (1) holds, too. Now let p be continuous on $\mathfrak{A}$, and $\mathfrak{A}$ be not a σ-regular Boolean subalgebra of $\mathfrak{B}$; then there exists a monotonically decreasing sequence $a_\nu \in \mathfrak{A}$, $\nu = 1, 2, \ldots$, with

$$(\mathfrak{A}) \bigwedge_{\nu=1}^{\infty} a_\nu = \emptyset,$$

and such that either

$$(\mathfrak{B}) \bigwedge_{\nu=1}^{\infty} a_\nu = a \neq \emptyset, \qquad a \in \mathfrak{B},$$

or

$$(\mathfrak{B}) \bigwedge_{\nu=1}^{\infty} a_\nu$$

does not exist in $\mathfrak{B}$.

In the first case

$$\lim_{\nu \to \infty} p(a_\nu) = p(a)$$

(see exercise Section 6), because of the continuity of p on $\mathfrak{B}$, and we have $p(a) > 0$. In the second case there exists an element $b \in \mathfrak{A}$, $b \neq \emptyset$ and $\emptyset < b \leqslant a_\nu$, $\nu = 1, 2, \ldots$. But then we must have $p(a_\nu) \geqslant p(b) > 0$. Hence we have in both cases $\lim_{\nu \to \infty} p(a_\nu) > 0$. But this is a contradition to the continuity of p on $\mathfrak{A}$, because $\lim p(a_\nu)$ must be equal to 0, since $(\mathfrak{A}) \bigwedge_{\nu=1}^{\infty} a_\nu = \emptyset$.

7.2. We shall introduce now another kind of σ-additivity (continuity) of the probability *relative* to a Boolean over-algebra, in which the Boolean algebra of the pr algebra is embedded. Let $(\mathfrak{B}, p)$ be a pr algebra; we suppose that the Boolean algebra $\mathfrak{B}$ is a Boolean subalgebra of another Boolean algebra $\mathfrak{B}^*$. Then the probability p is said to be σ-additive (continuous) on $\mathfrak{B}$ relative to $\mathfrak{B}^*$ if and only if the following condition holds:

(I_r) For every sequence of pairwise disjoint elements $a_\nu \in \mathfrak{B}$ for which

$$(\mathfrak{B}^*) \bigvee_{\nu=1}^{\infty} a_\nu = a \in \mathfrak{B},$$

we have

$$p(a) = \sum_{\nu=1}^{\infty} p(a_\nu).$$

Equivalently:

(II_r) For every decreasing sequence $a_\nu \in \mathfrak{B}$ with

$$(\mathfrak{B}^*) \bigwedge_{\nu=1}^{\infty} a_\nu = \emptyset,$$

we have

$$\lim_{\nu \to \infty} p(a_\nu) = 0.$$

The equivalence of conditions (I_r) and (II_r) can be proved in the same way as in Theorem 6.1.

We notice now that if the probability p is continuous (equivalently σ-additive) on the Boolean algebra $\mathfrak{B}$ and $(\mathfrak{A}, p)$ is any pr subalgebra of

$(\mathfrak{B}, p)$, then the probability p is always continuous (equivalently σ-additive) on $\mathfrak{A}$ relative to $\mathfrak{B}$, even if the Boolean algebra $\mathfrak{A}$ is not a σ-regular Boolean subalgebra of $\mathfrak{B}$.

8. QUASI-PROBABILITY ALGEBRAS

8.1. Let $\mathfrak{Q}$ be a Boolean algebra and v a real valued function on $\mathfrak{Q}$ which is additive and normed as a probability (see Section 1.1), but not strictly positive. We suppose instead that v is non-negative on $\mathfrak{Q}$; i.e.,

8.1.1. $v(x) \geqslant 0$ for every $x \in \mathfrak{Q}$.

We then say that v is a *quasi-probability* on $\mathfrak{Q}$ and we call $(\mathfrak{Q}, v)$ a *quasi-pr algebra.*

Properties 1.2.1, 1.2.3, 1.2.4, 1.2.5, 1.2.7 on probabilities hold also for quasi-probabilities and property 1.2.2 holds for a quasi-probability in the weaker form:

8.1.2. If $x \leqslant y$, then $v(x) \leqslant v(y)$.

We notice that the quasi-probability v can assume, for certain elements $x \neq \emptyset$, the value $v(x) = 0$ and, for certain elements $x \neq e$, the value $v(x) = 1$. We call an element $x \neq \emptyset$ with $v(x) = 0$ an *almost impossible* event and an element $x \neq e$ with $v(x) = 1$ an *almost sure* event.

The concepts σ-additivity, continuity, quasi-pr σ-algebra, σ-additivity and continuity relative to a Boolean over-algebra can be introduced for quasi-probabilities in the same way as for probabilities.

8.2. Let $(\mathfrak{Q}, v)$ be a quasi-pr algebra; then the class $\mathfrak{N}$ of all almost impossible events of $\mathfrak{Q}$ is an ideal in $\mathfrak{Q}$. Let now $\mathfrak{A} = \mathfrak{Q}/\mathfrak{N}$ be the quotient Boolean algebra $\mathfrak{Q}$ mod $\mathfrak{N}$ and write $p(x/\mathfrak{N}) = v(x)$ for every element $x/\mathfrak{N} \in \mathfrak{Q}/\mathfrak{N}$. Then the so defined function p on $\mathfrak{A}$ is a probability and $(\mathfrak{A}, p)$ is a probability algebra. Let $\mathfrak{Q}$ be a Boolean σ-algebra and v a σ-additive quasi-probability on $\mathfrak{Q}$; then $\mathfrak{N}$ is a σ-ideal in $\mathfrak{Q}$, hence $\mathfrak{A} = \mathfrak{Q}/\mathfrak{N}$ is a Boolean σ-algebra and p is σ-additive; i.e., $(\mathfrak{A}, p)$ is a pr σ-algebra.

8.3. A quasi-probability v is called a *two-valued* quasi-probability on the Boolean algebra $\mathfrak{Q}$ if and only if v assumes only the values 0 and 1. If the quasi-probability v is two-valued, then $\mathfrak{A} = \mathfrak{Q}/\mathfrak{N}$ is isomorphic to the Boolean algebra $\mathfrak{U} = \{\emptyset, e\}$ of two elements and $(\mathfrak{A}, p)$ is isometric to the improper pr algebra $(\mathfrak{U}, p)$ (*cf.* Section 2.3).

8.4. Every Boolean algebra $\mathfrak{Q}$ can be endowed with a quasi-probability or with a two-valued quasi-probability (*cf.* Kappos, [8] Nr. 4).

On the contrary there exist Boolean algebras that cannot be endowed with a probability or with a σ-additive quasi-probability. (*cf.* Kappos [8], Nr. 8.)

9. PROBABILITY SPACES

9.1. Let $(\mathfrak{K}, p)$ be a probability or quasi-pr algebra, where $\mathfrak{K}$ is a Boolean subalgebra of the Boolean algebra $\mathfrak{P}(\Omega)$ of all subsets of a non-empty set Ω; i.e., $\mathfrak{K}$ is a field (in the classical sense) of subsets of the set Ω with $\Omega \in \mathfrak{K}$. Let the probability or quasi-probability p be σ-additive (equivalently continuous) on $\mathfrak{K}$ relative to $\mathfrak{P}(\Omega)$; i.e., σ-additive (equivalently continuous) on $\mathfrak{K}$ in the classical set-theoretical meaning. Then $(\mathfrak{K}, p)$ is called, after Kolmogorov (Kolmogorov [1]), a *probability field* and $(\Omega, \mathfrak{K}, P)$ a *probability space.* If, moreover, $\mathfrak{K}$ is a Boolean σ-subalgebra of $\mathfrak{P}(\Omega)$, i.e., $\mathfrak{K}$ is a σ-field of subsets of the set Ω, then $(\mathfrak{K}, P)$ is called, after Kolmogorov, a *Borel probability field* and $(\Omega, \mathfrak{K}, P)$ a *Borel probability space.* The probability space $(\Omega, \mathfrak{K}, P)$ is called *P-complete* or a *Lebesgue probability space*, if $\mathfrak{K}$ is a σ-field and if the following condition is satisfied:

(K) If $Y \in \mathfrak{K}$ with $P(Y) = 0$, then every subset $X \subseteq Y$ belongs to $\mathfrak{K}$.

It is well known that every probability space $(\Omega, \mathfrak{K}, P)$ can be extended to a Borel probability space $(\Omega, \mathbf{B}\mathfrak{K}, P)$, where $\mathbf{B}\mathfrak{K}$ is the smallest Boolean σ-subalgebra of $\mathfrak{P}(\Omega)$ containing $\mathfrak{K}$. Moreover, the probability space $(\Omega, \mathbf{B}\mathfrak{K}, P)$ can be extended to a Lebesgue probability space $(\Omega, \mathbf{L}\mathfrak{K}, P)$, where $\mathbf{L}\mathfrak{K}$ is the smallest Boolean σ-subalgebra of $\mathfrak{P}(\Omega)$ containing the Boolean σ-algebra $\mathbf{B}\mathfrak{K}$, which satisfies the condition (K); i.e., $Y \in \mathbf{L}\mathfrak{K}$ with $P(Y) = 0$ implies $X \in \mathbf{L}\mathfrak{K}$, for every $X \subseteq Y$. All $X \in \mathbf{B}\mathfrak{K}$ with $P(X) = 0$ and all $Y \in \mathbf{L}\mathfrak{K}$ with $P(Y) = 0$ form a σ-ideal $\mathfrak{N}$ in $\mathbf{B}\mathfrak{K}$ and a σ-ideal $\mathfrak{N}^*$ in $\mathbf{L}\mathfrak{K}$ respectively. Then according to Section 8.2 there corresponds to the quasi-pr σ-algebra $(\mathbf{B}\mathfrak{K}, P)$ and to the quasi-pr σ-algebra $(\mathbf{L}\mathfrak{K}, P)$ a pr σ-algebra $(\mathfrak{B}, p)$ and $(\mathfrak{B}^*, p^*)$ respectively, where $\mathfrak{B} = \mathbf{B}\mathfrak{K}/\mathfrak{N}$ and $\mathfrak{B}^* = \mathbf{L}\mathfrak{K}/\mathfrak{N}^*$ and $p(X/\mathfrak{N}) = P(X)$, $X/\mathfrak{N} \in \mathfrak{B}$ and $p^*(X/\mathfrak{N}^*) = P(X)$, $X/\mathfrak{N}^* \in \mathfrak{B}^*$. It is easy to prove that the pr σ-algebras $(\mathfrak{B}, p)$ and $(\mathfrak{B}^*, p^*)$ are isometric.

9.2. The σ-additivity (equivalently continuity) of a probability or a quasi-probability P on $\mathfrak{K}$ in the set-theoretical meaning, i.e., relative to

$\mathfrak{P}(\Omega)$, does not in general imply the σ-additivity (equivalently continuity) of P on $\mathfrak{K}$ considered as an abstract Boolean algebra. We shall demonstrate this fact by an example.

Example. Let $\mathfrak{A}$ be the interval Boolean algebra (*cf.* Section 4.5). Further, let f be a strictly increasing real-valued function defined on the interval $[0, 1] = \{\xi \in R: 0 \leq \xi \leq 1\}$ with $f(0) = 0$, $f(1) = 1$. We set $P(L_\beta) = f(\beta)$, for every, $L_\beta \in \mathfrak{D}$ and then for every $a \in \mathfrak{A}$ with the representation (*cf.* Section 4.5 (I))

$$a = \bigcup_{i=0}^{n-1} (L_{\beta_{2i+1}} \cap L_{\beta_{2i}}^c),$$

$$m(a) = \sum_{i=0}^{n-1} [f(\beta_{2i+1}) - f(\beta_{2i})].$$

The so defined function m on $\mathfrak{A}$ is a probability and $(\mathfrak{A}, m)$ is a pr algebra. If we suppose, moreover, that the function f is continuous from the left on $[0, 1]$, then $(\Omega, \mathfrak{A}, m)$ where $\Omega = \{\xi \in R: 0 \leq \xi < 1\}$, is a pr space; i.e., m is continuous (σ-additive) on $\mathfrak{A}$ relative to $\mathfrak{P}(\Omega)$. Let us now suppose that f is discontinuous from the right at some point $\gamma \in [0, 1)$. Then m is continuous on $\mathfrak{A}$ relative to $\mathfrak{P}(\Omega)$, but not continuous on $\mathfrak{A}$ considered as an abstract Boolean algebra. In fact, let $[\gamma, \xi_\nu)$ with

$$1 > \xi_\nu > \xi_{\nu+1} > \gamma \quad (\nu = 1, 2, \ldots)$$

be half open subintervals of $[0, 1)$; i.e.,

$$[\gamma, \xi_\nu) = L_{\xi_\nu} \wedge L_\gamma^c \in \mathfrak{A},$$

with
$$\lim_{\nu \to \infty} \xi_\nu = \gamma;$$

obviously
$$(\mathfrak{A}) \bigwedge_{\nu=1}^{\infty} [\gamma, \xi_\nu) = \emptyset,$$

but
$$\lim_{\nu \to \infty} m(L_{\xi_\nu} \wedge L_\gamma^c) = \lim_{\nu \to \infty} (f(\xi_\nu) - f(\gamma)) = \eta > 0$$

because of the discontinuity from the right of the function f at the point γ.

The set-theoretical intersection

$$\bigcap_{\nu=1}^{\infty} [\gamma, \xi_\nu) \equiv (\mathfrak{P}(\Omega)) \bigwedge_{\nu=1}^{\infty} [\gamma, \xi_\nu)$$

is not empty but equal to the one-point set $\{\gamma\}$ and this set does not belong to $\mathfrak{A}$. Let $(\Omega, \mathbf{B}\mathfrak{A}, P)$ be the Borel pr space corresponding to the pr space $(\Omega, \mathfrak{A}, m)$; then $\{\gamma\} \in \mathbf{B}\mathfrak{A}$ and we have

$$P(\{\gamma\}) = \eta > 0;$$

i.e.,
$$P(\{\gamma\}) = \lim_{\nu \to \infty} m([\gamma, \xi_\nu)).$$

9.3. Every Boolean algebra $\mathfrak{B}$ can be mapped isomorphically into the Boolean algebra $\mathfrak{P}(\Omega)$ of all subsets of a set Ω (*cf.* Kappos, [8], Nr. 6). The set Ω is defined as the set of all the two-valued quasi-probabilities on $\mathfrak{B}$. The isomorphic map of $\mathfrak{B}$ into $\mathfrak{P}(\Omega)$ is defined as follows:

$$\mathfrak{B} \ni x \to X = \{v \in \Omega\colon\ v(x) = 1\} \in \mathfrak{P}(\Omega). \tag{1}$$

Let $\mathfrak{K}$ be the image of $\mathfrak{B}$ in $\mathfrak{P}(\Omega)$ under the isomorphic map (1). Then $\mathfrak{K}$ is a Boolean subalgebra of $\mathfrak{P}(\Omega)$; i.e., a field of subsets of Ω with $\Omega \in \mathfrak{K}$. Let now p be a probability or quasi-probability on $\mathfrak{B}$; we can then define a probability or quasi-probability P on $\mathfrak{K}$ as follows:

$$P(X) = p(x) \quad \text{for every} \quad x \in \mathfrak{B};$$

X is here the image of x under the map (1). We can then prove (*cf.* Kappos [8], Nr. 9) that $(\Omega, \mathfrak{K}, P)$ is a pr space; i.e., P is σ-additive on $\mathfrak{K}$ relative to $\mathfrak{P}(\Omega)$. Hence the following theorem holds:

Theorem 9.1.

Let $(\mathfrak{B}, p)$ *be any pr algebra or quasi-pr algebra; then there always exists a pr space* $(\Omega, \mathfrak{K}, P)$ *such that* $(\mathfrak{B}, p)$ *is isometric to* $(\mathfrak{K}, P)$. *The pr space* $(\Omega, \mathfrak{K}, p)$ *is called a* representation pr space *for the pr algebra or quasi-pr algebra* $(\mathfrak{B}, p)$.

We shall not explain this well-known representation theory for pr algebras, because we intend to introduce later (see Chapter 3 Section 4) another representation theory for pr σ-algebras, which is more important in probability theory.

II

EXTENSION OF PROBABILITY ALGEBRAS

1. THE PROBABILITY ALGEBRA AS A METRIC SPACE

1.1. Let $(\mathfrak{A}, p)$ and $(\mathfrak{B}, \pi)$ be two pr algebras. Let ϕ be an isomorphic map of $\mathfrak{A}$ into $\mathfrak{B}$ with the property $\pi(\phi(a)) = p(a)$ for every $a \in \mathfrak{A}$. Then we say that the pr algebra $(\mathfrak{B}, \pi)$ is an *extension* of the pr algebra $(\mathfrak{A}, p)$ and that the pr algebra $(\mathfrak{A}, p)$ is a *restriction* of the pr algebra $(\mathfrak{B}, \pi)$. Moreover, if $(\mathfrak{B}, \pi)$ is a pr σ-algebra (i.e., $\mathfrak{B}$ is a Boolean σ-algebra and π is continuous on $\mathfrak{B}$) with the property: The smallest Boolean σ-subalgebra of $\mathfrak{B}$ containing $\mathfrak{A}_0$ is the Boolean σ-algebra $\mathfrak{B}$ itself, i.e., if $b_\sigma(\mathfrak{A}_0) = \mathfrak{B}$, where $\mathfrak{A}_0$ is the image of $\mathfrak{A}$ under ϕ, then $(\mathfrak{B}, \pi)$ is called a *minimal* σ-extension of $(\mathfrak{A}, p)$. The pr subalgebra $(\mathfrak{A}_0, \pi)$ of $(\mathfrak{B}, \pi)$ is then isometric to the pr algebra $(\mathfrak{A}, p)$ and is a σ-basis of $(\mathfrak{B}, \pi)$; i.e., $(\mathfrak{A}_0, \pi)$ σ-generates $(\mathfrak{B}, \pi)$. The extension of pr algebras can be stated as follows: Find extensions of a given pr algebra $(\mathfrak{A}, p)$ and, in particular, find mimimal σ-extensions of $(\mathfrak{A}, p)$. The solution of this problem requires the definition of a metric in the pr algebra $(\mathfrak{A}, p)$. This is done in the following section.

1.2. Let $(\mathfrak{B}, p)$ be a probability algebra. We define a real valued function

$$\rho(a, b) = p(a+b) \text{ for every } (a, b) \in \mathfrak{B} \times \mathfrak{B}.$$

This function is a metric distance between pairs of elements of $\mathfrak{B}$; i.e., the function ρ has the following properties of a distance:

1. $\rho(a, b) \geqslant 0$ and $(a, b) = 0$ if and only if $a = b$.
2. $\rho(a, b) = (b, a)$.
3. $\rho(a, b) \leqslant \rho(a, c) + \rho(c, b)$.

Obviously 1 and 2 are true. We shall prove the property 3. We have

$$a+b = a+c+c+b = (a+c)+(c+b) \leqslant (a+c) \vee (c+b).$$

Hence

$$\rho(a, b) = p(a+b) \leqslant p(a+c)+p(c+b) = \rho(a, c)+\rho(c, b);$$

i.e., property 3 is true. Hence, the Boolean algebra $\mathfrak{B}$ can be considered as a metric topological space and the concept of metric convergence, equivalently, ρ-convergence or *p-convergence* can be introduced in $\mathfrak{B}$ in the usual way; namely, a sequence $a_\nu \in \mathfrak{B}$, $\nu = 1, 2, \ldots$, is said to be *p-convergent* to the element $a \in b$ if and only if

$$\lim_{\nu \to \infty} \rho(a_\nu, a) = \lim_{\nu \to \infty} p(a_\nu + a) = 0.$$

Then we write $p\text{-}\lim_{\nu \to \infty} a_\nu = a$.

A p-convergent sequence $a_\nu \in \mathfrak{B}$, $\nu = 1, 2, \ldots$, satisfies the p-Cauchy condition (criterion); i.e., for every $\varepsilon > 0$, there exists a natural number $N(\varepsilon)$ such that $p(a_\nu + a_\mu) < \varepsilon$ for every $\nu \geqslant N(\varepsilon)$, and for every $\mu \geqslant N(\varepsilon)$. A sequence $x_\nu \in \mathfrak{B}$; $\nu = 1, 2, \ldots$, which satisfies the p-Cauchy condition is said to be a *p-Cauchy* or, equivalently, a *p-fundamental* sequence in $\mathfrak{B}$, and will be denoted by $\{x_\nu\}$.

The metric space $\mathfrak{B}$ is called *p-complete* if and only if every p-fundamental sequence is a p-convergent sequence in $\mathfrak{B}$. It is easy to prove that $\mathfrak{B}$ is in general not p-complete. For example, if the probability p is continuous and $\mathfrak{B}$ is not a Boolean σ-algebra, then there exists in $\mathfrak{B}$ a decreasing sequence a_ν, $\nu = 1, 2, \ldots$, such that the meet $\bigwedge_{\nu=1}^{\infty} a_\nu$ does not exist in $\mathfrak{B}$; this sequence is p-fundamental, but not p-convergent. It is well known from the topology that each metric space $\mathfrak{B}$ determines a complete metric space, the so-called metric closure $\tilde{\mathfrak{B}}$ of $\mathfrak{B}$ into which $\mathfrak{B}$ can be mapped isometrically in such a way that the image $\mathfrak{B}_0$ of $\mathfrak{B}$ in $\tilde{\mathfrak{B}}$ is metrically dense in $\tilde{\mathfrak{B}}$.

We shall now construct the metric closure $\tilde{\mathfrak{B}}$ of $\mathfrak{B}$ and at the same time we shall pay close attention to the algebraic structure of the Boolean algebra $\tilde{\mathfrak{B}}$. By this process, the Boolean algebra $\mathfrak{B}$ will be extended to a Boolean σ-algebra and the probability p to a σ-additive probability.

2. CONSTRUCTION OF A σ-EXTENSION OF A PR ALGEBRA

2.1. Let $(\mathfrak{B}, p)$ be a pr algebra. We denote by $\mathfrak{M}$, $\mathfrak{K}$, and $\mathfrak{N}$ respectively, the class of all p-fundamental, p-convergent, and p-null sequences in $\mathfrak{B}$. We then have

$$\mathfrak{M} \supseteq \mathfrak{K} \supseteq \mathfrak{N}.$$

Obviously, $\mathfrak{N}$ is non-empty. We define equality in $\mathfrak{M}$ as follows:

$$\{a_\nu\} = \{b_\nu\} \quad \text{if and only if} \quad a_\nu = b, \ \nu = 1, 2, \ldots$$

and two operations, as follows:

$$\{a_\nu\} + \{b_\nu\} = \{a_\nu + b_\nu\}$$

$$\{a_\nu\} . \{b_\nu\} = \{a_\nu \, b_\nu\}.$$

The sequences $\{a_\nu + b_\nu\}$ and $\{a_\nu \, b_\nu\}$ are obviously p-fundamental and if $\{a_\nu\}$ and $\{b_\nu\}$ are p-convergent, then $\{a_\nu + b_\nu\}$ and $\{a_\nu \, b_\nu\}$ are also p-convergent. With respect to the operations $+$ and $.$, $\mathfrak{M}$ is an idempotent ring with unit, i.e., a Boolean ring, and $\mathfrak{K}$ is a Boolean subring of $\mathfrak{M}$. $\mathfrak{N}$ is an ideal in $\mathfrak{M}$, hence in $\mathfrak{K}$, for if $\{a_\nu\} \in \mathfrak{N}$ and $\{b_\nu\} \in \mathfrak{N}$, then $\{a_\nu + b_\nu\} \in \mathfrak{N}$ and if $\{a_\nu\} \in \mathfrak{N}$ and $\{b_\nu\} \in \mathfrak{M}$ or $\{b_\nu\} \in \mathfrak{K}$ then $\{a_\nu \, b_\nu\} \in \mathfrak{N}$. The unit element of $\mathfrak{M}$ is the sequence $\{x_\nu\}$ with $x_\nu = e$, $\nu = 1, 2, \ldots$, denoted by $\{e\}$ and the zero element is the sequence $\{x_\nu\}$ with $x_\nu = \emptyset$, $\nu = 1, 2, \ldots$, denoted by $\{\emptyset\}$.

We now define the lattice operations $\vee$ and $\wedge$ in $\mathfrak{M}$ in the usual way, i.e.,

$$\{a_\nu\} \vee \{b_\nu\} = \{a_\nu\} + \{b_\nu\} + \{a_\nu\}\{b_\nu\}$$

$$\{a_\nu\} \wedge \{b_\nu\} = \{a_\nu\} . \{b_\nu\}.$$

Clearly, we have

$$\{a_\nu\} \vee \{b_\nu\} = \{a_\nu \vee b_\nu\}.$$

The complement of $\{a_\nu\}$ is then defined by

$$\{a_\nu\}^c = \{e\} + \{a_\nu\}$$

and we have

$$\{a_\nu\}^c = \{a_\nu{}^c\}.$$

Thus $\mathfrak{M}$ can be considered as Boolean algebra. Now let $\tilde{\mathfrak{B}} = \mathfrak{M}/\mathfrak{N}$ and $\mathfrak{B}_0 = \mathfrak{K}/\mathfrak{N}$ be the quotient Boolean algebras $\mathfrak{M}$ mod $\mathfrak{N}$ and $\mathfrak{K}$ mod $\mathfrak{N}$ respectively. Then $\mathfrak{B}_0$ is a Boolean subalgebra of $\tilde{\mathfrak{B}}$ and is isomorphic to $\mathfrak{B}$, for the map

$$\mathfrak{B} \ni a \Rightarrow \{a_\nu\}/\mathfrak{N} \quad \text{with} \quad a_\nu = a, \quad \nu = 1, 2, \ldots$$

is an isomorphism of $\mathfrak{B}$ onto $\mathfrak{B}_0$. We define a real-valued function $\tilde{p}$ on $\mathfrak{B}$ as follows:

$$\tilde{p}(\{a_\nu\}/\mathfrak{N}) = \lim_{\nu \to \infty} p(a_\nu).$$

The function $\tilde{p}$ is independent of the particular sequence $\{a_\nu\}$ of the class $\{a_\nu\}/\mathfrak{N}$, hence $\tilde{p}$ is a single-valued function. Moreover $\tilde{p}$ possesses the properties of a probability on $\tilde{\mathfrak{B}}$. In fact, $\tilde{p}$ is obviously strictly positive and normed. Hence, it is sufficient to show that $\tilde{p}$ is additive. Let $\{a_\nu\}/\mathfrak{N} \in \tilde{\mathfrak{B}}$ and $\{b_\nu\}/\mathfrak{N} \in \tilde{\mathfrak{B}}$ with $\{a_\nu\}/\mathfrak{N} \wedge \{b_\nu\}/\mathfrak{N} = \{\emptyset\}/\mathfrak{N}$, i.e., $\{a_\nu b_\nu\} \in \mathfrak{N}$. Then we have

$$\begin{aligned}
\tilde{p}(\{a_\nu\}/\mathfrak{N} \vee \{b_\nu\}/\mathfrak{N}) &= \tilde{p}(\{a_\nu \vee b_\nu\}/\mathfrak{N}) \\
&= \lim_{\nu \to \infty} p(a_\nu \vee b_\nu) \\
&= \lim_{\nu \to \infty} p((a_\nu + a_\nu \wedge b_\nu) + b_\nu) \\
&= \lim_{\nu \to \infty} (p(a_\nu + a_\nu \wedge b_\nu) + p(b_\nu)) \\
&= \lim_{\nu \to \infty} \{p(a_\nu) + p(b_\nu) - p(a_\nu \wedge b_\nu)\} \\
&= \lim_{\nu \to \infty} p(a_\nu) + \lim_{\nu \to \infty} p(b_\nu) \\
&= \tilde{p}(\{a_\nu\}/\mathfrak{N}) + \tilde{p}(\{b_\nu\}/\mathfrak{N});
\end{aligned}$$

i.e., $\tilde{p}$ is additive; hence, $(\tilde{\mathfrak{B}}, \tilde{p})$ is a pr algebra and $(\mathfrak{B}_0, \tilde{p})$ a pr subalgebra of $(\tilde{\mathfrak{B}}, \tilde{p})$. The Boolean algebra $\mathfrak{B}$ is isomorphic to $\mathfrak{B}_0$ and we have $\tilde{p}(\{a\}/\mathfrak{N}) = p(a)$ for every $a \in \mathfrak{B}$; i.e., $(\mathfrak{B}_0, \tilde{p})$ is isometric to $(\mathfrak{B}, p)$ and hence $(\tilde{\mathfrak{B}}, \tilde{p})$ is an extension of $(\mathfrak{B}, p)$. We shall prove that $(\tilde{\mathfrak{B}}, \tilde{p})$ is a minimal σ-extension of $(\mathfrak{B}, p)$; i.e., $\tilde{\mathfrak{B}}$ is a Boolean σ-algebra with $b_\sigma(\mathfrak{B}_0) = \tilde{\mathfrak{B}}$ and the probability $\tilde{p}$ is continuous on $\tilde{\mathfrak{B}}$.

2.2. We shall prove first:

Theorem 2.1.

The Boolean algebra $\tilde{\mathfrak{B}}$, as a $\tilde{p}$-metric space, is $\tilde{p}$-complete and $\tilde{\mathfrak{B}}$ is the $\tilde{p}$-metric closure of $\mathfrak{B}_0$, i.e., $\mathfrak{B}_0$ is $\tilde{p}$-dense in $\tilde{\mathfrak{B}}$.

We denote the elements of $\tilde{\mathfrak{B}}$ by Greek letters. In order to prove Theorem 2.1, we require the following lemmas:

Lemma 2.1.

Let $\alpha \in \tilde{\mathfrak{B}}$ and $\{a_\nu\} \in \mathfrak{M}$ with $\{a_\nu\} \in \alpha$. Let ϕ be the isomorphic map of $\mathfrak{B}$ onto $\mathfrak{B}_0$ defined by $\mathfrak{B} \ni a \Rightarrow \phi(a) = \{a\}/\mathfrak{N} \in \mathfrak{B}_0$. Then we have

$$\alpha = \tilde{p}\text{-}\lim \phi(a_\nu).$$

Proof. We have, for every fixed μ, $\{a_\mu + a_\nu\}_{\nu=1,2,\ldots} \in \phi(a_\mu)+\alpha$; hence,

$$\tilde{p}(\phi(a_\mu)+\alpha) = \lim_{\nu\to\infty} p(a_\mu + a_\nu).$$

But $\{a_\nu\} \in \mathfrak{M}$, hence, for every $\varepsilon > 0$, there exists $N(\varepsilon) > 0$ such that $p(a_\mu + a_\nu) < \varepsilon$, for every $\mu \geqslant N(\varepsilon)$ and $\nu \geqslant N(\varepsilon)$. We have

$$\lim_{\nu\to\infty} p(a_\mu + a_\nu) \leqslant \varepsilon \quad \text{for every} \quad \mu \geqslant N(\varepsilon),$$

i.e.,

$$\tilde{p}(\phi(a_\mu)+\alpha) \leqslant \varepsilon \quad \text{for every} \quad \mu \geqslant N(\varepsilon),$$

or

$$\lim_{\mu\to\infty} \tilde{p}(\phi(a_\mu)+\alpha) = 0, \quad \text{i.e.,} \quad \tilde{p}\text{-}\lim_{\mu\to\infty} \phi(a_\mu) = \alpha.$$

Lemma 2.2.

$\mathfrak{B}_0$ *is* $\tilde{p}$*-dense in* $\tilde{\mathfrak{B}}$.

Proof. Let $\alpha \in \tilde{\mathfrak{B}}$, then there exists $\{a_\nu\} \in \mathfrak{M}$ with $\{a_\nu\} \in \alpha$ and according to Lemma 2.1, we have $\tilde{p}\text{-}\lim \phi(a_\nu) = \alpha$ with $\phi(a_\nu) \in \mathfrak{B}_0$. Hence, for every $\varepsilon > 0$, there exists a $\phi(a_\nu) \in \mathfrak{B}_0$ with $\tilde{p}(\phi(a_\nu)+\alpha) < \varepsilon$, i.e., $\mathfrak{B}_0$ is $\tilde{p}$-dense in $\mathfrak{B}$.

Lemma 2.3.

$\tilde{\mathfrak{B}}$ *is* $\tilde{p}$*-complete.*

Proof. Let $\alpha_\nu \in \tilde{\mathfrak{B}}$, $\upsilon = 1, 2, \ldots$, be a $\tilde{p}$-fundamental sequence in $\tilde{\mathfrak{B}}$. Then for every $1/\nu$, there exists, according to Lemma 2.2, an element $\beta_\nu \in \mathfrak{B}_0$ such that $\tilde{p}(\alpha_\nu + \beta_\nu) < 1/\nu$. But we have

$$\begin{aligned} \tilde{p}(\beta_\nu + \beta_\mu) &\leqslant \tilde{p}(\beta_\nu + \alpha_\nu) + \tilde{p}(\alpha_\nu + \alpha_\mu) + \tilde{p}(\alpha_\mu + \beta_\mu) \\ &\leqslant \tilde{p}(\alpha_\nu + \alpha_\mu) + \frac{1}{\nu} + \frac{1}{\mu}. \end{aligned}$$

Hence,

$$\beta_\nu \in \mathfrak{B}_0, \qquad \nu = 1, 2, \ldots,$$

is a $\tilde{p}$-fundamental sequence. Then

$$\phi^{-1}(\beta_\nu) = b_\nu \in \mathfrak{B}, \qquad \nu = 1, 2, \ldots,$$

is a p-fundamental sequence in $\mathfrak{B}$, i.e.,

$$\{b_\nu\}/\mathfrak{N} = \beta \in \tilde{\mathfrak{B}}$$

and according to Lemma 2.1, we have

$$\tilde{p}\text{-}\lim \beta_\nu = \beta \in \tilde{\mathfrak{B}}.$$

But

$$\tilde{p}(\alpha_\nu + \beta_\nu) < \frac{1}{\nu}, \qquad \nu = 1, 2, \ldots .$$

Hence, we have also that

$$\tilde{p}\text{-lim}\, \alpha_\nu = \beta \in \tilde{\mathfrak{B}},$$

i.e., $\{\alpha_\nu\}$ is p-convergent.

Theorem 2.1 follows now from Lemmas 2.2 and 2.3.

Theorem 2.2.

The probability $\tilde{p}$ is continuous (equivalently σ-additive) on $\tilde{\mathfrak{B}}$.

Proof. Let

$$\alpha_\nu \in \tilde{\mathfrak{B}}, \qquad \nu = 1, 2, \ldots, \text{ with } \alpha_\nu \geqslant \alpha_{\nu+1}$$

and

$$(\tilde{\mathfrak{B}}) \bigwedge_{\nu=1}^{\infty} \alpha_\nu = \varnothing.$$

Then

$$\tilde{p}(\alpha_\nu) \downarrow \quad \text{and} \quad \lim_{\nu \to \infty} \tilde{p}(\alpha_\nu) \geqslant 0$$

exists. The sequence of the real numbers $\tilde{p}(\alpha_\nu)$, $\nu = 1, 2, \ldots$, therefore, satisfies the Cauchy criterion: for every $\varepsilon > 0$ there exists $N(\varepsilon) > 0$ such that

$$0 \leqslant \tilde{p}(\alpha_\nu) - \tilde{p}(\alpha_{\nu+k}) = \tilde{p}(\alpha_\nu + \alpha_{\nu+k}) < \varepsilon,$$

for every $\nu > N(\varepsilon)$, $k = 0, 1, 2, \ldots$, i.e.,

$$\tilde{p}(\alpha_\nu + \alpha_\mu) < \varepsilon, \ \nu \geqslant N(\varepsilon), \ \mu \geqslant N(\varepsilon).$$

Hence α_ν, $\nu = 1, 2, \ldots$, is a $\tilde{p}$-fundamental sequence and, according to Theorem 2.1,

(1) $\tilde{p}$-lim a_ν exists and is equal to an element $\alpha \in \tilde{\mathfrak{B}}$.

Now, for a fixed index μ, let $\beta_\nu = \alpha_\mu$, $\nu = 1, 2, \ldots$,. We have

(2) $\alpha_\mu = \tilde{p}\text{-}\lim_{\nu \to \infty} \beta_\nu$.

(1) and (2) imply:

$$\begin{aligned} \alpha_\mu \wedge \alpha &= \alpha_\mu \wedge \tilde{p}\text{-}\lim_{\nu \to \infty} \alpha_\nu \\ &= \tilde{p}\text{-}\lim_{\nu \to \infty} \beta_\nu \wedge \tilde{p}\text{-}\lim_{\nu \to \infty} \alpha_\nu \\ &= \tilde{p}\text{-}\lim_{\nu \to \infty} (\beta_\nu \wedge a_\nu) \\ &= \tilde{p}\text{-}\lim_{\nu \to \infty} \alpha_\nu = \alpha, \end{aligned}$$

i.e., $\alpha \leqslant \alpha_\mu$, $\mu = 1, 2, \ldots$.

But

$$(\tilde{\mathfrak{B}}) \bigwedge_{\mu=1}^{\infty} a_\mu = \emptyset,$$

hence $\alpha = \emptyset$, i.e., $\tilde{p}$-lim $\alpha_\nu = \emptyset$; thus $\lim \tilde{p}(\alpha_\nu) = 0$; hence the probability $\tilde{p}$ is continuous (equivalently σ-additive) on $\tilde{\mathfrak{B}}$.

From Theorem 2.2 we have the following:

Corollary 2.1.

If $(\tilde{\mathfrak{B}}) a_\nu \overset{o}{\downarrow} \alpha$ *or* $(\tilde{\mathfrak{B}}) \alpha_\nu \overset{o}{\uparrow} \alpha$, *then*

$$\tilde{p}\text{-}\lim_{\nu\to\infty} \alpha_\nu = \alpha$$

and generally if $(\tilde{\mathfrak{B}})$ o-$\lim \gamma_\nu = \gamma$ *then* $\tilde{p}$-$\lim \gamma_\nu = \gamma$; *i.e., the order convergence in* $\tilde{\mathfrak{B}}$ *implies the* $\tilde{p}$*-convergence.*

Theorem 2.3.

The Boolean algebra $\tilde{\mathfrak{B}}$ *is a Boolean* σ*-algebra.*

Proof. We prove first that if $\alpha_\nu \in \tilde{\mathfrak{B}}$, $\nu = 1, 2, \ldots$, and if the sequence α_ν, $\nu = 1, 2, \ldots$, is monotone, i.e., increasing $\alpha_\nu\uparrow$ or decreasing $\alpha_\nu\downarrow$, then $(\tilde{\mathfrak{B}}) \bigvee_{\nu=1}^{\infty} \alpha_\nu$ or $(\tilde{\mathfrak{B}}) \bigwedge_{\nu=1}^{\infty} \alpha_\nu$ exists respectively. Let us prove this in the case where $\alpha_\nu\downarrow$:

In the same way as in proof of Theorem 2.2, we can prove that α_ν, $\nu = 1, 2, \ldots$, is a $\tilde{p}$-fundamental sequence and there exists $\alpha \in \tilde{\mathfrak{B}}$ such that $\tilde{p}$-$\lim \alpha_\nu = \alpha$. Now, for a fixed μ, let $\beta_\nu = \alpha_\mu$, $\nu = 1, 2, \ldots$; we then have

$$\alpha_\mu = \tilde{p}\text{-}\lim_{\nu\to\infty} \beta_\nu.$$

Hence,

$$\begin{aligned}\alpha_\mu \wedge \alpha &= \alpha_\mu \wedge \tilde{p}\text{-}\lim_{\nu\to\infty} \alpha_\nu \\ &= \tilde{p}\text{-}\lim_{\nu\to\infty} \beta_\nu \wedge \tilde{p}\text{-}\lim_{\nu\to\infty} \alpha_\nu \\ &= \tilde{p}\text{-}\lim_{\nu\to\infty} (\beta_\nu \wedge \alpha_\nu) \\ &= \tilde{p}\text{-}\lim_{\nu\to\infty} \alpha_\nu = \alpha,\end{aligned}$$

i.e.,

$$\alpha_\mu \wedge \alpha = \alpha;$$

hence

$$\alpha \leqslant \alpha_\mu, \qquad \mu = 1, 2, \ldots.$$

Now let γ be an element in $\tilde{\mathfrak{B}}$ such that $\gamma \leqslant \alpha_\mu$, i.e.,

$$\gamma \wedge \alpha_\mu = \gamma \quad \text{for every} \quad \mu = 1, 2, \ldots .$$

Then we have

$$\gamma \wedge \alpha = \gamma \wedge \tilde{p}\text{-}\lim_{\mu\to\infty} \alpha_\mu = \tilde{p}\text{-}\lim_{\mu\to\infty} (\gamma \wedge \alpha_\mu) = \tilde{p}\text{-}\lim_{\mu\to\infty} \gamma = \gamma;$$

therefore, $\gamma \leqslant \alpha$, i.e.,

$$\alpha = (\tilde{\mathfrak{B}}) \bigvee_{\nu=1}^{\infty} \alpha_\nu = \alpha.$$

In the same way (dually) we can prove the existence of $\tilde{p}\text{-}\lim \alpha_\nu = \alpha$, in the case $\alpha_\nu\uparrow$ and therefore

$$(\tilde{\mathfrak{B}}) \bigvee_{\nu=1}^{\infty} \alpha_\nu = \alpha.$$

Let $\alpha_\nu \in \tilde{\mathfrak{B}}$, $\nu = 1, 2, \ldots$, be any sequence in $\tilde{\mathfrak{B}}$. We write

$$s_1 = \alpha_1, \ s_2 = \alpha_1 \vee \alpha_2, \ldots, s_\nu = \alpha_1 \vee \alpha_2 \vee \ldots \vee \alpha_\nu, \ldots$$

respectively, and

$$d_1 = \alpha_1, \ d_2 = \alpha_1 \wedge \alpha_2, \ldots, d_\nu = \alpha_1 \wedge \alpha_2 \wedge \ldots \wedge \alpha_\nu, \ldots .$$

Then $s_\nu\uparrow$ and $d_\nu\downarrow$; hence,

$$(\tilde{\mathfrak{B}}) \bigvee_{\nu=1}^{\infty} s_\nu = s \quad \text{and} \quad (\tilde{\mathfrak{B}}) \bigwedge_{\nu=1}^{\infty} d_\nu = d$$

exist in $\tilde{\mathfrak{B}}$ and we have

$$(\tilde{\mathfrak{B}}) \bigvee_{\nu=1}^{\infty} \alpha_\nu = s \quad \text{and} \quad (\tilde{\mathfrak{B}}) \bigwedge_{\nu=1}^{\infty} \alpha_\nu = d;$$

i.e., $\tilde{\mathfrak{B}}$ is a Boolean σ-algebra.

Theorems 2.2 and 2.3 imply

Corollary 2.2.

$(\tilde{\mathfrak{B}}, \tilde{p})$ is a probability σ-algebra.

Theorem 2.4.

Let $(\mathfrak{B}, p)$ be a probability σ-algebra and $a_\nu \in \mathfrak{B}$ a p-fundamental sequence in $\mathfrak{B}$. Then there exists a subsequence a_{k_ν}, $\nu = 1, 2, \ldots$, of this sequence, which is o-convergent. Let $a \in \mathfrak{B}$ be the order-limit of a_{k_ν}, i.e.,

$$(\mathfrak{B})o\text{-}\lim_{\nu\to\infty} a_{k_\nu} = a;$$

then also

$$a = p\text{-}\lim_{\nu \to \infty} a_\nu,$$

i.e., $\mathfrak{B}$ *is p-complete.*

Proof. We choose a subsequence a_{k_ν} of the sequence $a_\nu \in \mathfrak{B}$ which satisfies the condition

$$p(a_{k_\nu} + a_{k_{\nu+1}}) < \frac{1}{2^\nu}, \qquad \nu = 1, 2, \ldots.$$

We then set

$$s_\mu = \bigvee_{\nu=\mu}^{\infty} a_{k_\nu};$$

then

$$o\text{-}\limsup_{\nu \to \infty} a_{k_\nu} = \bigwedge_{\mu=1}^{\infty} s_\mu = \tilde{a} \in \mathfrak{B}.$$

Since $s_\mu \downarrow \tilde{a}$, it follows from the continuity of p that $p(s_\mu) \downarrow p(\tilde{a})$, i.e.,

$$\tilde{a} = p\text{-}\lim_{\mu \to \infty} s_\mu.$$

Now

$$s_\mu + a_{k_\mu} \leqslant \bigvee_{\nu=\mu}^{\infty} (a_{k_\nu} + a_{k_{\nu+1}}).$$

Hence,

$$p(s_\mu + a_{k_\mu}) \leqslant \sum_{\nu=\mu}^{\infty} p(a_{k_\nu} + a_{k_{\nu+1}}) \leqslant \sum_{\nu=\mu}^{\infty} \frac{1}{2^\nu} = \frac{1}{2^{\mu-1}}, \qquad \mu = 1, 2, \ldots,$$

i.e., $s_\mu + a_{k_\mu}$ is p-convergent to $\emptyset$. Now

$$p\text{-}\lim_{\mu \to \infty} s_\mu = \tilde{a} \in \mathfrak{B}$$

and a_{k_ν} is a p-fundamental sequence; thus a_{k_ν} must also be p-convergent to $\tilde{a}$; i.e.,

$$p\text{-}\lim_{\nu \to \infty} a_{k_\nu} = \tilde{a} \in \mathfrak{B}.$$

Since a_ν, $\nu = 1, 2, \ldots$, is p-fundamental we have

$$p\text{-}\lim_{\nu \to \infty} a_\nu = p\text{-}\lim_{\nu \to \infty} a_{k_\nu} = \tilde{a} \in \mathfrak{B}.$$

In the same way as before, we can prove that

$$o\text{-}\liminf_{\nu \to \infty} a_{k_\nu} = a$$

is equal to $p\text{-}\lim a_{k_\nu}$. Therefore, we have

$$o\text{-}\liminf_{\nu \to \infty} a_{k_\nu} = o\text{-}\limsup_{\nu \to \infty} a_{k_\nu};$$

i.e., the sequence a_{k_ν} is order convergent and we have

$$o\text{-}\lim a_{k_\nu} = p\text{-}\lim_{\nu\to\infty} a_{k_\nu} = p\text{-}\lim_{\nu\to\infty} a_\nu.$$

Corollary 2.3.

In a pr σ-algebra $(\mathfrak{B}, p)$ the p-convergence of a sequence implies the existence of a subsequence which is order convergent to the same limit.

Corollary 2.4.

A pr σ-algebra $(\mathfrak{B}, p)$ is always p-complete.

Theorem 2.5.

Let $(\mathfrak{B}, p)$ be a pr algebra, $(\tilde{\mathfrak{B}}, \tilde{p})$ be the extension of $(\mathfrak{B}, p)$ defined in Section 2.1, and $(\mathfrak{B}_0, \tilde{p})$ be the isometric image of $(\mathfrak{B}, p)$ defined in Section 2.1. Then $(\tilde{\mathfrak{B}}, \tilde{p})$ is a minimal σ-extension of $(\mathfrak{B}, p)$; i.e., $b_\sigma(\mathfrak{B}_0) = \tilde{\mathfrak{B}}$ and in fact,

$$\tilde{\mathfrak{B}} = \mathfrak{B}_0{}^{\sigma\delta} = \mathfrak{B}_0{}^{\delta\sigma} = \mathfrak{B}_0{}^{o} = \mathfrak{B}_0{}^{p}.$$

(See for the definition, Appendix one, Section 6. $\mathfrak{B}_0{}^p$ is the smallest subsystem of $\tilde{\mathfrak{B}}$ which contains $\mathfrak{B}_0$ and is closed for the p-convergence.)

Proof. According to Theorems 2.1–2.3 the pr algebra $(\tilde{\mathfrak{B}}, \tilde{p})$ is a pr σ-algebra which is p-complete and such that $\tilde{\mathfrak{B}}$ is the p-closure of $\mathfrak{B}_0$, i.e., every element $b \in \tilde{\mathfrak{B}}$ is the $\tilde{p}$-limit of a sequence $b_\nu \in \mathfrak{B}_0$, $\nu = 1, 2, \ldots$. Hence, we have

$$b = \tilde{p}\text{-}\lim_{\nu\to\infty} b_\nu,$$

where $b_\nu \in \mathfrak{B}_0$, $\nu = 1, 2, \ldots$; i.e., $\tilde{\mathfrak{B}} = B_0{}^p$.

According to Theorem 2.4, there exists a subsequence $b_{k_\nu} = a_\nu \in \mathfrak{B}_0$ of $b_\nu \in B_0$, $\nu = 1, 2, \ldots$, such that

$$b = o\text{-}\lim_{\nu\to\infty} a_\nu, \quad a_\nu \in \mathfrak{B}_0, \qquad \nu = 1, 2, \ldots.$$

Hence, there exists, for every element $b \in \tilde{\mathfrak{B}}$, a sequence

$$a_\nu \in \mathfrak{B}_0, \qquad \nu = 1, 2, \ldots,$$

such that $b = o\text{-}\lim_{\nu\to\infty} a_\nu$, i.e., we have $\mathfrak{B} = \mathfrak{B}_0{}^o$; since

$$b = \bigwedge_{\nu=1}^{\infty} \bigvee_{k=1}^{\infty} a_{\nu+k} = \bigvee_{\nu=1}^{\infty} \bigwedge_{k=1}^{\infty} a_{\nu+k},$$

we have

$$\tilde{\mathfrak{B}} = \mathfrak{B}_0{}^{\sigma\delta} = \mathfrak{B}_0{}^{\delta\sigma}.$$

It is now obvious that $(\tilde{\mathfrak{B}}, \tilde{p})$ is a minimal σ-extension of $(\mathfrak{B}, p)$.

Theorem 2.6.

Let $(\mathfrak{B}, p)$ be a pr σ-algebra and $(\mathfrak{A}, p)$ a pr subalgebra of $(\mathfrak{B}, p)$ such that $b_\sigma(\mathfrak{A}) = \mathfrak{B}$. Then

$$\mathfrak{B} = \mathfrak{A}^p = \mathfrak{A}^o = \mathfrak{A}^{\sigma\delta} = \mathfrak{A}^{\delta\sigma}.$$

Proof. According to Theorem 2.4, $\mathfrak{B}$ is p-complete. Let $\mathfrak{A}^p$ be the set of all p-limits of sequences of $\mathfrak{A}$ and $(\tilde{\mathfrak{A}}, \tilde{p})$ be the p-closure of $(\mathfrak{A}, p)$. Then $(\mathfrak{A}^p, p)$ and $(\tilde{\mathfrak{A}}, \tilde{p})$ are isometric, i.e., $\mathfrak{A}^p$ is a Boolean σ-algebra, also a Boolean subalgebra of $\mathfrak{B}$. But p is continuous on $\mathfrak{B}$, i.e., according to Theorem 7.1, Chapter 1, $\mathfrak{A}^p$ is a σ-regular Boolean subalgebra of $\mathfrak{B}$, i.e., also a Boolean σ-subalgebra of $\mathfrak{B}$. But $\mathfrak{A} \subseteq \mathfrak{A}^p \subseteq \mathfrak{B}$ and $b_\sigma(\mathfrak{A}) = \mathfrak{B}$, hence $\mathfrak{A}^p = \mathfrak{B}$. Now we can prove, in the same way as in the proof of Theorem 2.5, that

$$\mathfrak{B} = \mathfrak{A}^o = \mathfrak{A}^{\sigma\delta} = \mathfrak{A}^{\delta\sigma}.$$

Theorems 2.5 and 2.6 imply:

Theorem 2.7.

All minimal σ-extensions of a pr algebra $(\mathfrak{B}, p)$ are isometric to each other; i.e., isometric to the p-closure $(\tilde{\mathfrak{B}}, \tilde{p})$ of $(\mathfrak{B}, p)$ defined in Section 2.1.

The following important theorem is true:

Theorem 2.8.

The Boolean algebra $\mathfrak{B}$ of a pr σ-algebra $(\mathfrak{B}, p)$ is complete with respect to the lattice operations; i.e., $\bigwedge_{i \in I} a_i$ and $\bigvee_{i \in I} a_i$ exist in $\mathfrak{B}$ for every family of elements $a_i \in \mathfrak{B}$, $i \in I$.

Proof. Let $a_i \in \mathfrak{B}$, $i \in I$, be an arbitrary family in $\mathfrak{B}$. Let $\mathbf{E}$ denote the class of all finite subsets of I. We write

$$s_E = \bigvee_{i \in E} a_i$$

for every $E \in \mathbf{E}$; then

$$0 \leqslant \sup_{E \in \mathbf{E}} p(s_E) = \xi \leqslant 1.$$

According to the definition of the supremum, there exists a sequence

$E_\nu \in \mathbf{E}$, $\nu = 1, 2, \ldots$, such that

$$p(s_{E_\nu}) \geqslant \xi - \frac{1}{2^\nu};$$

moreover, E_ν, $\nu = 1, 2, \ldots$, can be assumed to be increasing; i.e., $E_\nu \subseteq E_{\nu+1}$, $\nu = 1, 2, \ldots$. Now, since

$$\begin{aligned} p(s_{E_\nu}+s_{E_\mu}) &\leqslant p(s_{E_\mu}+s_{E_\nu\cup E_\mu})+p(s_{E_\nu\cup E_\mu}+s_{E_\nu}) \\ &= p(s_{E_\nu\cup E_\mu})-p(s_{E_\mu})+p(s_{E_\nu\cup E_\mu})-p(s_{E_\nu}) \\ &\leqslant \frac{1}{2^\mu}+\frac{1}{2^\nu} \end{aligned}$$

i.e.,

$$p(s_{E_\nu}+s_{E_\mu}) \leqslant \frac{1}{2^\mu}+\frac{1}{2^\nu},$$

the sequence s_{E_ν}, $\nu = 1, 2, \ldots$, is p-fundamental in $\mathfrak{B}$. Hence $p\text{-}\lim s_{E_\nu} = s$ exists in $\mathfrak{B}$ and

$$s = \bigvee_{\nu=1}^{\infty} s_{E_\nu}.$$

Obviously, we have $p(s) = \xi$. Further,

$$p(a_i \vee s)-p(s) = \lim_{\nu\to\infty} [p(a_i \vee s_{E_\nu})-p(s_{E_\nu})] \leqslant \lim \frac{1}{2^\nu} = 0 \qquad (1)$$

since

$$\xi \geqslant p(a_i \vee s_{E_\nu}) \geqslant p(s_{E_\nu}) \geqslant \xi - \frac{1}{2^\nu}$$

implies

$$0 \leqslant p(a_i \vee s_{E_\nu})-p(s_{E_\nu}) \leqslant \frac{1}{2^\nu},$$

for $i \in I$ and $\nu = 1, 2, \ldots$. Now, (1) implies $p((a_i \vee s)+s) = 0$, $a_i \vee s+s = \emptyset$, $a_i \vee s = s$, i.e.,

$$a_i \leqslant s \quad \text{for every} \quad i \in I. \qquad (2)$$

Let y be any element in $\mathfrak{B}$, such that $a_i \leqslant y$, for every $i \in I$; then $s_{E_\nu} \leqslant y$ for every $\nu = 1, 2, \ldots$, i.e.,

$$y \wedge s_{E_\nu} = s_{E_\nu},$$

and

$$y \wedge s = y \wedge \bigvee_{\nu=1}^{\infty} s_{E_\nu} = \bigvee_{\nu=1}^{\infty} (y \wedge s_{E_\nu}) = \bigvee_{\nu=1}^{\infty} s_{E_\nu} = s,$$

hence

$$s \leqslant y, \quad \text{if} \quad a_i \leqslant y, \qquad i \in I. \qquad (3)$$

Now (2) and (3) imply $\bigvee_{i \in I} a_i = s$. Similarly, one can prove that $\bigwedge_{i \in I} a_i$ exists in $\mathfrak{B}$.

Remarks. We proved previously that every Boolean σ-algebra $\mathfrak{B}$, which may carry a σ-additive probability, is complete with respect to the lattice operations. We remark that $\mathfrak{B}$ is in general not completely distributive, but it satisfies the so-called *weak σ-distributivity* condition, namely: For every double sequence $a_{i,j} \in \mathfrak{B}$, $i = 1, 2, \ldots$, $j = 1, 2, \ldots$, with $a_{i,j} \leqslant a_{i,j+1}$, we have

$$\bigwedge_{i=1}^{\infty} \bigvee_{j=1}^{\infty} a_{i,j} = \bigvee_{n \in \mathbf{N}} \bigwedge_{i=1}^{\infty} a_{i,n_i}$$

where the letter $\mathbf{N}$ denotes the set of all infinite sequences $n = \{n_1, n_2, \ldots\}$ of natural numbers. Further we remark that Theorem 2.8 can be proved without assuming that p is σ-additive. This can be done as follows:

Let $a_i \in \mathfrak{B}$, $i \in I$, be an arbitrary family in $\mathfrak{B}$. Let $\mathbf{E}$, s_E and ξ be the same as in the Theorem 2.8. Then, there exists again a sequence $E_\nu \in \mathbf{E}$, $\nu = 1, 2, \ldots$, with $\lim_{\nu \to \infty} p(s_{E_\nu}) = \xi$; we write $s \equiv \bigvee_{\nu=1}^{\infty} s_{E_\nu}$. We shall show that $s = \bigvee_{i \in I} a_i$.

We observe first that $s \geqslant a_i$, for every $i \in I$, because, if $a_{i_0} \ngeqslant s$, for one $i_0 \in I$, then $a_{i_0} \wedge s^c \neq \emptyset$, hence

$$p(a_{i_0} \wedge s^c) \equiv \varepsilon > 0.$$

But, then:

$$p((a_{i_0} \wedge s^c) \vee s_{E_\nu}) = p(a_{i_0} \wedge s^c) + p(s_{E_\nu}) = \varepsilon + p(s_{E_\nu}),$$

for every $\nu = 1, 2, \ldots$, and

$$\lim_{\nu \to \infty} p((a_{i_0} \wedge s^c) \vee s_{E_\nu}) = \varepsilon + \xi.$$

Hence, there exists a ν_0 such that $p((a_{i_0} \wedge s^c) \vee s_{E_\nu}) > \xi$. But, then, we have $p(a_{i_0} \vee s_{E_{\nu_0}}) > \xi$, which is impossible, since $a_{i_0} \vee s_{E_{\nu_0}}$ is a finite union of elements of the family a_i, $i \in I$. Hence, in fact, $s \geqslant a_i$ for every $i \in I$. Besides, if $s^* \geqslant a_i$, for every $i \in I$, then $s^* \geqslant s_{E_\nu}$, for every $\nu = 1, 2, \ldots$, hence $s^* \geqslant s$. So, we have proved that $s = \bigvee_{i \in I} a_i$.

Similarly, one can prove that $\bigwedge_{i \in I} a_i$ exists in $\mathfrak{B}$.

The above statement, as well as Theorem 2.8, can also be derived from Theorem 3.6, Appendix 1, since a Boolean algebra, which may carry an additive strictly positive probability, satisfies the countable chain condition.

3. THE LINEAR LEBESGUE PROBABILITY σ-ALGEBRA

3.1. Let $(\mathfrak{A}, m)$ be the probability interval algebra (see Section 4.5 and 9.2, Chapter 1). Then the pr space $(\Omega, \mathfrak{A}, m)$ with

$$\Omega = \{\xi \in R\colon\ 0 \leqslant \xi < 1\}$$

is a representation pr space for the pr algebra $(\mathfrak{A}, m)$. We denote by $(\mathfrak{J}, \mu)$ a smallest σ-extension of $(\mathfrak{A}, m)$, i.e., the pr σ-algebra $(\mathfrak{J}, \mu)$ is defined as isometric to the m-closure $(\tilde{\mathfrak{A}}, \tilde{m})$ of $(\mathfrak{A}, m)$ and is said to be *the linear Lebesgue pr σ-algebra.* Let now $(\Omega, \mathbf{B}\mathfrak{A}, m)$ and $(\Omega, \mathbf{L}\mathfrak{A}, m)$ be the Borel pr space and the Lebesgue pr space, respectively, which correspond to the pr space $(\Omega, \mathfrak{A}, m)$. Then $\mathbf{L}\mathfrak{A}$ is the σ-field of all Lebesgue measurable subsets of Ω. Let $\mathfrak{N}$ be the σ-ideal of all sets $X \in \mathbf{L}\mathfrak{A}$ with $m(X) = 0$, i.e., of all the so-called Lebesgue-null subsets of Ω; then the Boolean σ-algebra $\mathbf{L}\mathfrak{A}/\mathfrak{N}$, is isomorphic to the Boolean σ-algebra $\mathfrak{J}$ and if we define $p(A/\mathfrak{N}) = m(\mathfrak{A})$ for each class $A/\mathfrak{N} \in \mathbf{L}\mathfrak{A}/\mathfrak{N}$, then $(\mathbf{L}\mathfrak{A}/\mathfrak{N}, p)$ is a pr σ-algebra isometric to $(\mathfrak{J}, \mu)$. We call $(\Omega, \mathbf{L}\mathfrak{A}, m)$ the linear Lebesgue pr space and $(\Omega, \mathbf{B}\mathfrak{A}, m)$ the linear Borel pr space.

3.2. Let now $(\mathfrak{B}, p)$ be any pr σ-algebra and $(\Omega, \mathbf{L}, P)$ a pr space, where Ω is any set and $\mathbf{L}$ is a σ-field of subsets of Ω satisfying condition (K) of Section 9.1, Chapter 1, i.e.,

(K) if $A \in \mathbf{L}$ with $P(A) = 0$ then every subset $X \subseteq A$ belongs to $\mathbf{L}$.

We then say that $(\Omega, \mathbf{L}, P)$ is a Lebesgue pr space. Let $\mathfrak{N}$ be the σ-ideal of all subsets $X \in \mathbf{L}$ with $P(X) = 0$. We say that $(\Omega, \mathbf{L}, P)$ is a *Lebesgue representation space* of the pr σ-algebra $(\mathfrak{B}, p)$ if and only if the pr σ-algebra $(\mathbf{L}/\mathfrak{N}, \pi)$, where $\pi(X/\mathfrak{N}) = P(X)$ for every class $X/\mathfrak{N} \in \mathbf{L}/\mathfrak{N}$, is isometric to $(\mathfrak{B}, p)$.

According to this definition $(\Omega, \mathbf{L}\mathfrak{A}, m)$ is a Lebesgue representation space of the linear Lebesgue pr σ-algebra $(\mathfrak{J}, \mu)$.

3.3. A pr σ-algebra $(\mathfrak{B}, p)$ is said to be *p-separable* if and only if there exists a p-separable pr subalgebra $(\mathfrak{A}, p)$ of $(\mathfrak{B}, p)$ which σ-generates $(\mathfrak{B}, p)$; i.e., with the property $\mathfrak{A}^{\sigma\delta} = \mathfrak{B}$. It is easy to prove:

Theorem 3.1.

A pr σ-algebra $(\mathfrak{B}, p)$ is p-separable if and only if there exists a Boolean subalgebra $\mathfrak{A}$ of $\mathfrak{B}$ with a countable number of elements such that $\mathfrak{B} = \mathfrak{A}^{\sigma\delta} = \mathfrak{A}^{\delta\sigma}$. The pr σ-algebra $(\mathfrak{J}, \mu)$ is obviously μ-separable because

$(\mathfrak{A}, m)$ is m-separable and there exists a pr subalgebra $(\mathfrak{A}_0, \mu)$ of $(\mathfrak{J}, \mu)$ isometric to $(\mathfrak{A}, m)$ which σ-generates $(\mathfrak{J}, \mu)$.

3.4. We shall say: a pr σ-algebra $(\mathfrak{B}, p)$ is *σ-generated* by a chain $\mathfrak{S}$ of elements of $\mathfrak{B}$ if and only if the smallest σ-subalgebra of $\mathfrak{B}$ containing the chain $\mathfrak{S}$ is $\mathfrak{B}$ itself. The following theorem is true:

Theorem 3.2.

Every pr σ-algebra $(\mathfrak{B}, p)$, which is σ-generated by a chain $\mathfrak{S} \subseteq \mathfrak{B}$ is isometric to a pr σ-subalgebra of $(\mathfrak{J}, \mu)$.

Proof. By virtue of Theorem 4.2, Section 4.5, Chapter 1, Theorem 3.1 implies Theorem 3.2, i.e. the pr σ-algebra $(\mathfrak{J}, \mu)$ is universal for the pr σ-algebras which are σ-generated by chains.

Now, we shall prove that $(\mathfrak{J}, \mu)$ is universal for every p-separable pr σ-algebra $(\mathfrak{B}, p)$. More precisely the following theorem is true:

Theorem 3.3.

Every p-separable pr σ-algebra $(\mathfrak{B}, p)$ is isometric to a pr σ-subalgebra of $(\mathfrak{J}, \mu)$. Moreover, if the Boolean σ-algebra $\mathfrak{B}$ is atomless, then $(\mathfrak{B}, p)$ is isometric to $(\mathfrak{J}, \mu)$.

Proof. Let $\mathfrak{M}$ be a countable subset of $\mathfrak{B}$ which σ-generates $\mathfrak{B}$. We can assume, without loss of generality, that all the atoms of $\mathfrak{B}$ belong to $\mathfrak{M}$ (because the set of atoms is at most countable) and, moreover, that $\emptyset \notin \mathfrak{M}$ and $e \in \mathfrak{M}$. Let $\{a_1, a_2, \ldots\} \subseteq \mathfrak{M}$ be the set of all atoms of $\mathfrak{B}$. We write

$$\{e, b_1, b_2, \ldots\} = \mathfrak{M} - \{a_1, a_2, \ldots\}$$

and we can assume, without loss of generality, that $b_j \wedge a_i = \emptyset$, $i = 1, 2, \ldots$, $j = 1, 2, \ldots$. If the Boolean σ-algebra $\mathfrak{B}$ is not atomic, then the set $\{b_1, b_2, \ldots\}$ is non-empty and countably infinite. If the Boolean σ-algebra $\mathfrak{B}$ is atomic, then $\mathfrak{B}$ is σ-generated by a chain, namely the set $\mathfrak{S}$ of all elements $s_\nu = a_1 \vee a_2 \vee \ldots \vee a_\nu$, $\nu = 1, 2, \ldots$, because $\{b_1, b_2, \ldots\}$ is empty. Hence, according to Theorem 3.2, our assertion is valid. Let $\{b_1, b_2, \ldots\}$ be non-empty. Then we can replace in $\mathfrak{M}$ the set $\{b_1, b_2, \ldots\}$ by a set of elements $b_{i,j}$ (without loss of the condition that σ-generates $\mathfrak{B}$), which we define as follows:

1. $b_{11} = b_1$.

2. $b_{21} = b_1 \wedge b_2, \quad b_{22} = b_1 - b_1 \wedge b_2, \quad b_{23} = b_2 - b_1 \wedge b_2.$

Here the elements b_{21}, b_{22}, b_{23} are pairwise disjoint and we have

$$b_1 = b_{21} \vee b_{22}, \quad b_2 = b_{21} \vee b_{23}.$$

Now let $b_{v1}, b_{v2}, \ldots, b_{vk_v}$ (with $k_v = 2^v - 1$) be the elements of $\mathfrak{B}$ defined by the v-th step; then the elements $b_{v+1,j}$ of the $(v+1)$-th step can be defined as follows:

$$b_{v+1,\,2m-1} = b_{v,\,m} \wedge b_{v+1} \qquad m = 1, 2, \ldots, k_v$$

$$b_{v+1,\,2m} = b_{v,\,m} - b_{v,\,m} \wedge b_{v+1} \qquad m = 1, 2, \ldots, k_v$$

and

$$b_{v+1,\,2k_v+1} = b_{v+1} - \{b_{v,\,1} \vee b_{v,\,2} \vee \ldots \vee b_{v,\,k_v}\} \wedge b_{v+1}.$$

Hence we defined at the $(v+1)$-th step $k_{v+1} = 2k_v + 1 = 2^{v+1} - 1$ elements in all. Obviously we have

$$b_{v,\,m} = b_{v+1,\,2m-1} \vee b_{v+1,\,2m} \qquad 1 \leqslant m \leqslant k_v;$$

i.e., every element $b_{v,m}$ of the v-th step which is not zero is, after the $(v+1)$-th step, either decomposed in two disjoint $\neq \emptyset$ elements, namely the elements $b_{v+1,\,2m-1}$ and $b_{v+1,\,2m}$, or equal to one of them, if the other is equal to $\emptyset$. Let now

$$b_{v,\,\rho_{1v}}, b_{v,\,\rho_{2v}}, \ldots, b_{v,\,\rho_{\lambda_v,\,v}}, \quad \text{where} \quad 1 \leqslant \rho_{1v} < \rho_{2v} < \ldots < \rho_{\lambda_v,\,v} \leqslant k_v, \qquad (1)$$

be all the non-zero elements defined by the v-th step. Then (1) are pairwise disjoint. We write

$$\mathfrak{M}_* = \{e, a_1, a_2, \ldots, b_{v,\,\rho_{iv}}, \ldots\}, \qquad 1 \leqslant i \leqslant \lambda_v.$$

$\mathfrak{M}_*$ σ-generates $\mathfrak{B}$, i.e., we have

$$\mathfrak{M}_*^{\wedge + \sigma\delta} = \mathfrak{B}.$$

We define now a map of $\mathfrak{M}_*$ into $\mathfrak{A} \subseteq \mathfrak{J}$ as follows:

I. $\mathfrak{M}_* \ni e \Rightarrow \Omega = [0, 1) \in \mathfrak{A}$.

II. $\mathfrak{M}_* \ni a_v \Rightarrow A_v = \left[\sum_{i=1}^{v-1} p(a_i),\ \sum_{i=1}^{v} p(a_i)\right) \in \mathfrak{A}$.

Let now

$$\xi = \lim_{v \to \infty} \sum_{i=1}^{v} p(a_i);$$

then $0 \leqslant \xi < 1$ and $\xi = 0$, if $\mathfrak{B}$ atomless.

III. $\mathfrak{M}_* \ni b_{v,\,\rho_{iv}} \Rightarrow B_{v,\,\rho_{iv}} = \left[\xi + \sum_{j=1}^{i-1} p(b_{v,\,\rho_{jv}}),\ \xi + \sum_{j=1}^{i} p(b_{v,\,\rho_{jv}})\right) \in \mathfrak{A}$,

for every $i = 1, 2, \ldots, \lambda_v$, $v = 1, 2, \ldots$.

Here we put

$$\sum_{i=1}^{0} = 0$$

and the interval $B_{\nu, p_{i\nu}}$ is a subinterval of $[\xi, 1)$. Let

$$\mathfrak{M} = \{\Omega, A_1, A_2, \ldots, B_{\nu, p_{i\nu}}, \ldots\};$$

then $\mathfrak{M}^{\wedge +}$ is a Boolean subalgebra of $\mathfrak{A}$, isomorphic to the Boolean subalgebra $\mathfrak{M}_*{}^{\wedge +}$, which σ-generates $\mathfrak{B}$. Moreover, the pr algebra $(\mathfrak{M}^{\wedge +}, m)$ is isometric to the pr algebra $(\mathfrak{M}_*{}^{\wedge +}, p)$, hence the m-closure of $(\mathfrak{M}^{\wedge +}, m)$, i.e., $(\tilde{\mathfrak{M}}^{\wedge +}, \tilde{m})$, is a pr σ-algebra isometric to a pr σ-subalgebra of $(\mathfrak{J}, \mu)$ and isometric also to the p-closure of $(\mathfrak{M}_*{}^{\wedge +}, p)$ in $(\mathfrak{B}, p)$, i.e., isometric to the pr σ-algebra $(\mathfrak{B}, p)$ itself. Hence the first assertion of Theorem 3.8 is proved.

Now, let $\mathfrak{B}$ be atomless. Let Ξ be the set of all the end-points of the intervals $B_{\nu, j}$. We assert Ξ is dense in the interval $[0, 1)$, and if this assertion is true, the system $\mathfrak{M} = \{\Omega, B_{\nu, p_{i\nu}}, \ldots\}$ defines a Boolean algebra $\mathfrak{M}^{\wedge +}$ o-dense in $\mathfrak{A}$. Thus $(\mathfrak{M}^{\wedge +}, m)$ is isometric to a pr subalgebra of $(\mathfrak{J}, \mu)$, which is p-dense in $(\mathfrak{J}, \mu)$. Obviously then the pr σ-algebra $(\mathfrak{B}, p)$ is isometric to $(\mathfrak{J}, \mu)$. The assertion that Ξ is not dense in $[0, 1)$ implies a contradiction. In fact, if Ξ is not dense in $[0, 1)$ then there exist two points of accumulation of the set Ξ in $[0, 1)$ with $\xi_1 < \xi_2$, such that $(\xi_1, \xi_2) \cap \Xi = \emptyset$. It is easy to prove that the interval $[\xi_1, \xi_2]$ belongs to the Lebesgue σ-field $\mathbf{L}\mathfrak{M}^{\wedge +}$, and that the class $[\xi_1, \xi_2] | \mathfrak{N}$ defines in $\mathbf{L}\mathfrak{M}^{\wedge +} | \mathfrak{N}$ an atom. Then, since the Boolean σ-algebra $\mathfrak{B}$ is isomorphic to the Boolean σ-algebra $\mathbf{L}\mathfrak{M}^{\wedge +} | \mathfrak{N}$, $\mathfrak{B}$ must posses an atom, a contradiction to the assumption that $\mathfrak{B}$ is atomless.

Corollary 3.1.

Let $(\mathfrak{B}, p)$ be a pr σ-algebra; then the following statements are equivalents:

1. $(\mathfrak{B}, p)$ *is p-separable.*

2. *The Boolean σ-algebra is σ-generated by a chain $\mathfrak{S} \subseteq \mathfrak{B}$.*

3. $(\mathfrak{B}, p)$ *is isometric to pr σ-subalgebra of* $(\mathfrak{J}, \mu)$.

4. *There exists a Lebesgue representation space* $(\Omega, \mathfrak{K}, P)$ *where* $\Omega = [0, 1)$, $\mathfrak{K}$ *a σ-subfield of the linear Lebesgue σ-field, satisfying the condition* (K) *of Section* 9.1, *Chapter* 1, *and* $P(X) = m(X)$, *for all* $X \in \mathfrak{K}$.

4. CLASSIFICATION OF THE p-SEPARABLE PR σ-ALGEBRAS

4.1. Let $(\mathfrak{J}, \mu)$ be the linear Lebesgue pr σ-algebra and let a be any element of $\mathfrak{J}$, $a \neq \emptyset$. Then the set

$$a\mathfrak{J} = \{x \in \mathfrak{J}: x \leqslant a\}$$

i.e., the principal ideal in $\mathfrak{J}$ determined by a, is itself a Boolean σ-algebra with a as unit element; $\mu/\mu(a)$ is a σ-additive probability on $a\mathfrak{J}$; i.e., $(a\mathfrak{J}, \mu/\mu(a))$ is a pr σ-algebra. This pr σ-algebra is obviously $\mu/\mu(\alpha)$-separable and atomless; i.e., isometric to $(\mathfrak{J}, \mu)$. The linear Lebesgue pr σ-algebra $(\mathfrak{J}, \mu)$ satisfies thus the so-called *homogeneity condition*:

(o) for every $a \in \mathfrak{J}$, $a \neq \emptyset$, the pr σ-algebra $(a\mathfrak{J}, \mu/\mu(a))$ is isometric to $(\mathfrak{J}, \mu)$ itself.

4.2. Let now $(\mathfrak{B}, p)$ be a p-separable pr σ-algebra; then

I. If $\mathfrak{B}$ is atomless, then $(\mathfrak{B}, p)$ is isometric to $(\mathfrak{J}, \mu)$, i.e., $(\mathfrak{B}, p)$ satisfies the homogeneity condition. We shall then say $(\mathfrak{B}, p)$ is a *continuous* pr σ-algebra.

II. If $\mathfrak{B}$ is not atomless, i.e., there exist atoms $a_1, a_2, \ldots,$ in $\mathfrak{B}$ then we have two cases:

Case 1. $e = a_1 \vee a_2 \vee \ldots$, i.e., $\sum_{\nu \geqslant 1} p(a_\nu) = 1$, whenever the Boolean σ-algebra $\mathfrak{B}$ and hence the pr σ-algebra $(\mathfrak{B}, p)$ is atomic. We shall then say $(\mathfrak{B}, p)$ is a finite or infinite *discrete* pr σ-algebra, according as the number of atoms is finite or infinite.

Case 2. $a_1 \vee a_2 \vee \ldots = a \neq e$, i.e., $\sum_{\nu \geqslant 1} p(a_\nu) < 1$ and $a^c \neq \emptyset$. Then the Boolean σ-algebra $a\mathfrak{B}$ and hence the pr σ-algebra $(a\mathfrak{B}, p/p(a))$ is discrete finite or infinite, while the Boolean σ-algebra $a^c\,\mathfrak{B}$ is atomless; hence the pr σ-algebra $(a^c\,\mathfrak{B}, p/p(a^c))$ is continuous. The pr σ-algebra $(\mathfrak{B}, p)$ can thus be decomposed in this case in two pr σ-algebras, one discrete and one continuous. We shall say that the pr σ-algebra $(\mathfrak{B}, p)$ is of *mixed* type. We proved thus the theorem:

Theorem 4.1.

A p-separable pr σ-algebra $(\mathfrak{B}, p)$ belongs to one and only one of the following three types:

1. *Continuous*; *i.e., isometric to* $(\mathfrak{J}, \mu)$.

2. *Discrete finite or infinite*; *i.e.,* $\mathfrak{B}$ *is an atomic Boolean σ-algebra with a finite or infinite number of atoms.*

3. *Mixed; i.e. there exist two disjoint elements* $a \in \mathfrak{B}$, $b \in \mathfrak{B}$ *with* $a \vee b = e$, $a \neq \emptyset$, $b \neq \emptyset$ *such that the pr σ-algebra* $(a\mathfrak{B}, p/p(a))$ *is atomic, while the pr σ-algebra* $(b\mathfrak{B}, p/p(b))$ *is continuous. This decomposition of* $(\mathfrak{B}, p)$ *in two pr σ-algebras* $(a\mathfrak{B}, p/p(a))$ *of atomic type and* $(b\mathfrak{B}, p/p(b))$ *of continuous type is uniquely determined.*

Exercise. Prove that a pr σ-subalgebra of a continuous pr σ-algebra may be of any one of the three types, while a pr σ-subalgebra of an algebra of mixed type is either discrete or mixed, and a pr σ-subalgebra of a discrete algebra is always discrete.

4.3. Any pr σ-algebra $(\mathfrak{B}, p)$ is not always p-separable; for example the pr algebra $(\mathfrak{B}, p)$, defined in Section 4.3, Chapter I, where $\mathfrak{B}$ is the so-called free Boolean algebra with $\aleph$ generators, if $\aleph > \aleph_0$, does not possess a countable subclass $\mathfrak{K}$ of events, which is p-dense in $\mathfrak{B}$. The pr algebra $(\mathfrak{B}, p)$ and therefore, its σ-extension $(\tilde{\mathfrak{B}}, \tilde{p})$ are not p-separable. In Section 3 of Chapter III, a classification of any pr σ-algebra will be stated. However, we notice that every pr σ-algebra $(\mathfrak{B}, p)$ can be characterized by a minimal cardinality of its σ-bases. In fact, let $\mathfrak{A}$ be a σ-base of $\mathfrak{B}$, i.e., let $\mathfrak{A}$ be a Boolean subalgebra of the Boolean σ-algebra $\mathfrak{B}$ such that $b_\sigma(\mathfrak{A}) = \mathfrak{A}^{\sigma\delta} = \mathfrak{A}^{\delta\sigma} = \mathfrak{B}$. If $|\mathfrak{A}|$ denotes the cardinality of $\mathfrak{A}$, then there exists the minimal cardinality of $\mathfrak{A}$ for all σ-bases $\mathfrak{A}$ of $\mathfrak{B}$, the so-called character of $\mathfrak{B}$, denoted by:

$$c_{\mathfrak{B}} \equiv \min |\mathfrak{A}|, \quad \text{for all} \quad \sigma\text{-bases } \mathfrak{A} \text{ of } \mathfrak{B}.$$

Obviously, the *character* of $\mathfrak{B}_\aleph$ is $\aleph$ for every $\aleph \geqslant \aleph_0$, while the character of $\mathfrak{B}$ of any p-separable pr σ-algebra $(\mathfrak{B}, p)$ is $\leqslant \aleph_0$ and especially $c_{\mathfrak{B}} = \aleph_0$, if $|\mathfrak{B}| \geqslant \aleph_0$ and $c_{\mathfrak{B}} = |\mathfrak{B}|$, if $\mathfrak{B}$ is a finite set.

III

CARTESIAN PRODUCT OF PROBABILITY ALGEBRAS

1. CARTESIAN PRODUCT OF BOOLEAN ALGEBRAS

1.1. Let $\mathfrak{A}_i$, $i \in I$, be a family of Boolean algebras, where the cardinality $|I|$ of the index set I is $\geqslant 1$; then there exists a Boolean algebra $\mathfrak{F}$, called the *cartesian product* of the Boolean algebras $\mathfrak{A}_i$, $i \in I$, and denoted by $\mathfrak{F} = \mathbf{P}_{i \in I} \mathfrak{A}_i$, which satisfies the following condition:

There exists a family $\mathfrak{F}_i$, $i \in I$, of Boolean subalgebras of $\mathfrak{F}$ such that:

(1) For every $i \in I$, $\mathfrak{F}_i$ is isomorphic to the Boolean algebra $\mathfrak{A}_i$.

(2) The set-theoretical union $\bigcup_{i \in I} \mathfrak{F}_i$ generates the Boolean algebra.

(3) If, for every $i \in I$, h_i is a homomorphic map of $\mathfrak{F}_i$ into another Boolean algebra $\mathfrak{B}$, then there exists a homomorphic map h of $\mathfrak{F}$ into $\mathfrak{B}$ which is a common extension of all the homomorphic maps h_i, $i \in I$, i.e., we have $h_i(x) = h(x)$ for every $x \in \mathfrak{F}_i$, $i \in I$.

Remark. If, for every $i \in I$, the Boolean algebra $\mathfrak{A}_i$ is the atomic Boolean algebra with two atoms x_i and x_i^c, i.e., $\mathfrak{A}_i = \{\emptyset, x_i, x_i^c, e\}$, then

$$\mathfrak{F} = \mathbf{P}_{i \in I} \mathfrak{A}_i$$

is isomorphic to the free Boolean algebra $\mathfrak{B}$ with $\aleph = |I|$ generators x_i, $i \in I$ (see Section 4.3, Chapter 1).

1.2. Construction of the cartesian product $\underset{i \in I}{\mathbf{P}} \mathfrak{A}_i$

A family $\alpha = \{a_i \in \mathfrak{A}_i,\ i \in I\}$ with $a_i = e_i$† for all $i \in I$ except at most for a finite subset $\{i_1, i_2, \ldots, i_k\} \subseteq I$ is said to be a *product element* and is denoted by $\alpha = \underset{i \in I}{\mathbf{P}}\ a_i$ or briefly by $\alpha = \mathbf{P} a_i$. For every $i \in I$, the corresponding a_i is said to be the i-th factor of α.

Let now $\mathfrak{K}$ be the set of all product elements. We shall say the product element $\alpha = \underset{i \in I}{\mathbf{P}}\, a_i$ is *disjoint* to the product element $\beta = \underset{i \in I}{\mathbf{P}}\, b_i$ if and only if there exists (at least) an index $i \in I$, such that $a_i \wedge b_i = \emptyset_i$. Two product elements α and β are defined as equal, i.e., $\alpha = \beta$, if and only if either $a_i = \emptyset_i$ and $b_j = \emptyset_j$ at least for one index i and an index j or $a_i = b_i$ for every $i \in I$. The product element $\varepsilon = \underset{i \in I}{\mathbf{P}}\, e_i$ is said to be the unit and every product element $\underset{i \in I}{\mathbf{P}}\, a_i$ with (at least) one factor $a_i = \emptyset_i$ is said to be zero and is denoted by $\emptyset$. We define the operation

$$\alpha \wedge \beta = \underset{i \in I}{\mathbf{P}}\, a_i \wedge \underset{i \in I}{\mathbf{P}}\, b_i \equiv \underset{i \in I}{\mathbf{P}}\, (a_i \wedge b_i)$$

and, moreover, the binary relation

$$a \leqslant \beta \quad \text{if and only if} \quad \alpha \wedge \beta = \alpha.$$

$\mathfrak{K}$ is, with respect to the relation $\leqslant$, a *po*-set and $\alpha \wedge \beta$ is the infimum (meet) of α and β. There exists also, for each pair α, β, the supremum (join) $a \vee \beta$. The so-defined algebraic structure is hence the structure of a lattice, but not of a distributive lattice or of a Boolean algebra. We shall prove below that the *po*-set $\mathfrak{K}$ can however be embedded into a Boolean algebra $\mathfrak{F}$ such that ordering relation, meets and distributive joins are preserved.

1.3. We consider the set $\mathfrak{H}$ of all the finite subsets, the so-called *aggregates* $\{\alpha_1, \alpha_2, \ldots, \alpha_n\}$ of $\mathfrak{K}$, where the product elements $\alpha_1, \alpha_2, \ldots, \alpha_n$ are pairwise disjoint.

Let now $\{\alpha_1, \alpha_2, \ldots, \alpha_n\}$, where $\alpha_j = \underset{i \in I}{\mathbf{P}}\, a_{ji}$, $j = 1, 2, \ldots, n$, be an element of $\mathfrak{H}$. We shall call this element a *net-like aggregate*, if and only if, for every $(j, k) \in \{1, 2, \ldots, n\} \times \{1, 2, \ldots, n\}$ and each $i \in I$, we have either

$$a_{ji} \wedge a_{ki} = \emptyset_i \quad \text{or} \quad a_{ji} = a_{ki}.$$

† By e_i and $\emptyset_i$ are denoted the unit and zero of $\mathfrak{A}_i$ respectively.

Let $\{\alpha\} \in \mathfrak{H}$ where $\alpha = \mathbf{P}_{i \in I} a_i$; let, moreover, for a finite subset

$$\{i_1, i_2, \ldots, i_k\} \subseteq I, \quad \text{and} \quad \nu = 1, 2, \ldots, k,$$

$$a_{i_\nu} = a_{1,\, i_\nu} \vee a_{2,\, i_\nu} \vee \ldots \vee a_{\rho_\nu,\, i_\nu} \quad \text{in} \quad \mathfrak{A}_{i_\nu} \tag{1}$$

be a partition of the element $a_{i_\nu} \in \mathfrak{A}_{i_\nu}$ into ρ_ν pairwise disjoint elements' with $1 \leqslant \rho_\nu < +\infty$. We form the expression

$$(a_{1,\, i_1} \vee \ldots \vee a_{\rho_1,\, i_1}) \times (a_{1,\, i_2} \vee \ldots \vee a_{\rho_2,\, i_2}) \times \ldots \times (a_{1,\, i_k} \vee \ldots \vee a_{\rho_k,\, i_k})$$

and apply, formally, the distributive law. In this way, a number of $\rho = \rho_1 \rho_2 \ldots \rho_k$ expressions of the form

$$a_{\lambda_1,\, i_1} \times a_{\lambda_2,\, i_2} \times \ldots \times a_{\lambda_k,\, i_k}, \quad 1 \leqslant \lambda_j \leqslant \rho_j, \quad j = 1, 2, \ldots, k, \tag{2}$$

may be obtained. For every expression (2), let now $\alpha_{\lambda_1,\, \lambda_2,\, \ldots,\, \lambda_k}$ be a product element $\mathbf{P}_{i \in I} x_i$, in which $x_{i_j} = a_{\lambda_j,\, i_j}$, $j = 1, 2, \ldots, k$, and if $i \neq i_j$, $j = 1, 2, \ldots, k$, then $x_i = a_i$. The so defined product elements

$$\alpha_{\lambda_1,\, \lambda_2,\, \ldots,\, \lambda_k},$$

for all λ_j with $j = 1, 2, \ldots, k$, are pairwise disjoint and form a net-like aggregate with a number of $\rho = \rho_1 \rho_2 \ldots \rho_k$ product elements. Let the elements of this aggregate be enumerated and denoted by

$$\{\beta_1, \beta_2, \ldots, \beta_\rho\} \in \mathfrak{H}.$$

We shall call the previously stated process: a decomposition of $\{\alpha\} \in \mathfrak{H}$ into a net-like aggregate $\{\beta_1, \beta_2, \ldots, \beta_\rho\} \in \mathfrak{H}$.

1.4. Let now $\{\alpha_1, \alpha_2, \ldots, \alpha_n\}$ be any aggregate in $\mathfrak{H}$ then, for every $\nu = 1, 2, \ldots, n$, we may decompose, by the previously stated process, the one product element aggregate $\{\alpha_\nu\}$ in a suitable net-like aggregate $\{\beta_{\nu,\, 1}, \beta_{\nu,\, 2}, \ldots, \beta_{\nu,\, \sigma_\nu}\}$ such that the aggregate:

$$\{\beta_{1,\, 1}, \beta_{1,\, 2}, \ldots, \beta_{1,\, \sigma_1}, \beta_{2,\, 1}, \beta_{2,\, 2}, \ldots, \beta_{2,\, \sigma_2}, \ldots, \beta_{n,\, 1}, \beta_{n,\, 2}, \ldots, \beta_{n,\, \sigma_n}\}$$

is net-like. There exist obviously, for every $\{\alpha_1, \alpha_2, \ldots, \alpha_n\} \in \mathfrak{H}$, several net-like aggregates which may be obtained by the previously stated decomposition process. We shall say briefly that these net-like aggregates belong to the aggregate $\{\alpha_1, \alpha_2, \ldots, \alpha_n\} \in \mathfrak{H}$. We shall define in $\mathfrak{H}$ a binary relation $\leqslant$ as follows:

$$\{\alpha_1, \alpha_2, \ldots, \alpha_n\} \leqslant \{\beta_1, \beta_2, \ldots, \beta_m\}$$

if and only if there exist two net-like aggregates: $\{\alpha_1^*, \alpha_2^*, \ldots, \alpha_t^*\}$ belonging to $\{\alpha_1, \alpha_2, \ldots, \alpha_n\}$ and $\{\beta_1^*, \beta_2^*, \ldots, \beta_s^*\}$ belonging to

$\{\beta_1, \beta_2, \ldots, \beta_m\}$ such that, for every α_j^*, $j = 1, 2, \ldots, t$ there exists a β_i^* with $\alpha_j^* = \beta_i^*$.

It is easy to prove that this relation is reflexive and transitive; i.e., $\mathfrak{H}$ is, with respect to this relation, a quasi *po*-set. An equivalence relation $\sim$ may hence be defined in $\mathfrak{H}$, as follows:

$$\{\alpha_1, \alpha_2, \ldots, \alpha_n\} \sim \{\beta_1, \beta_2, \ldots, \beta_m\}$$

if and only if

$$\{\alpha_1, \alpha_2, \ldots, \alpha_n\} \leqslant \{\beta_1, \beta_2, \ldots, \beta_m\}$$

and
$$\{\beta_1, \beta_2, \ldots, \beta_m\} \leqslant \{\alpha_1, \alpha_2, \ldots, \alpha_n\}.$$

Let now $\mathfrak{F}$ be the set of all equivalence classes in $\mathfrak{H}$ with respect to $\sim$; then $\mathfrak{F}$ becomes a *po*-set if we define for a pair $\bar{a} \in \mathfrak{F}$ and $\bar{b} \in \mathfrak{F}$:

$\bar{a} \leqslant \bar{b}$ if and only if there exist $\{\alpha_1, \alpha_2, \ldots, \alpha_n\} \in \bar{a}$ and $\{\beta_1, \beta_2, \ldots, \beta_m\} \in \bar{b}$ such that

$$\{\alpha_1, \alpha_2, \ldots, \alpha_n\} \leqslant \{\beta_1, \beta_2, \ldots, \beta_m\}.$$

Let now $\bar{a} \in \mathfrak{F}$ with $\{\alpha_1, \alpha_2, \ldots, \alpha_n\} \in \bar{a}$ and $\bar{b} \in \mathfrak{F}$ with $\{\beta_1, \beta_2, \ldots, \beta_m\} \in \bar{b}$; then there exists $\bar{a} \wedge \bar{b} \in \mathfrak{F}$, namely the class which is defined by the aggregate $\{\alpha_i \wedge \beta_j | i = 1, 2, \ldots, n;\ j = 1, 2, \ldots, m\}$.

1.5. In order to prove that $\mathfrak{F}$ is a Boolean algebra we remark that a mapping h of $\mathfrak{K}$ into $\mathfrak{F}$ can be defined as follows: for each $\alpha \in \mathfrak{K}$ let $h(\alpha)$ be in $\mathfrak{F}$ the class of all aggregates equivalent to the one-element aggregate $\{\alpha\}$. This mapping h is an isomorphic, relative to the ordering relation, mapping of $\mathfrak{K}$ into $\mathfrak{F}$, preserving meets. Let now the class of all aggregates equivalent to the one-element aggregate $\{\alpha\}$ (i.e., $h(\alpha) \in \mathfrak{F}$) be identified with α (i.e., $\alpha \in \mathfrak{F}$ is defined as identical to $h(\alpha) \in \mathfrak{F}$) and $\mathfrak{K}$ with $h(\mathfrak{K})$ (i.e., $\mathfrak{K} \subseteq \mathfrak{F}$ is defined as identical to $h(\mathfrak{K}) \subseteq \mathfrak{F}$). First, we shall prove: if $\alpha \in \mathfrak{K} \subseteq \mathfrak{F}$ and $\beta \in \mathfrak{K} \subseteq \mathfrak{F}$ then there exists $\bar{s} \in \mathfrak{F}$ which is the supremum of α and β in $\mathfrak{F}$, i.e., $\alpha \vee \beta = \bar{s}$ and generally if $\alpha_\nu \in \mathfrak{K} \subseteq \mathfrak{F}$, $\nu = 1, 2, \ldots, k$, then there exists an element $\bar{s} \in \mathfrak{F}$ such that $\bar{s} = \alpha_1 \vee \alpha_2 \vee \ldots \vee \alpha_k$. In order to prove this, set

$$\alpha_\nu = \mathbf{P}_{i \in I}\, a_{\nu i}, \quad \nu = 1, 2, \ldots, k;$$

there exists then a smallest finite subset J of I such that, for every $i \in (I - J)$, we have

$$a_{\nu i} = e_i, \quad \nu = 1, 2, \ldots, k.$$

We consider now, for every $j \in J$, the elements of the Boolean algebra $\mathfrak{A}_j$:

$$a_{1j}, a_{2j}, \ldots, a_{kj};$$

there exists then a finite set of pairwise disjoint elements

$$b_{1j}, b_{2j}, \ldots, b_{\lambda_j j}$$

such that every $a_{\nu j}$ can be represented as a join of the form

$$a_{\nu j} = \bigvee_{\rho \in J_{\nu j}} b_{\rho j} \quad \text{where} \quad J_{\nu j} \subseteq \{1, 2, \ldots, \lambda_j\}.$$

For every $\nu = 1, 2, \ldots, k$, the aggregate $\{\alpha_\nu\}$ can now be decomposed into a net-like aggregate as follows: we note that

$$a_{\nu j} = \bigvee_{\rho \in J_{\nu j}} b_{\rho j}$$

is a partition of the factor $a_{\nu j} \in \mathfrak{A}_j$ of α_ν for every $j \in J$ and the decomposition process of Section 1.3 can be applied to these partitions.

Let now

$$\{\beta_{\nu 1}, \beta_{\nu 2}, \ldots, \beta_{\nu, \sigma_\nu}\} \sim \{\alpha_\nu\}$$

be the corresponding, by this decomposition process, net-like aggregate to $\{\alpha_\nu\}$. Note now that every $\beta_{\nu j}$ is either equal or disjoint to one $\beta_{\lambda i}$, for $\lambda \neq \nu$ and any $i = 1, 2, \ldots, \sigma_\lambda$, and consider the set $\{\beta_1, \beta_2, \ldots, \beta_\sigma\}$ of all product elements β_j such that for (at least) one $\nu \in \{1, 2, \ldots, k\}$ we have $\beta_j \in \{\beta_{\nu 1}, \beta_{\nu 2}, \ldots, \beta_{\nu, \sigma_\nu}\}$; then the product elements $\beta_1, \beta_2, \ldots, \beta_\sigma$ are pairwise disjoint, namely $\{\beta_1, \beta_2, \ldots, \beta_\sigma\}$ is an aggregate and defines an equivalence class $\bar{s}$ in $\mathfrak{F}$, for which we have $\bar{s} = \alpha_1 \vee \alpha_2 \vee \ldots \vee \alpha_k$, because, obviously, $\alpha_\nu \leqslant \bar{s}$, for every $\nu = 1, 2, \ldots, k$, and, if $\bar{x} \in \mathfrak{F}$ with $\alpha_\nu \leqslant \bar{x}$, then $\bar{s} \leqslant \bar{x}$.

1.6. Let now $\alpha_1 \in \mathfrak{R} \subseteq \mathfrak{F}$, $\alpha_2 \in \mathfrak{R} \subseteq \mathfrak{F}$ and let us apply the previous process for $k = 2$. Consider

$$\{\beta_{11}, \beta_{12}, \ldots, \beta_{1, \sigma_1}\} \sim \{\alpha_1\}$$

$$\{\beta_{21}, \beta_{22}, \ldots, \beta_{2, \sigma_2}\} \sim \{\alpha_2\};$$

then the set theoretical difference

$$\{\beta_{11}, \beta_{12}, \ldots, \beta_{1, \sigma_1}\} - \{\beta_{21}, \beta_{22}, \ldots, \beta_{2, \sigma_2}\} = \{\gamma_1, \gamma_2, \ldots, \gamma_\tau\},$$

where for every $i = 1, 2, \ldots, \tau$,

$$\gamma_i \in \{\beta_{11}, \beta_{12}, \ldots, \beta_{1, \sigma_1}\}$$

but $$\gamma_i \notin \{\beta_{21}, \beta_{22}, \ldots, \beta_{2, \sigma_2}\},$$

defines an aggregate $\{\gamma_1, \gamma_2, \ldots, \gamma_\tau\}$ and the corresponding equivalence class $\bar{d}$ to this aggregate in $\mathfrak{F}$ is obviously the difference $\alpha_1 - \alpha_2$; i.e., we have $\bar{d} = \alpha_1 - \alpha_2$ in $\mathfrak{F}$. It is easy now to prove:

(α) For every equivalent class $\bar{s}$ with $\{\alpha_1, \alpha_2, \ldots, \alpha_n\} \in \bar{s}$, we have $\bar{s} = \alpha_1 \vee \alpha_2 \vee \ldots \vee \alpha_n$ in $\mathfrak{F}$.

(β) If $\bar{s} \in \mathfrak{F}$ and $\bar{t} \in \mathfrak{F}$ with $\{\alpha_1, \alpha_2, \ldots, \alpha_n\} \in \bar{s}$, $\{\beta_1, \beta_2, \ldots, \beta_k\} \in \bar{t}$ then

$$\bar{s} \vee \bar{t} = \alpha_1 \vee \alpha_2 \vee \ldots \vee \alpha_n \vee \beta_1 \vee \beta_2 \vee \ldots \vee \beta_k.$$

(γ) There exists an element $\bar{d} \in \mathfrak{F}$ such that $\bar{d} = \bar{s} - \bar{t}$. Moreover it can easily be proved that $\mathfrak{F}$ is a distributive lattice, hence $\mathfrak{F}$ is a Boolean algebra. Obviously the smallest Boolean subalgebra of $\mathfrak{F}$ containing $\mathfrak{K}$ is $\mathfrak{F}$ itself and moreover we have $\mathfrak{K}^+ = \mathfrak{F}$ where $+$ is the addition of $\mathfrak{F}$ considered as a Boolean ring. For every $j \in I$, there exists an isomorphic mapping of $\mathfrak{A}_j$ into $\mathfrak{F}$, namely: for every $\alpha \in \mathfrak{A}_j$, assign as image the product element $\alpha = \mathbf{P}_{i \in I} a_i \in \mathfrak{K} \subseteq \mathfrak{F}$ with $a_j = a$ and $a_i = e_i$ for every $i \neq j$, $i \in I$. Let now $\mathfrak{F}_j$ be the image of $\mathfrak{A}_j$ in $\mathfrak{F}$ by this isomorphic mapping; then the set theoretical union $\mathfrak{S} = \bigcup_{j \in I} \mathfrak{F}_j$ generates $\mathfrak{F}$ because we have $\mathfrak{S}^\wedge = \mathfrak{K}$, hence $\mathfrak{S}^{\wedge +} = \mathfrak{K}^+ = \mathfrak{F}$. The family $\mathfrak{F}_j$, $j \in I$, satisfies thus the conditions (1) and (2) of Section 1.1. It is easy to prove that the condition (3) is also satisfied. Let namely h_j be a homomorphism of $\mathfrak{F}_j$ into another Boolean algebra $\mathfrak{B}$, $j \in I$; then the mapping h, which is defined as follows:

(α) $h(\alpha) = h_j(\alpha)$ for every $\alpha \in \mathfrak{F}_j$, $j \in I$.

(β) $h(\alpha_{j_1} \wedge \alpha_{j_2} \wedge \ldots \wedge \alpha_{j_k}) = h_{j_1}(\alpha_{j_1}) \wedge h_{j_2}(\alpha_{j_2}) \wedge \ldots \wedge h_{j_k}(\alpha_{j_k})$

for every $\qquad \alpha_{j_1} \wedge \alpha_{j_2} \wedge \ldots \wedge \alpha_{j_k} \in \mathfrak{S}^\wedge = \mathfrak{K}$

with $\qquad \alpha_{j_\rho} \in \mathfrak{F}_j, \quad \rho = 1, 2, \ldots, k.$

(γ) $h(\beta_{j_1} + \beta_{j_2} + \ldots + \beta_{j_m}) = h(\beta_{j_1}) + h(\beta_{j_2}) + \ldots + h(\beta_{j_m})$

for every $\qquad \beta_{j_1} + \beta_{j_2} + \ldots + \beta_{j_m} \in \mathfrak{S}^{\wedge +} = \mathfrak{F}$

with $\qquad \beta_{j_k} \in \mathfrak{K}, \quad k = 1, 2, \ldots, m,$

is a homomorphism of $\mathfrak{F}$ into $\mathfrak{B}$ and, moreover, a common extension of all homomorphisms h_j, $j \in I$. It is, hence, proved that a cartesian product $\mathfrak{F} = \mathbf{P}_{i \in I} \mathfrak{A}_i$ exists.

2. PRODUCT PROBABILITY ALGEBRAS

2.1. Let $(\mathfrak{A}_i, p_i)$, $i \in I$, be a family of pr algebras with $|I| \geqslant 1$; then the product algebra $\mathfrak{F} = \mathbf{P}_{i \in I} \mathfrak{A}_i$ can be endowed with a probability π as follows:

(I) If $\alpha \in \mathfrak{K}$, then $\alpha = \mathbf{P}_{i \in I} a_i$, where $a_i = e_i$ for every $i \in (I - J)$ and J a finite subset of I. We can hence define

$$\pi(\alpha) = \prod_{j \in J} p_j(a_j) \in R.$$

(II) Consider now any $\alpha \in \mathfrak{F}$; then there exists a finite number of pairwise disjoint elements $\alpha_\nu \in \mathfrak{K} \subseteq \mathfrak{F}$, $\nu = 1, 2, \ldots, k$, such that

$$\alpha = \alpha_1 \vee \alpha_2 \vee \ldots \vee \alpha_k.$$

We can, hence, define:

$$\pi(\alpha) = \pi(\alpha_1) + \pi(\alpha_2) + \ldots + \pi(\alpha_k).$$

In this way π is defined for every $\alpha \in \mathfrak{F}$ and it is easy to prove that the value $\pi(\alpha)$ is uniquely determined for every $\alpha \in \mathfrak{F}$. We have $\pi(e) = 1$ and $\pi(\emptyset) = 0$, and, moreover, π is strictly positive and additive, i.e., π is a probability on $\mathfrak{F}$. Thus $(\mathfrak{F}, \pi)$ is a pr algebra which we shall call *product pr algebra* of the pr algebras $(\mathfrak{A}_i, p_i)$, $i \in I$, and denote also by

$$(\mathfrak{F}, \pi) = \underset{i \in I}{\mathbf{P}} (\mathfrak{A}_i, p_i).$$

Note that every $(\mathfrak{F}_j, \pi)$ is a pr subalgebra of $(\mathfrak{F}, \pi)$ isometric to $(\mathfrak{A}_j, p_j)$ and, if $(\mathfrak{A}_j, p_j)$ is a pr σ-algebra, then $(\mathfrak{F}_j, \pi)$ is a pr σ-subalgebra of $(\mathfrak{F}, \pi)$ with $\mathfrak{F}_j$ a σ-regular Boolean σ-subalgebra of $\mathfrak{F}$.

2.2. Let now

$$(\mathfrak{F}, \pi) = \underset{i \in I}{\mathbf{P}} (\mathfrak{A}_i, p_i)$$

be a product pr algebra; then we shall denote by

$$(\tilde{\mathfrak{F}}, \pi) = \underset{i \in I}{\tilde{\mathbf{P}}} (\mathfrak{A}_i, p_i)$$

the σ-extension of $(\mathfrak{F}, \pi)$ and we will call it a *product pr σ-algebra* of the pr algebras $(\mathfrak{A}_i, p_i)$, $i \in I$. Note that the product pr σ-algebra $\underset{i \in I}{\tilde{\mathbf{P}}} (\tilde{\mathfrak{A}}_i, p_i)$, where $(\tilde{\mathfrak{A}}_i, p_i)$ is the σ-extension of $(\mathfrak{A}_i, p_i)$, $i \in I$, is isometric to the product pr σ-algebra $\underset{i \in I}{\tilde{\mathbf{P}}} (\mathfrak{A}_i, p_i)$. We shall hence write

$$(\tilde{\mathfrak{F}}, \pi) = \underset{i \in I}{\tilde{\mathbf{P}}} (\mathfrak{A}_i, p_i) = \underset{i \in I}{\tilde{\mathbf{P}}} (\tilde{\mathfrak{A}}_i, p_i).$$

There exists, for every factor $(\mathfrak{A}_i, p_i)$ respectively $(\tilde{\mathfrak{A}}_i, p_i)$, an isometric image in $(\tilde{\mathfrak{F}}, \pi)$, namely the pr subalgebra $(\mathfrak{F}_i, \pi)$ resp the pr σ-subalgebra $(\tilde{\mathfrak{F}}_i, \pi)$ of $(\tilde{\mathfrak{F}}, \pi)$. The pr subalgebras $(\mathfrak{F}_i, \pi)$ resp pr σ-subalgebras $(\tilde{\mathfrak{F}}_i, \pi)$, $i \in I$, are π-independent in $(\tilde{\mathfrak{F}}, \pi)$. (See for this concept Section 5 below.)

2.3. Let
$$(\tilde{\mathfrak{F}}, \pi) = \tilde{\mathbf{P}}_{i \in I} (\mathfrak{A}_i, p_i)$$
be a product pr σ-algebra, in which every factor $(\mathfrak{A}_i, p_i)$ is a pr σ-algebra. Let the character $c_{\mathfrak{A}_i}$ be $\geqslant \aleph_0$ for every $i \in I$; then the character $c_{\tilde{\mathfrak{F}}}$ is also $\geqslant \aleph_0$ and, moreover, if $|I| \leqslant \aleph_0$, then $c_{\tilde{\mathfrak{F}}} = \max_{i \in I} c_{\mathfrak{A}_i}$. If $c_{\mathfrak{A}_i} = \aleph_0$, for every $i \in I$, and $|I| \leqslant \aleph_0$, then $c_{\tilde{\mathfrak{F}}} = \aleph_0$. If $|I| > \aleph_0$, then the Boolean σ-algebra $\tilde{\tilde{\mathfrak{F}}}$ is always atomless. The Boolean σ-algebra $\tilde{\tilde{\mathfrak{F}}}$ can possess atoms if $|I| \leqslant \aleph_0$.

Let now $|I| = \aleph_0$ and suppose α is an atom in $\tilde{\tilde{\mathfrak{F}}}$. We have $\tilde{\tilde{\mathfrak{F}}} = \mathfrak{R}^{+\delta\sigma}$. An atom $\alpha \in \tilde{\tilde{\mathfrak{F}}}$ must obviously belong to $\mathfrak{R}^{+\delta}$; i.e., α may be represented
$$\alpha = \bigwedge_{\nu=1}^{\infty} \alpha_\nu,$$
where $\alpha_\nu \downarrow$, and $\alpha_\nu \in \mathfrak{R}^+$, i.e., $\alpha_\nu = \alpha_{\nu 1} \vee \alpha_{\nu 2} \vee \ldots \vee \alpha_{\nu k}$ where $\alpha_{\nu\rho} \in \mathfrak{R}$, $\rho = 1, 2, \ldots, k_\nu$ and pairwise disjoint. We have thus
$$\alpha = \bigwedge_{\nu=1}^{\infty} \{\alpha_{\nu 1} \vee \alpha_{\nu 2} \vee \ldots \vee \alpha_{\nu, k_\nu}\} \tag{3}$$
where
$$\alpha_{\nu\rho} \in \mathfrak{R}, \qquad \rho = 1, 2, \ldots, k_\nu,$$
are pairwise disjoint and
$$\alpha_\nu = \alpha_{\nu 1} \vee \alpha_{\nu 2} \vee \ldots \vee \alpha_{\nu, k_\nu} \downarrow,$$
i.e., a non-increasing sequence. We can obviously suppose that, for every $\alpha_{\nu+1, \rho}$, there exists an element $\alpha_{\nu, \tau}$ such that $\alpha_{\nu+1, \rho} \leqslant \alpha_{\nu, \tau}$ and moreover we can suppose that $\alpha_\nu \downarrow$ is strictly decreasing, for the sequence α_ν cannot be eventually constant; because if $\alpha_{\nu_0 + k} = \alpha_{\nu_0}$, $k = 1, 2, \ldots,$ then $\alpha_{\nu_0} = \alpha \in \mathfrak{F}$ and α an atom in $\mathfrak{F}$. This however is impossible; for, if any $\alpha \in \mathfrak{F}$ is $\neq \emptyset$, then we have $\alpha = \beta_1 \vee \beta_2 \vee \ldots \vee \beta_k$ with $\beta_\rho \in \mathfrak{R}$, $\beta_\rho \neq \emptyset$, $\rho = 1, 2, \ldots, k$, and pairwise disjoint; if α is, moreover, an atom, then we must have $k = 1$, i.e., $\alpha \in \mathfrak{R}$. Let now $\alpha = \mathbf{P}_{i \in I} a_i$; then there exist

(α) an index $i_0 \in I$ such that $a_{i_0} = e_{i_0}$ and

(β) an element $b_{i_0} \in \mathfrak{A}_{i_0}$ such that $\emptyset_{i_0} < b_{i_0} < e_{i_0}$;

consider now the element $\delta = \mathbf{P}_{i \in I} d_i$, with $d_i = a_i$, for $i \neq i_0$, and $d_{i_0} = b_{i_0}$; then we have $\delta < \alpha$, i.e., a contradiction to the assumption "α is an atom".

Consider an arbitrary natural number ν. Since
$$\alpha \leqslant \alpha_\nu = \alpha_{\nu 1} \vee \alpha_{\nu 2} \vee \ldots \vee \alpha_{\nu k_\nu}$$

and α is an atom in $\tilde{\mathfrak{F}}$, there exists exactly one index t_ν with $1 \leqslant t_\nu \leqslant k_\nu$, such that $\alpha \leqslant \alpha_{\nu, t_\nu}$. From this inequality, it follows that:

$$\alpha = \bigwedge_{\nu=1}^{\infty} \alpha_{\nu, t_\nu}.$$

On the other hand

$$\bigwedge_{\nu=1}^{\infty} \alpha_{\nu, t_\nu} \leqslant \bigwedge_{\nu=1}^{\infty} \alpha_\nu = \alpha.$$

Therefore:

$$\alpha = \bigwedge_{\nu=1}^{\infty} \alpha_{\nu, t_\nu}.$$

The sequence α_{ν, t_ν}, $\nu = 1, 2, \ldots$, is decreasing, because of the assumption that, for every $\alpha_{\nu+1, \rho}$, there exists $\alpha_{\nu, \tau}$ with $\alpha_{\nu+1, \rho} \leqslant \alpha_{\nu, \tau}$. Moreover, this sequence can be assumed to be strictly decreasing, since, as it has been proved above, it cannot be eventually constant. We proved thus:

If

$$(\tilde{\mathfrak{F}}, \pi) = \mathbf{P}_{i \in I} (\mathfrak{A}_i, p_i)$$

is a product pr σ-algebra where each factor $(\mathfrak{A}_i, p_i)$ is a pr σ-algebra and $|I| = \aleph_0$, then, if an element $\alpha \in \tilde{\mathfrak{F}}$ is an atom, there exists a strictly decreasing sequence $\alpha_\nu \in \mathfrak{R}$, $\nu = 1, 2, \ldots$, such that

$$\alpha = \bigwedge_{\nu=1}^{\infty} \alpha_\nu.$$

Let now the index set I be enumerated, i.e., $I = \{i_1, i_2, i_3, \ldots\}$ or equivalently $I = \{1, 2, 3, \ldots\}$; then we can prove (see Kappos [8], Nr. 11.5)

Theorem 2.1.

If

$$(\tilde{\mathfrak{F}}, \pi) = \tilde{\mathbf{P}}_{\nu=1}^{\infty} (\mathfrak{A}_\nu, p_\nu)$$

is a product pr σ-algebra, where each factor $(\mathfrak{A}_\nu, p_\nu)$ is a pr σ-algebra and if α is an atom in $\tilde{\mathfrak{F}}$, then there exists a uniquely determined sequence $\alpha_\nu \in \mathfrak{R}$ with

$$\alpha_\nu = \mathbf{P}_{j=1}^{\infty} a_{j\nu},$$

where $a_{j\nu} = e_j$ for all $\nu \neq j$ and the element a_{jj} is an atom in the algebra $\mathfrak{A}_j$, for every $j = 1, 2, \ldots$, such that

$$\alpha = \bigwedge_{\nu=1}^{\infty} \alpha_\nu.$$

We have obviously:

$$a_n = \bigwedge_{\nu=1}^{n} \alpha_\nu = a_{11} \times a_{22} \times \ldots \times a_{nn} \times e_{n+1} \times e_{n+2} \times \ldots$$

where $a_{nn} \in \mathfrak{A}_n$ and a_{nn} an atom in $\mathfrak{A}_n$ for every $n = 1, 2, \ldots$.

We shall denote it as follows:

$$\alpha = o\text{-lim}\, a_n = o\text{-lim} \left(\mathop{\mathbf{P}}_{\nu=1}^{n} a_{\nu\nu} \times \mathop{\mathbf{P}}_{\nu=n+1}^{\infty} e_\nu \right) = \mathop{\mathbf{P}}_{\nu=1} a_{\nu\nu}$$

where $a_{\nu\nu}$ is an atom in $\mathfrak{A}_\nu$ for every $\nu = 1, 2, \ldots$. We will call

$$\mathop{\mathbf{P}}_{\nu=1}^{\infty} a_{\nu\nu}$$

a *product limit element*. We have obviously:

$$\pi(\alpha) = \prod_{\nu=1}^{\infty} p_\nu(a_{\nu\nu}).$$

We can now conversely prove that if a_ν is an atom in the algebra $\mathfrak{A}_\nu$ for every $\nu = 1, 2, \ldots$ and the infinite product

$$\prod_{\nu=1}^{\infty} p_\nu(a_\nu)$$

converges to a real number $\xi \neq 0$, then there exists an atom $\alpha \in \mathfrak{F}$, such that

$$\alpha = \bigwedge_{\nu=1}^{\infty} \alpha_n$$

where

$$\alpha_n = \mathop{\mathbf{P}}_{\nu=1}^{n} a_\nu \times \mathop{\mathbf{P}}_{\nu \geq n+1} e_\nu,$$

i.e., such that

$$\alpha = \mathop{\mathbf{P}}_{\nu=1}^{\infty} a_\nu,$$

and moreover we have $\pi(\alpha) = \xi$. Hence the following theorem holds:

Theorem 2.2.

An element $\alpha \neq 0$ of the Boolean algebra $\tilde{\mathfrak{F}}$, where

$$(\tilde{\mathfrak{F}}, \pi) = \mathop{\tilde{\mathbf{P}}}_{\nu=1}^{\infty} (\mathfrak{A}_\nu, p_\nu)$$

is a product pr σ-algebra with $(\mathfrak{A}_\nu, p_\nu)$ pr σ-algebras, is an atom in $\tilde{\mathfrak{F}}$ if *and only if every algebra* $\mathfrak{A}_\nu$ *possesses an atom* a_ν, $\nu = 1, 2, \ldots$, *such that*

$$\alpha = \mathop{\mathbf{P}}_{\nu=1}^{\infty} a_\nu.$$

Then we have

$$\prod_{\nu=1}^{\infty} p_\nu(a_\nu) = \pi(\alpha).$$

The following theorem can be proved (see author's [8], Nr. 11, Theorem 6):

Theorem 2.3.

Let every Boolean σ*-algebra* $\mathfrak{A}_\nu$ *be atomic and let* $\{a_{\nu 0}, a_{\nu 1}, \ldots\}$ *be the set of all atoms of* $\mathfrak{A}_\nu$ *enumerated so that* $p_\nu(a_{\nu 0}) \geqslant p_\nu(a_{\nu\rho})$, $\rho = 1, 2, \ldots$, *for every* $\nu = 1, 2, \ldots$. *Then the following statements are equivalent*:

(I) *The product pr* σ*-algebra*

$$(\tilde{\mathfrak{F}}, \pi) = \mathop{\tilde{\mathbf{P}}}_{\nu=1}^{\infty} (\mathfrak{A}_\nu, p_\nu)$$

is atomic, i.e., $\tilde{\mathfrak{F}}$ *is isomorphic to the Boolean algebra of all subsets of the set* $\mathbf{N} = \{1, 2, \ldots\}$ *of all natural numbers.*

(II) *The double series*

$$\sum_{\nu=1}^{\infty} \sum_{\rho \geqslant 1} p_\nu(a_{\nu\rho})$$

converges to a real number.

Corollary 2.1.

A product pr σ*-algebra*

$$(\tilde{\mathfrak{F}}, \pi) = \mathop{\tilde{\mathbf{P}}}_{\nu=1}^{\infty} (\mathfrak{A}_\nu, p_\nu)$$

where every $\mathfrak{A}_\nu$ *is an atomic Boolean* σ*-algebra with a set* $\{a_{\nu\rho}, \rho \geqslant 0\}$ *of atoms is either atomic or atomless (and in the latter case, isometric to the Lebesgue pr* σ*-algebra* $(\mathfrak{J}, \mu)$*) according as the double series*

$$\sum_{\nu=1}^{\infty} \sum_{\rho \geqslant 1} p_\nu(a_{\nu\rho})$$

converges to a real number or to $+\infty$.

3. CLASSIFICATION OF PROBABILITY σ-ALGEBRAS

3.1. In Section 4, Chapter II, the p-separable pr σ-algebras are classified; we proved that there exist three types of separable pr σ-algebras, namely type III, *atomic*, type I, *continuous*, i.e. isometric to the Lebesgue pr σ-algebra, and type II, *mixed*. In the latter type, there exist atoms but the sum of the probabilities of all atoms is < 1; there exists, therefore, a continuous part in the pr algebra. The theory of product pr algebras allows us now to state a general classification of all pr σ-algebras, which was first investigated by Dorothy Maharam [1].

3.2. A Boolean σ-algebra $\mathfrak{B}$ is said to be *homogeneous*, if and only if, for every $b \in \mathfrak{B}$, $b \neq \emptyset$, the character of the principal ideal

$$b\mathfrak{B} = \{x \in \mathfrak{B}\colon\ x \leqslant b\},$$

considered as a Boolean σ-algebra with unit the element b, is equal to the character of $\mathfrak{B}$, i.e.,

$$c_{\mathfrak{B}} = c_{b\mathfrak{B}} \quad \text{for every} \quad b \in \mathfrak{B}, \quad b \neq \emptyset.$$

An example of a homogeneous Boolean σ-algebra is the Boolean σ-algebra $\mathfrak{J}$ of the linear Lebesgue pr σ-algebra $(\mathfrak{J}, \mu)$. The character of $\mathfrak{J}$ is $\aleph_0$. The linear Lebesgue pr σ-algebra $(\mathfrak{J}, \mu)$ possesses the following property:

for every $b \neq \emptyset$, $b \in \mathfrak{J}$, the pr σ-algebra $(b\mathfrak{J}, \mu/\mu(b))$ is isometric to the pr σ-algebra $(\mathfrak{J}, \mu)$.

It is easy to prove that the Boolean σ-algebra $\mathfrak{B}_\gamma$ of the product pr σ-algebra

$$(\mathfrak{B}_\gamma, \pi_\gamma) = \mathop{\widetilde{\mathbf{P}}}_{0 \leqslant \xi < \omega_\gamma} (\mathfrak{J}_\xi, \mu_\xi)$$

where ξ and γ are ordinal numbers and $(\mathfrak{J}_\xi, \mu_\xi) = (\mathfrak{J}, \mu)$, i.e., equal to the linear Lebesgue pr σ-algebra, for every $\xi \in [0, \omega_\gamma)$, is also homogeneous and of character $\aleph_\gamma$ = the power of the set $W(\omega_\gamma)$ of all ordinal numbers $< \omega_\gamma$. The pr σ-algebra $(\mathfrak{B}_\gamma, \pi_\gamma)$ possesses also the property:

for every $b \in \mathfrak{B}$, $b \neq \emptyset$, the pr σ-algebra $(b\mathfrak{B}, \pi_\gamma/\pi_\gamma(b))$ is isometric to $(\mathfrak{B}_\gamma, \pi_\gamma)$.

We shall say a pr σ-algebra $(\mathfrak{B}, p)$ is homogeneous if and only if the Boolean σ-algebra $\mathfrak{B}$ is homogeneous. Moreover, if the character of $\mathfrak{B}$ is $\aleph_\beta$, where β is an ordinal number $\geqslant 0$, then we shall say the pr σ-algebra $(\mathfrak{B}, p)$ is *β-homogeneous*. According to this definition the pr σ-algebra $(\mathfrak{J}, \mu)$ is 0-homogeneous and for every ordinal number $\gamma \geqslant 0$ the product pr σ-algebra $(\mathfrak{B}_\gamma, \pi_\gamma)$ is γ-homogeneous. Note that the product pr σ-algebra

$$(\mathfrak{B}_0, \pi_0) = \widetilde{\mathbf{P}}_{0 \leqslant \xi < \omega_0} (\mathfrak{J}_\xi, \mu_\xi)$$

is π_0-separable (with character $\aleph_0$) without atoms, hence isometric to $(\mathfrak{J}, \mu)$, i.e., 0-homogeneous. Moreover, for every ordinal number η with $\omega_\beta \leqslant \eta < \omega_{\beta+1}$ the product pr σ-algebra

$$(\widetilde{\mathfrak{F}}_\eta, \pi) = \widetilde{\mathbf{P}}_{0 \leqslant \xi < \eta} (\mathfrak{J}_\xi, \mu_\xi)$$

is isometric to the product pr σ-algebra $(\mathfrak{B}_\beta, \pi_\beta)$, i.e., β-homogeneous. Dorothy Maharam has proved [1] that every homogeneous pr σ-algebra of character $\aleph_\beta$, $\beta \geqslant 0$, is isometric to the product pr σ-algebra $(\mathfrak{B}_\beta, \pi_\beta)$. Hence the following definition is equivalent to the definition of homogeneous pr σ-algebras given previously: a pr σ-algebra $(\mathfrak{B}, p)$ of character $\aleph_\beta$, $\beta \geqslant 0$, is homogeneous if and only if $(\mathfrak{B}, p)$ is isometric to the product pr σ-algebra $(\mathfrak{B}_\beta, \pi_\beta)$.

We shall consider every atomic pr σ-algebra also as homogeneous and that as minus one-homogeneous ((-1)-homogeneous).

Now, it is easy to see that

$$(\mathfrak{B}_1, \pi_1) = \widetilde{\mathbf{P}}_{0 \leqslant \xi < \omega_1} (\mathfrak{J}_\xi, \mu_\xi)$$

satisfies

$$(\mathfrak{B}_1, \pi_1) = \widetilde{\mathbf{P}}_{0 \leqslant \xi < \omega_1} (\mathfrak{B}_{0\xi}, \pi_{0\xi}) = \widetilde{\mathbf{P}}_{0 \leqslant \xi < \omega_1} (\mathfrak{B}_{1\xi}, \pi_{1\xi}),$$

where

$$(\mathfrak{B}_{0\xi}, \pi_{0\xi}) = (\mathfrak{B}_0, \pi_0) \quad \text{and} \quad (\mathfrak{B}_{1\xi}, \pi_{1\xi}) = (\mathfrak{B}_1, \pi_1)$$

for every ξ.

Generally, for every ordinal number $\beta \geqslant 0$, we have

$$(\mathfrak{B}_\beta, \pi_\beta) = \widetilde{\mathbf{P}}_{0 \leqslant \xi < \omega_\beta} (\mathfrak{B}_{\beta\xi}, \pi_{\beta\xi}) \quad \text{with} \quad (\mathfrak{B}_{\beta\xi}, \pi_{\beta\xi}) = (\mathfrak{B}_\beta, \pi_\beta)$$

for every ξ. In general we have

$$(\Gamma) \qquad (\mathfrak{B}_\beta, \pi_\beta) = \widetilde{\mathbf{P}}_{0 \leqslant \xi < \omega_\beta} (\mathfrak{B}_{\gamma\xi}, \pi_{\gamma\xi}),$$

where

$$(\mathfrak{B}_{\gamma\xi}, \pi_{\gamma\xi}) = (\mathfrak{B}_\gamma, \pi_\gamma)$$

for every ξ, γ being any ordinal number with $0 \leqslant \gamma \leqslant \beta$.

According to the previous statement, the following theorem is true:

Theorem 3.1.

Every β-homogeneous pr σ-algebra with $\beta \geqslant 0$ is universal for all γ-homogeneous pr σ-algebras with $-1 \leqslant \gamma \leqslant \beta$, i.e., a γ-homogeneous pr σ-algebra $(\mathfrak{B}_\gamma, \pi_\gamma)$ can be embedded isometrically in the β-homogeneous pr σ-algebra $(\mathfrak{B}_\beta, \pi_\beta)$.

Proof. For $\gamma \geqslant 0$ our assertion follows from (Γ) and from the proposition that every factor $(\mathfrak{B}_{\gamma\xi}, \pi_{\gamma\xi})$ is isometric to a pr σ-subalgebra of the product pr σ-algebra $(\mathfrak{B}_\beta, \pi_\beta)$. For $\gamma = -1$ we have some atomic σ-algebra $(\mathfrak{A}, p)$. Let $\{a_\nu, \nu \geqslant 1\}$ be the set of all the atoms of $(\mathfrak{A}, p)$; then we have

$$\sum_{\nu \geqslant 1} p(a_\nu) = 1.$$

Since now $(\mathfrak{B}_\beta, \pi_\beta)$ is non-atomic, there exists an element $b_1 \in \mathfrak{B}_\beta$ such that $\pi_\beta(b_1) = p(a_1)$ and an element $b_2 \leqslant b_1{}^c$ such that $\pi_\beta(b_2) = p(a_2)$, and, generally, an element

$$b_k \leqslant (b_1 \vee b_2 \vee \ldots \vee b_{k-1})^c = \bigwedge_{\nu=1}^{k-1} b_\nu{}^c$$

such that $\pi_\beta(b_k) = p(a_k)$; then $b_1, b_2, \ldots$ are pairwise disjoint and

$$\sum_{k \geqslant 1} \pi_\beta(b_k) = 1,$$

i.e.,

$$\bigvee_{k \geqslant 1} b_k = e_\beta = \text{unit of } \mathfrak{B}_\beta.$$

The set $\{b_1, b_2, \ldots\} \subseteq \mathfrak{B}_\beta$ σ-generates a Boolean σ-subalgebra $\mathfrak{A}^*$ of $\mathfrak{B}_\beta$ isomorphic to $\mathfrak{A}$, i.e., $(\mathfrak{A}^*, \pi_\beta)$ is then a pr σ-subalgebra of $(\mathfrak{B}_\beta, \pi_\beta)$ isometric to $(\mathfrak{A}, p)$.

3.3. Let now $(\mathfrak{B}, p)$ be any pr σ-algebra with character $\aleph_\beta \geqslant \aleph_0$. Assume, moreover, the Boolean σ-algebra $\mathfrak{B}$ is atomless. For every $b \in \mathfrak{B}$, $b \neq \emptyset$, the character of the element b is by definition the same as the character of the principal ideal $b\mathfrak{B}$, and is denoted by $c(b)$. Obviously there exists the

$$\min_{b \in \mathfrak{B},\, b \neq \emptyset} c(b) = m(\mathfrak{B})$$

and is equal to $\aleph_{\beta_1}$ for a certain ordinal number $\beta_1 \geqslant 0$. Moreover there exist elements $x \in \mathfrak{B}$, $x \neq \emptyset$, with $c(x) = m(\mathfrak{B})$. Let now

$$a_1 = \bigvee \{x \in \mathfrak{B};\ x \neq \emptyset \text{ and } c(x) = m(\mathfrak{B}) = \aleph_{\beta_1}\};$$

then $c(a_1) = \aleph_{\beta_1}$ and the pr σ-algebra $(a_1 \mathfrak{B}, p/p(a_1))$ is obviously β_1-homogeneous, because for every $x \leqslant a_1$, $x \neq \emptyset$, we have

$$\aleph_{\beta_1} = c(a_1) = c(x) = c_{x\mathfrak{B}}.$$

If $a_1 = e$ then our pr σ-algebra is classified as a β_1-homogeneous.

Suppose $a_1 \neq e$ and set $b_2 = e - a_1$; then $m(b_2) = \aleph_{\beta_2}$ with $\beta_2 > \beta_1$ obviously. Let now

$$a_2 = \bigvee\{x \in b_2 \mathfrak{B};\ x \neq \emptyset \text{ and } c(x) = \aleph_{\beta_2}\};$$

then we have

$$c(a_2) = \aleph_{\beta_2}$$

and the pr σ-algebra $(a_2 \mathfrak{B}, p/p(a_2))$ is obviously β_2-homogeneous. If $a_1 \vee a_2 = e$ then our process is finished and our pr σ-algebra $(\mathfrak{B}, p)$ is decomposed in two homogeneous pr σ-algebras $(a_1 \mathfrak{B}, p/p(a_1))$ and $(a_2 \mathfrak{B}, p/p(a_2))$ such that the smallest Boolean σ-subalgebra of $\mathfrak{B}$ containing the principal ideals $a_1 \mathfrak{B}$ and $a_2 \mathfrak{B}$ is $\mathfrak{B}$ itself.

Suppose $a_1 \vee a_2 \neq e$; then our process may be continued, if we set $b_3 = e - (a_1 \vee a_2)$, until we have $\bigvee a_j = e$; but the number of steps needed to finish this decomposition process is always countable, for the elements $a_1, a_2, \ldots$ are pairwise disjoint and every set of pairwise disjoint elements in $\mathfrak{B}$ is countable. The following theorem is thus proved:

Theorem 3.2.

If a pr σ-algebra $(\mathfrak{B}, p)$ is atomless and with a character $\aleph_\beta \geqslant \aleph_0$, then there exist, and are uniquely determined,

(1) *an ordinal number N with $1 \leqslant N < \omega_1$,*

(2) *ordinal numbers β_ν with $0 \leqslant \beta_\nu < \beta_{\nu+1}$,*

and (3) *pairwise disjoint elements $a_\nu \in \mathfrak{B}$, for all ν with $1 \leqslant \nu < N$, such that*

(α) *we have*
$$e = \bigvee_{1 \leqslant \nu < N} a_\nu,$$

and (β) *for every ν with $1 \leqslant \nu < N$, the pr σ-algebra $(a_\nu \mathfrak{B}, p/p(a_\nu))$ is β_ν-homogeneous.*

3.4. Assume now the pr σ-algebra $(\mathfrak{B}, p)$ of Section 3.3 possesses atoms, and let $\{d_1, d_2, \ldots\}$ be the set of all atoms in $\mathfrak{B}$ with $p(\bigvee d_\nu) \leqslant 1$. We set $a_1 = d_1 \vee d_2 \vee \ldots$ and we consider the pr σ-algebra $(a_1 \mathfrak{B}, p/p(a_1))$, which is atomic, i.e., (-1)-homogeneous. If $a_1 = e$, then $(\mathfrak{B}, p)$ is classified as a (-1)-homogeneous pr σ-algebra. Suppose $a_1 \neq e$ and set $b_2 = e - a_1$; then the pr σ-algebra $(b_2 \mathfrak{B}, p/p(b_2))$ is atomless and with character $\aleph_\beta \geqslant \aleph_0$. We can therefore apply the decomposition process of Section 3.3 for this pr σ-algebra $(b_2 \mathfrak{B}, p/p(b_2))$. In this way a decomposition for the pr σ-algebra $(\mathfrak{B}, p)$ can be obtained in which the first member is a (-1)-homogeneous pr σ-algebra. Theorem 3.2 can therefore be stated in its general form as follows:

Theorem 3.3. (of D. Maharam).

Let $(\mathfrak{B}, p)$ be any pr σ-algebra with character $\aleph_\beta \geqslant \aleph_0$; then there exist, and are uniquely determined,

(1) *an ordinal number N with $1 \leqslant N \leqslant \omega_1$,*

(2) *ordinal numbers β_ν with $-1 \leqslant \beta_\nu < \beta_{\nu+1}$, for every ν with $1 \leqslant \nu < \nu+1 < N$,*

and (3) *pairwise disjoint elements a_ν, for every ν with $1 \leqslant \nu < N$, such that*

$$e = \bigvee_{1 \leqslant \nu < N} a_\nu$$

and every principal ideal $a_\nu \mathfrak{B}$, considered as a Boolean algebra with a_ν as a unit, is β_ν-homogeneous.

Remark. Obviously, (α) the set-theoretical union $\bigcup_{1 \leqslant \nu < N} a_\nu \mathfrak{B}$ σ-generates the Boolean algebra $\mathfrak{B}$; and (β) $\sup_{1 \leqslant \nu < N} \beta_\nu = \beta$.

We shall write, according to Theorem 3.3,

$$(\mathfrak{B}, p) = \left(a_1 \mathfrak{B}, \frac{p}{p(a_1)}\right) \otimes \left(a_2 \mathfrak{B}, \frac{p}{p(a_2)}\right) \otimes \ldots$$

and say that the pr σ-algebra $(\mathfrak{B}, p)$ is decomposed into the homogeneous pr σ-algebras $(a_\nu \mathfrak{B}, p/p(a_\nu))$, $1 \leqslant \nu < N$, which classify uniquely the pr σ-algebra $(\mathfrak{B}, p)$.

The following theorem can now be proved:

Theorem 3.4.

The β-homogeneous pr σ-algebra $(\mathfrak{B}_\beta, \pi_\beta)$ is universal for every pr σ-algebra $(\mathfrak{B}, p)$ with a character $c_{\mathfrak{B}} \leqslant \aleph_\beta$, i.e., every pr σ-algebra with a character $c_{\mathfrak{B}} \leqslant \aleph_\beta$ can be embedded isometrically in the β-homogeneous pr σ-algebra $(\mathfrak{B}_\beta, \pi_\beta)$.

Proof. Let $(a_\nu \mathfrak{B}, p/p(a_\nu))$ be the β_ν-homogeneous pr σ-algebras with $-1 \leqslant \beta_\nu < \beta_{\nu+1}$, $1 \leqslant \nu < \nu+1 < N$, into which $(\mathfrak{B}, p)$ is decomposed according to Theorem 3.3. Then there exists a decomposition

$$e_\beta = \bigvee_{1 \leqslant \nu < N} b_\nu$$

of the unit $e_\beta \in \mathfrak{B}_\beta$ in pairwise disjoint elements $b_1, b_2, \ldots$ such that we have:

$$\pi_\beta(b_\nu) = p(a_\nu), \qquad 1 \leqslant \nu < N.$$

We consider now the pr σ-algebra $(b_\nu \mathfrak{B}_\beta, \pi_\beta/\pi_\beta(b_\nu))$, which is β-homogeneous. Since $\max_{1 \leqslant \nu < N} \beta_\nu \leqslant \beta$, we have $\beta_\nu \leqslant \beta$, i.e., the β_ν-homogeneous pr σ-algebra $(a_\nu \mathfrak{B}, p/p(a_\nu))$ can be embedded isometrically into the β-homogeneous pr σ-algebra $(b_\nu \mathfrak{B}_\beta, \pi_\beta/\pi_\beta(b_\nu))$, $1 \leqslant \nu < N$. Let now $\mathfrak{A}_\nu$ be the isometric image of $a_\nu \mathfrak{B}$ into $b_\nu \mathfrak{B}_\beta$ by this isometric mapping of $a_\nu \mathfrak{B}$ into $b_\nu \mathfrak{B}_\beta$; then there exists a smallest Boolean σ-subalgebra $\mathfrak{A}$ of $\mathfrak{B}_\beta$ containing all images $\mathfrak{A}_\nu$, $1 \leqslant \nu < N$. The pr σ-subalgebra $(\mathfrak{A}, \pi_\beta)$ of $(\mathfrak{B}_\beta, \pi_\beta)$ is now isometric to $(\mathfrak{B}, p)$. Our theorem is proved.

4. REPRESENTATION OF PR σ-ALGEBRAS BY PROBABILITY SPACES

4.1. We proved in Section 3.4, Chapter II, Corollary 3.1, that, if a pr σ-algebra $(\mathfrak{B}, p)$ is separable, i.e., with a character $c_\mathfrak{B} \leqslant \aleph_0$, then there exists a σ-field $\mathfrak{K}$ of subsets of the set $\Omega \equiv \{\xi \in R\colon\ 0 \leqslant \xi < 1\}$, which is a σ-subfield of the linear Lebesgue σ-field such that the pr space $(\Omega, \mathfrak{K}, m)$, where m is the Lebesgue measure, is a set-theoretical representation of the pr σ-algebra $(\mathfrak{B}, p)$; i.e., if $\mathfrak{N}$ is the σ-ideal of all $X \in \mathfrak{K}$ with $m(X) = 0$, then the quotient Boolean σ-algebra $\mathfrak{K}/\mathfrak{N}(\mathfrak{K} \bmod \mathfrak{N})$ is isometric to $\mathfrak{B}$, and, if we define $\tilde{p}(A/\mathfrak{N}) = m(A)$, for every class $A/\mathfrak{N}$ in $\mathfrak{K}/\mathfrak{N}$ with $A \in A/\mathfrak{N}$, then $(\mathfrak{K}/\mathfrak{N}, \tilde{p})$ is isometric to $(\mathfrak{B}, p)$. Moreover, we can choose the σ-field $\mathfrak{K}$ so that it is m-complete, i.e., if X is any subset of Ω with $X \subseteq A$, where $A \in \mathfrak{N}$, then $X \in \mathfrak{N}$. We shall now prove that, for any pr σ-algebra, with arbitrary character $c_\mathfrak{B} \geqslant \aleph_0$, there exists a set-theoretical representation by a pr space.

4.2. We shall first introduce some concepts from the classical measure theory.

Let

$$\Omega_\beta = \mathbf{P}_{0 \leqslant \xi < \omega_\beta} E_\xi$$

be the set-theoretical product, where every factor

$$E_\xi = \{\rho \in R\colon\ 0 \leqslant \rho \leqslant 1\},$$

i.e., the set Ω_β of all families: ρ_ξ, $0 \leqslant \xi < \omega_\beta$, where ρ_ξ a real number with $0 \leqslant \rho_\xi \leqslant 1$. A subset $Z \subseteq \Omega_\beta$ is called a Lebesgue *cylinder*, if and only if there exists a representation of Z as a cartesian product,

$$Z = \mathop{\mathbf{P}}_{0 \leqslant \xi < \omega_\beta} Z_\xi$$

where $Z_\xi = E_\xi$ for all ξ except a finite set $\xi_1, \xi_2, \ldots, \xi_n$, for which Z_{ξ_ν} is a Lebesgue-measurable subset of $E = \{\rho \in R: 0 \leqslant \rho \leqslant 1\}$, $\nu = 1, 2, \ldots, n$. Let now $\mathfrak{S}$ be the set of all Lebesgue cylinders, $\mathfrak{A}_\beta$ be the smallest subfield of subsets of Ω_β containing $\mathfrak{S}$; then we can define a measure (the so-called product measure) m_β on $\mathfrak{A}_\beta$ as follows:

$$m_\beta(Z) = m(Z_{\xi_1})\, m(Z_{\xi_2}) \ldots m(Z_{\xi_n}) \quad \text{for every } Z \in \mathfrak{S},$$

where $m(Z_{\xi_j})$ is the Lebesgue measure of Z_{ξ_j}, $j = 1, 2, \ldots, n$.

m_β can be extended in a well-known way to $\mathfrak{A}_\beta$ and to $\mathfrak{K}_\beta$, i.e., to the smallest σ-field $\mathfrak{K}_\beta$ of subsets of Ω_β containing $\mathfrak{A}_\beta$. $\mathbf{B}_\beta$ is then the smallest σ-field of subsets of Ω_β containing $\mathfrak{K}_\beta$ which satisfies the condition:

If $X \subseteq \Omega_\beta$ and $X \subseteq A$, where $A \in \mathbf{B}_\beta$ with $m_\beta(A) = 0$, then $X \in \mathbf{B}_\beta$.

Note that $(\Omega_\beta, \mathbf{B}_\beta, m_\beta)$ is a pr space and obviously the β-homogeneous pr σ-algebra $(\mathfrak{B}_\beta, \pi_\beta)$ can be represented set-theoretically by this pr space $(\Omega_\beta, \mathbf{B}_\beta, m_\beta)$.

4.3. Let now $(\mathfrak{B}, p)$ be any pr σ-algebra with character $c_\mathfrak{B} = \aleph_\beta$. Then $(\mathfrak{B}, p)$ can be embedded isometrically into the β-homogeneous pr σ-algebra $(\mathfrak{B}_\beta, \pi_\beta)$. There exists therefore a σ-subfield $\mathfrak{K}$ of $\mathbf{B}_\beta$ such that the pr subspace $(\Omega_\beta, \mathfrak{K}, m_\beta)$ of the space $(\Omega_\beta, \mathbf{B}_\beta, m_\beta)$ is a set-theoretical representation of the image of $(\mathfrak{B}, p)$ in $(\mathfrak{B}_\beta, \pi_\beta)$, hence also of $(\mathfrak{B}, p)$ itself. Moreover, we can choose $\mathfrak{K}$ so that the condition:

(K) If $X \subseteq \Omega_\beta$ and $X \subseteq A$, where $A \in \mathfrak{K}$ with $m_\beta(A) = 0$, then $X \in \mathfrak{K}$, is satisfied. The following theorem is hence true:

Theorem 4.1.

If $(\mathfrak{B}, p)$ is any pr σ-algebra with the character $c_\mathfrak{B} = \aleph_\beta$, then there exists a pr space $(\Omega_\beta, \mathfrak{K}, m_\beta)$, where $\mathfrak{K}$ is a σ-subfield of the Lebesgue product σ-field $\mathbf{B}_\beta$, satisfying the condition (K), *which represents $(\mathfrak{B}, p)$ set-theoretically; i.e., if $\mathfrak{N}$ is the σ-ideal of all $A \in \mathfrak{K}$ with $m_\beta(A) = 0$, then the quotient σ-algebra $\mathfrak{K}/\mathfrak{N}$ is isomorphic to $\mathfrak{B}$ and if we define*

$$\tilde{p}(B/\mathfrak{N}) = m_\beta(B)$$

for every $B/\mathfrak{N} \in \mathfrak{K}/\mathfrak{N}$, then the pr σ-algebra $(\mathfrak{K}/\mathfrak{N}, \tilde{p})$ is isometric to $(\mathfrak{B}, p)$.

5. INDEPENDENCE IN PROBABILITY

Related to the concept of product probability algebras is the concept of probability independence. The connection between these concepts will be stated here. Details one can find in the author's [8], Chapters IV and V.

5.1. Let $(\mathfrak{B}, p)$ be any pr algebra and $a_i \in \mathfrak{B}$, $i \in I$, any family of events, $a_i \neq \emptyset$, $i \in I$. Then the events a_i, $i \in I$, are said to be *independent* in $(\mathfrak{B}, p)$ (briefly p-independent), if and only if, for every finite subset $\{i_1, i_2, \ldots, i_n\} \subseteq I$

$$p\left(\bigwedge_{k=1}^{n} a_{i_k}\right) = \prod_{k=1}^{n} p(a_{i_k}). \tag{1}$$

Classes $\mathfrak{A}_i$ of events, $\mathfrak{A}_i \subseteq \mathfrak{B}$, $i \in I$, are said to be p-independent if and only if their events are p-independent, exactly, if for every finite subset $\{i_1, i_2, \ldots, i_n\} \subseteq I$ and every arbitrarily selected $a_{i_k} \in \mathfrak{A}_{i_k}$, $a_{i_k} \neq \emptyset$, $k = 1, 2, \ldots, n$, we have (1).

Clearly, if $J \subseteq I$ and $\mathfrak{A}_j' \subseteq \mathfrak{A}_j$, for every $j \in J$, then the p-independence of $\mathfrak{A}_i$, $i \in I$, implies the p-independence of $\mathfrak{A}_j'$, $j \in J$. In other words: *Subclasses of p-independent classes are p-independent.*

5.2. Let now $(\mathfrak{B}, p)$ be a pr algebra and $\mathfrak{A}_i \subseteq \mathfrak{B}$, $i \in I$, be a family of p-independent and $\wedge$-closed classes $\mathfrak{A}_i$, each of which contains an element different from the unit and zero. Let $b(\mathfrak{A}_i)$, resp $\mathfrak{A}$,denotes the smallest Boolean subalgebra of $\mathfrak{B}$ containing $\mathfrak{A}_i$, $i \in I$, respectively, the smallest Boolean subalgebra of $(\mathfrak{B}, p)$ containing the set-theoretical union $\mathfrak{S} = \bigcup_{i \in I} \mathfrak{A}_i$; then the following statements are true (see author's [8] Chapter V, Theorems 3 and 4):

(1) The Boolean subalgebras $b(\mathfrak{A}_i)$, $i \in I$, are p-independent,

(2) The probability algebra $(\mathfrak{A}, p)$ is isometric to the product pr algebra

$$(\mathfrak{F}, \pi) = \mathbf{P}_{i \in I} (b(\mathfrak{A}_i), p_i)$$

with $p_i = p$, $i \in I$.

Particularly, by this isometricity the image of every event of the form $a_{i_1} \wedge a_{i_2} \wedge \ldots \wedge a_{i_n}$ with $\{i_1, i_2, \ldots, i_n\} \subseteq I$ and $a_{i_k} \in b(\mathfrak{A}_{i_k})$, $a_{i_k} \neq \emptyset$, is the product event $\mathbf{P}_{i \in I} x_i$ with $x_i = a_i$, if $i = i_1, i_2, \ldots, i_n$, and $x_i = e_i = e$, if $i \in (I - \{i_1, i_2, \ldots, i_n\})$.

(3) If $(\mathfrak{B}, p)$ is a pr σ-algebra and $b_\sigma(\mathfrak{A}_i)$ denotes the smallest Boolean σ-subalgebra of $\mathfrak{B}$ containing $\mathfrak{A}_i$, $i \in I$, then the Boolean σ-subalgebras $b_\sigma(\mathfrak{A}_i)$, $i \in I$, are also p-independent and the pr σ-subalgebra $(b_\sigma(\mathfrak{A}), p)$, where $b_\sigma(\mathfrak{A})$ is the smallest Boolean σ-subalgebra of $\mathfrak{B}$ containing $\mathfrak{A}$, is isometric to the product pr σ-algebra

$$(\tilde{\tilde{\mathfrak{F}}}, \tilde{\pi}) = \tilde{\mathbf{P}}_{i \in I} (b(\mathfrak{A}_i), p_i)$$

where $p_i = p$, $i \in I$.

5.3. Let

$$(\mathfrak{F}, \pi) = \mathbf{P}_{i \in I} (\mathfrak{B}_i, p_i)$$

be a product pr algebra; then the class of product elements:

$$\mathfrak{F}_j \equiv \left\{ \alpha_j = \mathbf{P}_{i \in I} x_i, \quad \text{where} \quad x_j = a_j \in \mathfrak{B}_j \quad \text{and} \right.$$
$$\left. x_i = e_i, \quad \text{for every} \quad i \in I, \quad \text{with} \quad i \neq j \right\}$$

forms a Boolean subalgebra of $\mathfrak{F}$, which is isomorphic to the Boolean algebra $\mathfrak{B}_j$, $j \in I$. Clearly the Boolean subalgebras $\mathfrak{F}_j$, $j \in I$, are π-independent.

It follows easily now that:

5.3.1. If a pr algebra $(\mathfrak{B}, p)$ or a pr subalgebra $(\mathfrak{B}^*, p)$ of $(\mathfrak{B}, p)$ is isometric to any product pr algebra

$$(\mathfrak{F}, \pi) = \mathbf{P}_{i \in I} (\mathfrak{B}_i, p_i),$$

then and only then there are classes $\mathfrak{A}_i \subseteq \mathfrak{B}$, $i \in I$, which are p-independent.

5.3.2. Every β-homogeneous pr σ-algebra $(\mathfrak{B}, p)$ always possesses classes $\mathfrak{A}_i \subseteq \mathfrak{B}$, $i \in I$, with $|I| \leqslant \aleph_\beta$, which are p-independent. However, it is impossible in this case, or generally if the character of $(\mathfrak{B}, p)$ is $\aleph_\beta$ to exist classes $\mathfrak{A}_i \subseteq \mathfrak{B}$, $i \in I$, with $|I| > \aleph_\beta$, which are p-independent.

IV

STOCHASTIC SPACES

1. EXPERIMENTS (TRIALS) IN PROBABILITY ALGEBRAS

1.1. Let $(\mathfrak{B}, p)$ be a pr σ-algebra. An *experiment* (or *trial*) $\mathbf{a}$ in $\mathfrak{B}$ is a class $\mathbf{a}$ of pairwise disjoint and different from $\emptyset$ events of $\mathfrak{B}$, whose join $\bigvee_{i \in I} a_i$ is the sure event (unit) e. The events a_i, $i \in I$, are said to be the *outcomes* of the trial $\mathbf{a}$. Because of the strict positivity of the pr p the class $\mathbf{a}$ defining an experiment is at most countable. Hence, we will denote an experiment $\mathbf{a}$ in $\mathfrak{B}$, by its outcomes, as follows:

$$\mathbf{a} = \{a_1, a_2, \ldots, a_k\} \quad \text{or} \quad \mathbf{a} = \{a_1, a_2, \ldots\},$$

if $\mathbf{a}$ is a finite or an infinite experiment respectively. We shall denote by $\mathscr{T}(\mathfrak{B})$ the set of all experiments and by $\mathscr{T}_0(\mathfrak{B})$ the set of all finite experiments in $\mathfrak{B}$. Then we have:

$$\mathscr{T}_0(\mathfrak{B}) \subseteq \mathscr{T}(\mathfrak{B}).$$

1.2. In $\mathscr{T}(\mathfrak{B})$ we can define an order relation, denoted by $[$, as follows:

$\mathbf{a} \,[\, \mathbf{b}$, read $\mathbf{a}$ *finer* than $\mathbf{b}$, if and only if each outcome $a_i \in \mathbf{a}$ is contained in some outcome $b_j \in \mathbf{b}$. The relation $[$ is reflexive, antisymmetric and transitive. Moreover, there always exists the infimum $\mathbf{a} \wedge \mathbf{b}$ of two experiments $\mathbf{a}$ and $\mathbf{b}$, namely the experiment:

$$\mathbf{a} \wedge \mathbf{b} = \{a_i \wedge b_j \neq \emptyset,\ a_i \in \mathbf{a},\ b_j \in \mathbf{b}\}.$$

The po-set $(\mathscr{T}(\mathfrak{B}), [)$ so defined satisfies the Moore–Smith condition, i.e., given a pair $\mathbf{a}$, $\mathbf{b}$ of experiments, there always exists an experiment $\mathbf{d}$ finer than each of the given experiments.

1.3. The *norm* of an experiment $\mathbf{a} = \{a_j,\ j \geqslant 1\} \in \mathscr{T}(\mathfrak{B})$ is denoted by $v(\mathbf{a})$ and is defined as the supremum of the probabilities $p(a_j)\ j \geqslant 1$, i.e., $v(\mathbf{a}) = \sup_{j \geqslant 1} p(a_j)$.

It is easy to prove:

Theorem 1.1.

Let $\mathbf{a}_n = \{a_{nj}, j \geqslant 1\} \in \mathscr{T}(\mathfrak{B})$, $n = 1, 2, \ldots$. *Then the infimum* $\bigwedge_{n=1}^{\infty} \mathbf{a}_n$ *exists in* $\mathscr{T}(\mathfrak{B})$ *if and only if the smallest Boolean* σ*-subalgebra of* $\mathfrak{B}$ *containing all the experiments* $\mathbf{a}_n$, $n = 1, 2, \ldots$, *is atomic.*

Theorem 1.2.

Assumptions as in Theorem 1.1 *and, moreover, that the experiments* $\mathbf{a}_n$, $n = 1, 2, 3, \ldots$, *are p-independent, i.e., for every*

$$\{n_1, n_2, \ldots, n_k\} \subseteq \{1, 2, 3, \ldots\}$$

and every $a_{n_i j_i} \in \mathbf{a}_{n_i}$, *we have*

$$p(a_{n_1 j_1} \wedge a_{n_2 j_2} \wedge \ldots \wedge a_{n_k j_k}) = p(a_{n_1 j_1}) \cdot p(a_{n_2 j_2}) \ldots p(a_{n_k j_k}).$$

Then, if the outcomes of $\mathbf{a}_n$ *are enumerated so that* $p(a_{n1}) \geqslant p(a_{nj})$, *for every* $j \geqslant 2$ *and* $n = 1, 2, \ldots$, *the infimum* $\bigwedge_{n=1}^{\infty} \mathbf{a}_n$ *exists in* $\mathscr{T}$ *if and only if the double series* $\sum_{n \geqslant 1} \sum_{j \geqslant 2} p(a_{nj})$ *converges to a positive number* $< +\infty$.

1.4. Let $(\mathfrak{A}, p)$ be a pr subalgebra of the pr σ-algebra $(\mathfrak{B}, p)$ such that $\mathfrak{B}$ is σ-generated by $\mathfrak{A}$. Then $v(\mathbf{a})$, for all $\mathbf{a} \in \mathscr{T}_0(\mathfrak{A})$ (i.e., for all finite experiments in $\mathfrak{A}$), is a net (a directed family) of real numbers $v(\mathbf{a})$ with respect to the order $[$, and, further, isotone and bounded, i.e.,

$$0 \leqslant v(\mathbf{a}_1) \leqslant v(\mathbf{a}_2) \quad \text{if} \quad \mathbf{a}_1 \,[\, \mathbf{a}_2;$$

hence the $\lim_{\mathbf{a} \in \mathscr{T}_0(\mathfrak{A})} v(\mathbf{a})$ exists and is $\geqslant 0$. If this limit is equal to zero, then we can prove that the Boolean σ-algebra $\mathfrak{B}$ is atomless. In fact, assume $\mathfrak{B}$ is not atomless; then an atom x exists in $\mathfrak{B}$ and we have $p(x) = \delta > 0$. Now, let $\mathbf{a} \in \mathscr{T}_0(\mathfrak{A})$; then an outcome a_i exists in $\mathbf{a}$ such that $a_i \geqslant x$, hence $p(a_i) \geqslant \delta$ and $v(\mathbf{a}) \geqslant \delta$ for arbitrary $\mathbf{a} \in \mathscr{T}_0(\mathfrak{A})$. This implies the contradiction $\lim_{\mathbf{a} \in \mathscr{T}_0(\mathfrak{A})} v(\mathbf{a}) \geqslant \delta > 0$.

1.5. We shall say a family of experiments $\mathbf{a}_i \in \mathscr{T}_0(\mathfrak{B})$, $i \in I$, *is p-dense* in $\mathfrak{B}$ if and only if, for every $b \in \mathfrak{B}$ and every positive real number $\varepsilon > 0$, there exists an index $i = i(\varepsilon, b)$ and outcomes $a_{i, k_1}, a_{i, k_2}, \ldots, a_{i, k_p}$ of the corresponding experiment $\mathbf{a}_i$ such that

$$p(b + (a_{i, k_1} \vee a_{i, k_2} \vee \ldots \vee a_{i, k_p})) < \varepsilon.$$

The following theorem is true:

Theorem 1.3.

If $(\mathfrak{B}, p)$ *is an atomless pr* σ*-algebra and* $\mathbf{a}_n = \{a_{nj}, j \geqslant 1\} \in \mathscr{T}_0(\mathfrak{B})$, $n = 1, 2, \ldots$, *a decreasing (i.e.,* $\mathbf{a}_{n+1} [\, \mathbf{a}_n$, $n = 1, 2, \ldots$) *sequence of finite experiments, which is p-dense in* $\mathfrak{B}$, *then* $\lim_{n \to \infty} \nu(\mathbf{a}_n) = 0$.

Proof. Suppose $\lim_{n \to \infty} \nu(\mathbf{a}_n) = \delta > 0$; further let the outcomes of every experiment

$$\mathbf{a}_n = \{a_{n1}, a_{n2}, \ldots, a_{n, \rho_n}\}$$

be enumerated so that

$$p(a_{nj}) \geqslant p(a_{n, j+1}), \qquad j = 1, 2, \ldots, \rho_n - 1;$$

then there exists a greatest index $j_n \leqslant \rho_n$ such that $p(a_{n, j_n}) \geqslant \delta$, i.e., $p(a_{n, j}) \geqslant \delta$, for $j \leqslant j_n$, and $p(a_{nj}) < \delta$, for $j > j_n$. Obviously we have

$$s_n \equiv \bigvee_{j=1}^{j_n} a_{nj} \geqslant \bigvee_{j=1}^{j_{n+1}} a_{n+1, j} \equiv s_{n+1},$$

i.e.,

$$s_n \geqslant s_{n+1},$$

and every $a_{n+1, j}$, $j \leqslant j_{n+1}$ is contained in one $a_{n, i}$ for an $i \leqslant j_n$; hence, there exists at least one sequence: $a_{1, n_1}, a_{2, n_2}, \ldots,$ with $n_k \leqslant j_k$, $k = 1, 2, \ldots,$ such that

$$a_{k, n_k} \geqslant a_{k+1, n_{k+1}}, \qquad k = 1, 2, \ldots;$$

then we have

$$\bigwedge_{k=1}^{\infty} a_{k, n_k} = s \in \mathfrak{B}, \quad \text{but} \quad p(a_{k, n_k}) \geqslant \delta,$$

i.e.,

$$\lim_{k \to \infty} p(a_{k, n_k}) = p(s) \geqslant \delta;$$

hence $s \neq \emptyset$. The Boolean σ-algebra is assumed to be atomless, hence there exists an element $b \in B$ such that $\emptyset < b < s$ and $0 < p(b) < p(s)$. Now we have $b < a_{k, n_k}$ and $b \wedge a_{k, i} = \emptyset$ for every $i \neq n_k$, $k = 1, 2, \ldots$. It follows that if $\varepsilon > 0$ and $\varepsilon < \min \{p(b), p(s) - p(b)\}$, then there exists no experiment $\mathbf{a}_k$ such that there exist outcomes

$a_{k,j_i} \in \mathbf{a}_k$, $i = 1, 2, \ldots, \rho$, with

$$p\left(b + \bigvee_{i=1}^{\rho} a_{k,j_i}\right) < \varepsilon.$$

Since this contradicts the p-density of the sequence $\mathbf{a}_k$, $k = 1, 2, \ldots$, in $\mathfrak{B}$, the proof of the theorem is complete.

2. ELEMENTARY RANDOM VARIABLES (ELEMENTARY STOCHASTIC SPACE)

2.1. Let $(\mathfrak{B}, p)$ be a pr σ-algebra and $\mathbf{a} = \{a_j, j \geqslant 1\}$ be an experiment in $\mathfrak{B}$, i.e., $\mathbf{a} \in \mathscr{T}(\mathfrak{B})$. We shall define an *elementary random variable* (briefly erv) in $\mathfrak{B}$ as a real-valued function X on any experiment $\mathbf{a} \in \mathscr{T}(\mathfrak{B})$, i.e., X is defined as a map:

$$X\colon \mathbf{a} \ni a_j \Rightarrow X(a_j) = \xi_j \in R, \qquad j = 1, 2, \ldots .$$

We shall denote by $\mathscr{E}(\mathfrak{B})$ or briefly by $\mathscr{E}$ the set of all erv's in $\mathfrak{B}$. An erv defined on the experiment $\mathbf{e} = \{e\}$, i.e., $X\colon \mathbf{e} \ni e \Rightarrow X(e) = \xi \in R$ is said to be a *constant rv* and is denoted by its value ξ. An erv defined on an experiment of two outcomes $\{a, a^c\}$ and valued as follows:

$$X\colon \quad \begin{matrix} a \Rightarrow 1 \\ a^c \Rightarrow 0 \end{matrix}$$

is said to be the *indicator* of $a \in \mathfrak{B}$ and is denoted by I_a. We shall denote by $\mathscr{J}(\mathfrak{B})$ or briefly by $\mathscr{J}$ the set of all indicators I_a, $a \in \mathfrak{B}$. The constant rv's 1 and 0 will be considered as the indicators of e and $\varnothing$ respectively, i.e., we consider $I_e = 1$ and $I_\varnothing = 0$. An erv defined on a finite experiment $\mathbf{a} = \{a_1, a_2, \ldots, a_k\}$ is said to be a *simple* rv (briefly srv). We shall denote by $\mathscr{S}(\mathfrak{B})$ or briefly by $\mathscr{S}$ the set of all srv's. We have obviously

$$\mathscr{J}(\mathfrak{B}) \subseteq \mathscr{S}(\mathfrak{B}) \subseteq \mathscr{E}(\mathfrak{B}).$$

2.2. We can define in $\mathscr{E}(\mathfrak{B})$ an order relation " $\leqslant$ " as follows: Let $X \in \mathscr{E}(\mathfrak{B})$ and $Y \in \mathscr{E}(\mathfrak{B})$, i.e.,

$$X\colon \mathbf{a} \ni a_j \Rightarrow X(a_j) = \xi_j \in R, \qquad j \geqslant 1$$

$$Y\colon \mathbf{b} \ni b_i \Rightarrow Y(b_i) = \eta_i \in R, \qquad i \geqslant 1;$$

then we define:

$$X \leqslant Y$$

if and only if, for every (j, i) with $a_j \wedge b_i \neq \emptyset$, we have

$$\xi_j \leqslant \eta_i,$$

The relation " $\leqslant$ " so defined is reflexive and transitive, and if we define equality in $\mathscr{E}(\mathfrak{B})$ by: $X = Y$, if and only if, for every (j, i) with $a_j \wedge b_i \neq \emptyset$, we have $\xi_j = \eta_i$, then the relation " $\leqslant$ " is antisymmetric, i.e., with respect to this relation, $\mathscr{E}(\mathfrak{B})$ is a *po*-set. It is easy to prove that $X \wedge Y$ and $X \vee Y$ (i.e., the infimum and supremum of two elements) always exist in $\mathscr{E}(\mathfrak{B})$ and we have

$$X \wedge Y: \ \mathbf{a} \wedge \mathbf{b} \ni a_j \wedge b_i \Rightarrow \min\{X(a_j), \ Y(b_i)\}$$

for every (j, i) with $a_j \wedge \mathbf{b}_i \neq \emptyset$,

$$X \vee Y: \ \mathbf{a} \wedge \mathbf{b} \ni a_j \wedge b_i \Rightarrow \max\{X(a_j), \ Y(b_i)\}$$

for every (j, i) with $a_j \wedge b_i \neq \emptyset$. Hence, $\mathscr{E}(\mathfrak{B})$ is a lattice and, moreover, we can prove that $\mathscr{E}(\mathfrak{B})$ is a distributive lattice. The map

$$\mathfrak{B} \ni b \Rightarrow I_b \in \mathscr{I}(\mathfrak{B}) \tag{1}$$

is a Boolean (i.e., with respect to $\vee$, $\wedge$ and complement) isomorphism of $\mathfrak{B}$ onto $\mathscr{I}(\mathfrak{B})$. More specifically, we have

$$I_a \wedge I_b = I_{a \wedge b}, \quad I_a \vee I_b = I_{a \vee b}$$

$$I_{a^c} = 1 - I_a = I_e - I_a; \quad \text{i.e.,} \quad I_a \wedge I_{a^c} = 0$$

and

$$I_a \vee I_{a^c} = 1.$$

The Boolean σ-algebra $\mathfrak{B}$ is embedded by the map (1) completely regularly (invariantly) in $\mathscr{E}(\mathfrak{B})$. The image $\mathscr{I}(\mathfrak{B})$ of $\mathfrak{B}$ in $\mathscr{E}(\mathfrak{B})$ by the isomorphism (1) is also said to be the *Boolean kernel* of the lattice $\mathscr{E}(\mathfrak{B})$.

2.3. We can define in $\mathscr{E}(\mathfrak{B})$ the following operations:

Addition $\quad X + Y: \ \mathbf{a} \wedge \mathbf{b} \ni a_j \wedge b_i \Rightarrow X(a_j) + Y(b_i),$

for every $a_j \wedge b_i \neq \emptyset, \quad j \geqslant 1, \ i \geqslant 1.$

Multiplication of X by a real number λ:

$$\lambda X: \quad \mathbf{a} \ni a_j \Rightarrow \lambda X(a_j), \qquad j \geqslant 1.$$

Multiplication $\quad XY: \ \mathbf{a} \wedge \mathbf{b} \ni a_j \wedge b_i \Rightarrow X(a_j) Y(b_i),$

for every $a_j \wedge b_i \neq \emptyset, \quad j \geqslant 1, \quad i \geqslant 1.$

It is easy to prove that the following properties hold in $\mathscr{E}(\mathfrak{B})$:

(1) $X + Y = Y + X$.

(2) $(X+Y)+Z = X+(Y+Z)$.

(3) $\lambda(X+Y) = \lambda X+\lambda Y$.

(4) $(\lambda+\mu) X = \lambda X+\mu X$.

(5) $\lambda(\mu X) = (\lambda\mu) X$.

(6) There exists an element $I_{\emptyset} = 0 \in \mathscr{E}(\mathfrak{B})$, such that $X+I_{\emptyset} = X$ for every $X \in \mathscr{E}(\mathfrak{B})$.

(7) $1.X = X$ for every $X \in \mathscr{E}(\mathfrak{B})$.

(8) For every $X \in \mathscr{E}(\mathfrak{B})$, ther exists an element $-X \in \mathscr{E}(\mathfrak{B})$ such that $X+(-X) = I_{\emptyset}$.

Hence $\mathscr{E}(\mathfrak{B})$ is, with respect to the operation $X+Y$ and λX, $X \in \mathscr{E}(\mathfrak{B})$, $Y \in \mathscr{E}(\mathfrak{B})$, $\lambda \in R$, a linear (vector) space.

We note some well-known, simple consequences of the previous statements (see R. Christescu [1] Chapter II):

(α) $0.X = I_{\emptyset} = 0$.

(β) $(-1).X = -X$.

(γ) If $\lambda \neq 0$ and $\lambda X = 0$, then $X = 0$.

(δ) If $\lambda \neq 0$ and $\lambda X = \lambda Y$, then $X = Y$.

(ε) If $X \neq 0$ and $\lambda X = 0$, then $\lambda = 0$.

The order relation " $\leqslant$ ", with respect to which $\mathscr{E}(\mathfrak{B})$ is a lattice, satisfies the conditions:

(O_1) If $X \leqslant Y$, then $X+Z \leqslant Y+Z$.

(O_2) If $X \leqslant Y$ and λ a positive real number $\geqslant 0$, then $\lambda X \leqslant \lambda Y$.

Hence $\mathscr{E}(\mathfrak{B})$ is an *ordered vector space* and specially a *vector lattice*. We note some simple consequences of this vector lattice structure of $\mathscr{E}(\mathfrak{B})$:

(O_3) $X \leqslant Y$ if and only if $Y-X \geqslant 0$.

(O_4) If $X \geqslant 0$ and $Y \geqslant 0$, then $X+Y \geqslant 0$.

(O_5) If $X \geqslant 0$ and $\lambda \geqslant 0$, then $\lambda X \geqslant 0$.

(O_6) If $X \leqslant Y$ and $X' \leqslant Y'$, then $X+X' \leqslant Y+Y'$.

(O_7) If $X \leqslant Y$ and $\lambda \leqslant 0$, then $\lambda X \geqslant \lambda Y$.

(O_8) If $\lambda \leqslant \mu$ and $X \geqslant 0$, then $\lambda X \leqslant \mu X$.

(O_9) $X \wedge Y = -[(-X) \vee (-Y)]$.

(O_{10}) $(X \vee Y)+Z = (X+Z) \vee (Y+Z)$,

$(X \wedge Y)+Z = (X+Z) \wedge (Y+Z)$,

and if $\lambda \geqslant 0$, then $\lambda(X \vee Y) = \lambda X \vee \lambda Y$.

(O_{11}) $(X \vee Y)+(X \wedge Y) = X+Y$.

(O_{12}) If $X \geqslant 0$, $Y \geqslant 0$, and $Z \geqslant 0$, then $(X+Y) \wedge Z \leqslant X \wedge Z + Y \wedge Z$.

(O_{13}) If there exists $\bigwedge_{i \in I} X_i$, then there exists $\bigvee_{i \in I} (-X_i)$, and we have

$$\bigwedge_{i \in I} X_i = -\bigvee_{i \in I} (-X_i),$$

and dually.

(O_{14}) If there exist $\bigvee_{i \in I} X_i$ and $\bigvee_{j \in J} Y_j$, then there exists

$$\bigvee_{(i,j) \in I \times J} (X_i + Y_j),$$

and we have

$$\bigvee_{(i,j) \in I \times J} (X_i + Y_j) = \bigvee_{i \in I} X_i + \bigvee_{j \in J} Y_j,$$

and dually.

(O_{15}) If there exists $\bigvee_{i \in I} X_i$, and $\lambda > 0$ (or $\lambda < 0$), then there exists

$$\bigvee_{i \in I} (\lambda X_i) \quad (\text{or} \bigwedge_{i \in I} (\lambda X_i)),$$

and we have

$$\bigvee_{i \in I} (\lambda X_i) = \lambda \bigvee_{i \in I} X_i \quad (\text{or} \bigwedge_{i \in I} (\lambda X_i) = \lambda \bigvee_{i \in I} X_i),$$

and dually.

(O_{16}) For every $X \neq 0$, the sequence nX, $n = 1, 2, \ldots$, is not bounded.

2.4. The constant random variables are identified in $\mathscr{E}(\mathfrak{B})$ with their real values on e. The set of all constant rv's is isomorphic to the ordered field R of real numbers, i.e., R may be considered as a completely regular (invariant) vector sublattice of $\mathscr{E}(\mathfrak{B})$. It is easy to show that the Boolean kernel $\mathscr{J}(\mathfrak{B})$ of $\mathscr{E}(\mathfrak{B})$ generates the vector sublattice $\mathscr{S}(\mathfrak{B})$ of $\mathscr{E}(\mathfrak{B})$, i.e., $\mathscr{S}(\mathfrak{B})$ is the smallest vector sublattice of $\mathscr{E}(\mathfrak{B})$ containing $\mathscr{J}(\mathfrak{B})$. The vector lattice $\mathscr{E}(\mathfrak{B})$ is, with respect to the addition $X+Y$, the multiplication XY, and the multiplication λX, $X \in \mathscr{E}(\mathfrak{B})$, $Y \in \mathscr{E}(\mathfrak{B})$, $\lambda \in R$, a commutative (real) algebra, with unit $1 = I_e$. In fact, we have $XY = YX$, $X(YZ) = (XY)Z$, $(X+Y)Z = XZ+YZ$, $(\lambda X)\, Y = \lambda(XY)$, and $XI_e = I_e X = X$ for every $X \in \mathscr{E}(\mathfrak{B})$.

Moreover, $\mathscr{E}(\mathfrak{B})$ possesses the so-called *Freudenthal property*, i.e.,

(F). There exists an element $I_e = 1$ (so-called *weak unit*) such that $X \wedge I_e > 0$ for every $X > 0$.

Note that I_e is a *strong unit* for the vector sublattice $\mathscr{S}(\mathfrak{B})$, i.e., if $X \in \mathscr{S}(\mathfrak{B})$, then there exists a positive integer n such that $nI_e > X$. $\mathscr{E}(\mathfrak{B})$ does not possess a strong unit in general.

2.4.1. We define *the positive part of* X : $X^+ \underset{\text{def}}{=} X \vee 0$

the negative part of X : $X^- \underset{\text{def}}{=} (-X) \vee 0$

the modulus (*absolute value*) of X : $|X| = X^+ + X^-$.

(1) We have $X^+ \geqslant 0$, $X^- \geqslant 0$, $|X| \geqslant 0$, and $|X| \geqslant X^+ \geqslant X$.

(2) $X = X^+ - X^-$.

(3) $X^+ \wedge X^- = 0$.

(4) $|X| = X \vee (-X)$.

(5) The maps $X \Rightarrow X^+$, $X \Rightarrow X^-$ and $X \Rightarrow |X|$ of $\mathscr{E}(\mathfrak{B})$ into itself possess the properties:

(α) $(X+Y)^+ \leqslant X^+ + Y^+$, $(X+Y)^- \leqslant X^- + Y^-$

$$|X+Y| \leqslant |X| + |Y|.$$

(β) If $\lambda \geqslant 0$, then $(\lambda X)^+ = \lambda X^+$, $(\lambda X)^- = \lambda X^-$ $|\lambda X| = \lambda |X|$, and if λ is any real number, then $|\lambda X| = |\lambda|\,|X|$.

(γ) $|XY| = |X|\,|Y|$.

(6) $$X \vee Y = Y + (X-Y)^+, \qquad X \wedge Y = X - (X-Y)^+.$$

Definition. Two elements $X \in \mathscr{E}(\mathfrak{B})$ and $Y \in \mathscr{E}(\mathfrak{B})$ are said to be *disjoint* (or *orthogonal*) if and only if $|X| \wedge |Y| = 0$.

Property (F) implies:

(7) The weak (Freudenthal) unit I_e of $\mathscr{E}(\mathfrak{B})$ is orthogonal (disjoint) only to the element 0.

Exercise. Equivalent to the orthogonality condition $|X| \wedge |Y| = 0$ are the conditions

$$|X| \vee |Y| = |X| + |Y|, \quad \text{and} \quad |X| = \{|X| - |Y|\}^+.$$

2.4.2. We note that $\mathscr{E}(\mathfrak{B})$ satisfies the Archimedean property, i.e.,

(A) If $nX \leqslant Y$, $n = 1, 2, \ldots$, then $X \leqslant 0$,

and is a completely distributive lattice, i.e.,

(D) If there exists $\bigvee_{i \in I} X_i$, then there exists $\bigvee_{i \in I} (X_i \wedge Y)$, and we have:

$$\bigvee_{i \in I} (X_i \wedge Y) = \left(\bigvee_{i \in I} X_i\right) \wedge Y$$

and dually.

(D_0) If $\lambda_i \in R$, $i \in I$, is upper, respectively lower, bounded and $X \in \mathscr{E}(\mathfrak{B})$, with $X \geqslant 0$, then there exists $\bigvee_{i \in I} (\lambda_i X)$, resp $\bigwedge_{i \in I} (\lambda_i X)$, and we have

$$\bigvee_{i \in I} (\lambda_i X) = \left(\bigvee_{i \in I} \lambda_i\right) X, \quad \text{resp} \quad \bigwedge_{i \in I} (\lambda_i X) = \left(\bigwedge_{i \in I} \lambda_i\right) X.$$

In general for every $X \in \mathscr{E}(\mathfrak{B})$ and $\lambda_i \in R$, $i \in I$, we have

$$\bigvee_{i \in I} (\lambda_i X) = \left(\bigvee_{i \in I} \lambda_i\right) X^+ - \left(\bigwedge_{i \in I} \lambda_i\right) X^-, \quad \text{resp,}$$

$$\bigwedge_{i \in I} (\lambda_i X) = \left(\bigwedge_{i \in I} \lambda_i\right) X^+ - \left(\bigvee_{i \in I} \lambda_i\right) X^-.$$

2.5. Let X be a srv, i.e., $X \in \mathscr{S}(\mathfrak{B})$; then there always exist, first, a finite experiment $\mathbf{a} = \{a_1, a_2, \ldots, a_n\}$ and second, pairwise different real numbers $\xi_1, \xi_2, \ldots, \xi_n$ such that

(r) $$X = \xi_1 I_{a_1} + \xi_2 I_{a_2} + \ldots + \xi_n I_{a_n}.$$

This representation of X is uniquely determined and is said to be the *reduced representation* of X *by indicators*.

Consider now an $X \in \mathscr{E}(\mathfrak{B})$ which is not a srv, i.e., $X \notin \mathscr{S}(\mathfrak{B})$; then, obviously, there exist first, an infinite experiment $\mathbf{a} = \{a_1, a_2, \ldots\}$ and second, a countable sequence of pairwise different real numbers $\xi_1, \xi_2, \ldots$, such that

$$X\colon \mathbf{a} \ni a_j \Rightarrow X(a_j) = \xi_j, \qquad j = 1, 2, \ldots.$$

We shall also write in this case

(R) $$X = \sum_{j=1}^{\infty} \xi_j I_{a_j},$$

and shall call (R) the reduced representation of X by indicators; the representation (R) is also unique. We shall justify (R) by a limit process.

In fact, we remark: If $b \in \mathfrak{B}$, then $I_b X$ is an erv, which may be defined as follows:

$$I_b X\colon\ b \wedge a_j \Rightarrow \xi_j, \quad \text{for every } j \text{ with} \quad b \wedge a_j \neq \emptyset,$$
$$b^c \wedge a_j \Rightarrow 0, \quad \text{for every } j \text{ with} \quad b^c \wedge a_j \neq \emptyset.$$

If the number of the j with $b \wedge a_j \neq \emptyset$ is finite, then $I_b X$ is a srv and may be represented as follows:

$$I_b X = \sum \xi_j I_{a_j \wedge b} \quad \text{for all } j \text{ with} \quad b \wedge a_j \neq \emptyset.$$

Now let $s_n = a_1 \vee a_2 \vee \ldots \vee a_n$; we then have $s_n \wedge a_j = a_j$ for all $j \leqslant n$ and $s_n \wedge a_j = \emptyset$ for all $j > n$; hence we have

$$X_n \underset{\text{def}}{=} I_{s_n} X = \xi_1 I_{a_1} + \xi_2 I_{a_2} + \ldots + \xi_n I_{a_n} \in \mathscr{S}(\mathfrak{B}),$$

for every $n = 1, 2, \ldots$; furthermore, we have

$$o\text{-}\lim_{n \to \infty} X_n = X \quad \text{in} \quad \mathscr{E}(\mathfrak{B})\dagger.$$

In fact, we first have

$$(\mathfrak{B})o\text{-}\lim s_n = (\mathfrak{B}) \bigvee_{n=1}^{\infty} s_n = e,$$

hence

$$(\mathscr{E}(\mathfrak{B}))\ o\text{-}\lim I_{s_n} = (\mathscr{E}(\mathfrak{B})) \bigvee_{n=1}^{\infty} I_{s_n} = I_e = 1,$$

because $\mathscr{J}(\mathfrak{B})$ is a completely regular sublattice of $\mathscr{E}(\mathfrak{B})$. Now we have

$$(\mathscr{E}(\mathfrak{B}))\ o\text{-}\lim I_{s_n} X = ((\mathscr{E}(\mathfrak{B}))\ o\text{-}\lim I_{s_n})\, X = I_e X = X.$$

Hence, we have in $\mathscr{E}(\mathfrak{B})$:

$$X = o\text{-}\lim_{n \to \infty} \{\xi_1 I_{a_1} + \xi_2 I_{a_2} + \ldots + \xi_n I_{a_n}\} = \sum_{n=1}^{\infty} \xi_n I_{a_n}.$$

We can, therefore, justify (R) by the order convergence in $\mathscr{E}(\mathfrak{B})$ of the series on the right to the erv X. Thus the following theorem is true:

Theorem 2.1.

Every $X \in \mathscr{E}(\mathfrak{B})$, *where* $(\mathfrak{B}, p)$ *is a pr* σ*-algebra, possesses a uniquely reduced representation by indicators*

(I) $$X = \sum_{n \geqslant 1} \xi_n I_{a_n}, \quad \text{where} \quad \mathbf{a} = \{a_n,\ n \geqslant 1\} \in \mathscr{T}(\mathfrak{B}),$$

---

† About o-convergence and properties of it compare below, Section 3.

and ξ_n, $n \geqslant 1$, *pairwise different real numbers. The right-hand side of* (I) *will be regarded as an o-convergent series* $\sum_{n \geqslant 1} \xi_n I_{a_n}$ *with limit* X *in* $\mathscr{E}(\mathfrak{B})$, *if the experiment* **a** *is infinite.*†

Exercise. Prove: If $b_1, b_2, \ldots$ are any pairwise disjoint events in $\mathfrak{B}$ and $\xi_1, \xi_2, \ldots$ any real numbers, then the series $\sum_{n=1}^{\infty} \xi_n I_{b_n}$ *o*-converges to an erv X in $\mathscr{E}(\mathfrak{B})$, i.e., there exists an element X in $\mathscr{E}(\mathfrak{B})$ such that

$$X = \sum_{n=1}^{\infty} \xi_n I_{b_n}.$$

2.6. Let $X \in \mathscr{E}(\mathfrak{B})$ and let

$$X = \sum_{j \geqslant 1} \xi_j I_{a_j}$$

be the reduced representation of X by indicators; we shall denote by $\mathfrak{B}_X$ the smallest Boolean-subalgebra of $\mathfrak{B}$ containing all $a_1, a_2, \ldots$ and call $(\mathfrak{B}_X, p)$ the *distribution pr σ-algebra* of X. The map

$$R \ni \xi \Rightarrow [X < \xi] \underset{\text{def}}{=} \bigvee_{X(a_j) < \xi} a_j \in \mathfrak{B}$$

may be considered as an order preserving map of the chain $\mathfrak{D}$ of all intervals $(-\infty, \xi)$, $\xi \in R$, into $\mathfrak{B}$. Hence $[X < \xi]$, $-\infty < \xi < +\infty$, is also a chain in the Boolean σ-algebra $\mathfrak{B}$, the so-called *spectral chain* of X in $\mathfrak{B}$ and possesses the following properties:

(1) $\underset{\xi \to -\infty}{o\text{-lim}} [X < \xi] = \emptyset$, $\underset{\xi \to +\infty}{o\text{-lim}} [X < \xi] = e$

(2) If $\xi_\nu \uparrow \xi$, then $\underset{\nu \to \infty}{o\text{-lim}} [X < \xi_\nu] = [X < \xi]$, i.e. $[X < \xi]$ considered as a function of ξ is continuous from the left. We call the real-valued function

(3) $\phi_X(\xi) \underset{\text{def}}{=} p([X < \xi])$, $-\infty < \xi < +\infty$

the *distribution function* of X; obviously ϕ_X is monotone, increasing, continuous from the left, and such that

$$\phi_X(-\infty) = \lim_{\xi \to -\infty} \phi_X(\xi) = 0, \quad \phi_X(+\infty) = \lim_{\xi \to +\infty} \phi_X(\xi) = 1.$$

The chain $\mathfrak{D} \equiv \{(-\infty, \xi), -\infty < \xi + \infty\}$ considered as a subset of the Boolean algebra $\mathfrak{P}(R)$ of all subsets of the real line $R = (-\infty, +\infty)$

† About *o*-convergence and properties of it compare below, Section 3.

σ-generates the Boolean σ-subalgebra (Borel σ-field) $\mathbf{B}$ of all Borel subsets of the real line R. Let now h_X be the map

(4) $\mathfrak{D} \ni (-\infty, \xi) \Rightarrow h_X((-\infty, \xi)) = [X < \xi] \in \mathfrak{B};$

h_X is an order preserving map of $\mathfrak{D}$ into $\mathfrak{B}$. It is easy to prove that this map h_X can be extended to a σ-homomorphism $\hat{h}_X$ of the Borel σ-field $\mathbf{B}$ into the Boolean σ-algebra $\mathfrak{B}$. The image $\hat{h}_X(\mathbf{B})$ of $\mathbf{B}$ in $\mathfrak{B}$ by this σ-homomorphism $\hat{h}_X$ is the Boolean σ-subalgebra $\mathfrak{B}_X$ of $\mathfrak{B}$, which was defined at the beginning of this section. We can define on $\mathbf{B}$ a quasi-probability as follows

$$P_X(K) = p(\hat{h}_X(K)) \quad \text{for every} \quad K \in \mathbf{B};$$

then P_X is σ-additive on $\mathbf{B}$, i.e. $(R, \mathbf{B}, P_X)$ is a Borel pr space, the so-called " distribution pr space " or " *the sample pr space* " of the erv X.

Let
$$\mathfrak{N}_X = \{A \in \mathbf{B}\colon \hat{h}_X(A) = \emptyset\} = \{A \in \mathbf{B}\colon P_X(A) = 0\};$$

then $\mathfrak{N}_X$ is a σ-ideal in $\mathbf{B}$ and the quotient Boolean σ-algebra $\mathbf{B}/\mathfrak{N}_X$ is isomorphic to the Boolean σ-subalgebra $\mathfrak{B}_X$ of $\mathfrak{B}$. We may define a probability π on $\mathbf{B}/\mathfrak{N}_X$ as follows

$$\pi(K/\mathfrak{N}_X) = P_X(K) = p(\hat{h}_X(K)),$$

for every class $K/\mathfrak{N}_X \in \mathbf{B}/\mathfrak{N}_X$; then the pr σ-algebra $(\mathbf{B}/\mathfrak{N}_X, \pi)$ is isometric to the pr σ-subalgebra $(\mathfrak{B}_X, p)$ of $(\mathfrak{B}, p)$.

We remark that the map (4), and, hence, the uniquely determined extension of it to a σ-homomorphism $\hat{h}_X$ of $\mathbf{B}$ into $\mathfrak{B}$, characterizes uniquely the erv X, i.e., $[X < \xi] = [Y < \xi]$ for every $\xi \in R$ (which implies $\hat{h}_X(K) - \hat{h}_Y(K)$ for every $K \in \mathbf{B}$) if and only if $X = Y$.

2.7. In the previous sections, we proved that, to any pr σ-algebra $(\mathfrak{B}, p)$, there corresponds a vector lattice $\mathscr{E}(\mathfrak{B})$ which possesses a sublattice $\mathscr{J}(\mathfrak{B})$ isomorphic to $\mathfrak{B}$, the so-called Boolean kernel of all indicators I_b, $b \in \mathfrak{B}$; $J(\mathfrak{B})$ generates a vector sublattice $\mathscr{S}(\mathfrak{B})$ of the vector lattice $\mathscr{E}(\mathfrak{B})$. We shall call $\mathscr{E}(\mathfrak{B})$ and $\mathscr{S}(\mathfrak{B})$ the *elementary stochastic space* and the *simple stochastic space* over $\mathfrak{B}$, respectively. In the definition of $\mathscr{E}(\mathfrak{B})$, $\mathscr{S}(\mathfrak{B})$ and $\mathscr{J}(\mathfrak{B})$, we did not use the concept of probability. To every $X \in \mathscr{E}(\mathfrak{B})$ there corresponds a uniquely determined representation

$$X = \sum_{j \geqslant 1} \xi_j I_{a_j}$$

by indicators, with ξ_j, $j \geqslant 1$, pairwise different.

A family $X_i \in \mathscr{E}(\mathfrak{B})$, $i \in I$, is called a *stochastic process*. The index set I of a stochastic process is usually the set of all integer numbers,

$$I \equiv \{\ldots, -2, -1, 0, +1, +2, \ldots\},$$

or of all natural numbers, $I \equiv \{0, 1, 2, \ldots\}$, or a time interval,

$$I = \{i \in R\colon \alpha \leqslant i < +\infty\}, \quad \text{or} \quad I = \{i \in R\colon -\infty < i < +\infty\}.$$

In general, any set I or any net (directed system) I of indices may be used to define a stochastic process X_i, $i \in I$. The different kinds of stochastic processes are characterized by the index set I and by other conditions. One is then interested in questions concerning convergence. There are several kinds of convergence in a stochastic space. We shall define below in $\mathscr{E}(\mathfrak{B})$ the more important kinds of convergence, and the elementary stochastic space $\mathscr{E}(\mathfrak{B})$ will be extended to a stochastic space which will be closed for all these convergences.

2.8. *Remark.* Let $(\mathfrak{A}, p)$ be a pr algebra (not necessarily a pr σ-algebra); then the class $\mathscr{S}(\mathfrak{A})$ of all simple random variables in $\mathfrak{A}$ can be defined, and it is a vector lattice. The class $\mathscr{E}(\mathfrak{A})$ of all elementary random variables in $\mathfrak{A}$ can also be considered, i.e., for every infinite experiment $\mathbf{a} = \{a_1, a_2, \ldots\}$ existing in $\mathfrak{A}$, one can define erv's. Furthermore, we notice, if $\mathbf{a}$ and $\mathbf{b}$ are two infinite experiments existing in $\mathfrak{A}$, then $\mathbf{a} \wedge \mathbf{b}$ exists also in $\mathfrak{A}$; hence, operations can be defined in $\mathscr{E}(\mathfrak{A})$, and $\mathscr{E}(\mathfrak{A})$ is a vector lattice containing the vector lattice $\mathscr{S}(\mathfrak{A})$. Let now $(\mathfrak{A}, p)$ be a pr subalgebra of a pr σ-algebra $(\mathfrak{B}, p)$; then $\mathscr{S}(\mathfrak{A}) \subseteq \mathscr{S}(\mathfrak{B}) \subseteq \mathscr{E}(\mathfrak{B})$, but $\mathscr{E}(\mathfrak{A})$ is not always a subset of $\mathscr{E}(\mathfrak{B})$, because $\mathbf{a} = \{a_1, a_2, \ldots\}$ can be an experiment in $\mathfrak{A}$ without also being an experiment in $\mathfrak{B}$. In general, we have

$$(\mathfrak{A}) \bigvee_{i=1}^{\infty} a_i = e \geqslant (\mathfrak{B}) \bigvee_{i=1}^{\infty} a_i.$$

$\mathscr{E}(\mathfrak{A})$ is a subset of $\mathscr{E}(\mathfrak{B})$, and indeed a σ-regular vector sublattice of $\mathscr{E}(\mathfrak{B})$, if and only if $\mathfrak{A}$ is a σ-regular Boolean subalgebra of $\mathfrak{B}$, equivalently if p is continuous on $\mathfrak{A}$. Hence it is recommended in the general case, to consider only the vector lattice $\mathscr{S}(\mathfrak{A})$.

3. CONVERGENCE IN STOCHASTIC SPACES

3.1. Let $(\mathfrak{B}, p)$ be a fixed (for this section) pr σ-algebra. We shall briefly denote by $\mathscr{E}$ the elementary stochastic space over $\mathfrak{B}$, by $\mathscr{S}$ the simple stochastic space over $\mathfrak{B}$, and by $\mathscr{J}$ the Boolean kernel of all indicators of $\mathfrak{B}$ in $\mathscr{E}$. We shall introduce in $\mathscr{E}$ the following kind of convergence:

Order convergence, briefly o-convergence.

We shall say the sequence $X_n \in \mathscr{E}$, $n = 1, 2, \ldots$, o-converges to $X \in \mathscr{E}$ and write $X_n \overset{o}{\underset{\mathscr{E}}{\to}} X$, or $(\mathscr{E})\, o\text{-}\lim X_n = X$, if and only if there exists a decreasing sequence $U_n \in \mathscr{E}$, $n = 1, 2, \ldots$, such that

$$\bigwedge_{n=1}^{\infty} U_n = 0$$

in $\mathscr{E}$, and $|X_n - X| \leqslant U_n$, $n = 1, 2, \ldots$.

One verifies easily the following statements:

(1) If $X_n = X \in \mathscr{E}$, $n = 1, 2, 3, \ldots$, then $X_n \overset{o}{\underset{\mathscr{E}}{\to}} X$.

(2) If $X_n \overset{o}{\underset{\mathscr{E}}{\to}} X$ and $X_n \overset{o}{\underset{\mathscr{E}}{\to}} Y$, then $X = Y$.

(3) If $X_{k_n} \in \mathscr{E}$, $n = 1, 2, \ldots$, is any subsequence of $X_n \in \mathscr{E}$, $n = 1, 2, \ldots$, then:

$$\text{If} \quad X_n \overset{o}{\underset{\mathscr{E}}{\to}} X, \quad \text{then} \quad X_{k_n} \overset{o}{\underset{\mathscr{E}}{\to}} X.$$

The following theorem is true:

Theorem 3.1.

The sequence $X_n \in \mathscr{E}$, $n = 1, 2, \ldots$, o-converges to $X \in \mathscr{E}$, if and only if there exists two sequences $D_n \in \mathscr{E}$, $V_n \in \mathscr{E}$, $n = 1, 2, \ldots$, such that

$$D_n \leqslant D_{n+1}, \quad V_n \geqslant V_{n+1}, \quad D_n \leqslant X_n \leqslant V_n, \qquad n = 1, 2, \ldots,$$

and

$$\bigvee_{n=1}^{\infty} D_n = \bigwedge_{n=1}^{\infty} V_n = X \quad \text{in} \quad \mathscr{E}.$$

Proof. (A) Let $U_n \in \mathscr{E}$, $U_n \geqslant U_{n+1}$, $|X_n - X| \leqslant U_n$, $n = 1, 2, \ldots$, and

$$\bigwedge_{n=1}^{\infty} U_n = 0,$$

i.e., $X_n \overset{o}{\underset{\mathscr{E}}{\to}} X$; then we have

$$-U_n \leqslant X_n - X \leqslant U_n, \qquad n = 1, 2, \ldots;$$

hence

$$D_n = X - U_n \leqslant X_n \leqslant X + U_n = V_n, \qquad n = 1, 2, \ldots .$$

Moreover,

$$D_n \leqslant D_{n+1}, \quad V_n \geqslant V_{n+1}, \qquad n = 1, 2, \ldots,$$

and

$$\bigvee_{n=1}^{\infty} D_n = \bigvee_{n=1}^{\infty} (X - U_n) = X + \bigvee_{n=1}^{\infty} (-U_n) = X - \bigwedge_{n=1}^{\infty} U_n = X$$

$$\bigwedge_{n=1}^{\infty} V_n = \bigwedge_{n=1}^{\infty} (X + U_n) = X + \bigwedge_{n=1}^{\infty} U_n = X$$

in $\mathscr{E}$.

(B) Now let two sequences $D_n \leqslant D_{n+1}$, $V_n \geqslant V_{n+1}$, $n = 1, 2, \ldots$, in E, such that

$$D_n \leqslant X_n \leqslant V_n, \qquad n = 1, 2, \ldots,$$

and

$$\bigvee_{n=1}^{\infty} D_n = \bigwedge_{n=1}^{\infty} V_n = X \quad \text{in} \quad \mathscr{E};$$

then we have

$$D_n \leqslant X_n \wedge X, \quad X_n \vee X \leqslant V_n, \qquad n = 1, 2, \ldots;$$

hence

$$|X_n - X| = X_n \vee X - X_n \wedge X \leqslant V_n - D_n = U_n$$

with

$$U_n \geqslant U_{n+1}, \qquad n = 1, 2, \ldots,$$

and

$$\bigwedge_{n=1}^{\infty} U_n = \bigwedge_{n=1}^{\infty} (V_n - D_n) = \bigwedge_{n=1}^{\infty} V_n + \bigwedge_{n=1}^{\infty} (-D_n) = \bigwedge_{n=1}^{\infty} V_n - \bigvee_{n=1}^{\infty} D_n = 0,$$

i.e.,

$$\bigwedge_{n=1}^{\infty} U_n = 0 \quad \text{in} \quad \mathscr{E};$$

hence

$$X_n \underset{\mathscr{E}}{\overset{o}{\to}} X.$$

It is easy to show that the following theorems are true:

Theorem 3.2.

If the sequence $X_n \in \mathscr{E}$, $n = 1, 2, \ldots$, is increasing or decreasing respectively, then:

$$X_n \underset{\mathscr{E}}{\overset{o}{\to}} X \textit{ if and only if } \bigvee_{n=1}^{\infty} X_n = X \quad \textit{or} \quad \bigwedge_{n=1}^{\infty} X_n = X$$

in $\mathscr{E}$ respectively. We write in this case:

$$X_n \overset{o}{\underset{\mathscr{E}}{\uparrow}} X \quad \text{or} \quad X_n \overset{o}{\underset{\mathscr{E}}{\downarrow}} X \quad \textit{respectively.}$$

Theorem 3.3.

If $X_n \in \mathscr{E}$, $n = 1, 2, \ldots$, and

$$V_n = \bigvee_{j \geqslant n} X_j, \qquad D_n = \bigwedge_{j \geqslant n} X_j$$

exist in $\mathscr{E}$, for every $n = 1, 2, \ldots$, then $X_n \overset{o}{\underset{\mathscr{E}}{\rightarrow}} X$ if and only if

$$\bigwedge_{n=1}^{\infty} \bigvee_{j \geqslant n} X_j = \bigvee_{n=1}^{\infty} \bigwedge_{j \geqslant n} X_j = X \quad \textit{in} \quad \mathscr{E}.$$

3.1.1. For any $X_n \in \mathscr{E}$, $n = 1, 2, \ldots$, if

$$V_n = \bigvee_{j \geqslant n} X_j \quad \text{and} \quad D_n = \bigwedge_{j \geqslant n} X_j$$

exist in $\mathscr{E}$ for every $n = 1, 2, \ldots$ and, moreover,

$$\bigwedge_{n=1}^{\infty} \bigvee_{j \geqslant n} X_j \quad \text{and} \quad \bigvee_{n=1}^{\infty} \bigwedge_{j \geqslant n} X_j$$

exist in $\mathscr{E}$, we define

$$(\mathscr{E})o\text{-}\lim\sup X_n = \bigwedge_{n=1}^{\infty} \bigvee_{j \geqslant n} X_j \equiv \tilde{X}$$

$$(\mathscr{E})o\text{-}\lim\inf X_n = \bigvee_{n=1}^{\infty} \bigwedge_{j \geqslant n} X_j \equiv \underset{\sim}{X}.$$

We have in this case:

$$(\mathscr{E})o\text{-}\lim\inf X_n \leqslant (\mathscr{E})o\text{-}\lim\sup X_n,$$

and $$(\mathscr{E})o\text{-}\lim X_n = X \quad \text{if and only if} \quad \tilde{X} = \underset{\sim}{X} = X.$$

Theorem 3.4.

(*Continuity of the operations*). *If $X_n \overset{o}{\underset{\mathscr{E}}{\rightarrow}} X$ and $Y_n \overset{o}{\underset{\mathscr{E}}{\rightarrow}} Y$ and if*

$$\lambda_n \in R, \qquad n = 1, 2, \ldots,$$

with $\lim_{n\to\infty} \lambda_n = \lambda$, then

$$X_n \vee Y_n \underset{\mathscr{E}}{\overset{o}{\to}} X \vee Y, \quad X_n \wedge Y_n \underset{\mathscr{E}}{\overset{o}{\to}} X \wedge Y$$

$$X_n \pm Y_n \underset{\mathscr{E}}{\overset{o}{\to}} X \pm Y, \quad X_n\, Y_n \underset{\mathscr{E}}{\overset{o}{\to}} XY$$

$$\lambda_n X_n \underset{\mathscr{E}}{\overset{o}{\to}} \lambda X, \quad |X_n| \underset{\mathscr{E}}{\overset{o}{\to}} |X|, \quad X_n^+ \underset{\mathscr{E}}{\overset{o}{\to}} X^+, \quad X_n^- \underset{\mathscr{E}}{\overset{o}{\to}} X^-.$$

Theorem 3.5.

(1) $|X_n| \underset{\mathscr{E}}{\overset{o}{\to}} 0$ *if and only if* $X_n \underset{\mathscr{E}}{\overset{o}{\to}} 0$.

(2) $X_n \underset{\mathscr{E}}{\overset{o}{\to}} X$ *if and only if* $|X_n - X| \underset{\mathscr{E}}{\overset{o}{\to}} 0$.

Theorem 3.6.

The simple stochatic space $\mathscr{S}$ is o-dense in $\mathscr{E}$, i.e., for every $X \in \mathscr{E}$, there exists a sequence $X_n \in \mathscr{S}$, $n = 1, 2, \ldots$, such that $X_n \underset{\mathscr{E}}{\overset{o}{\to}} X$.

Proof. If $X \in \mathscr{S} \subseteq \mathscr{E}$, then $X_n = X$, $n = 1, 2, \ldots$, satisfies our assertion. Let $X \in \mathscr{E}$ but $X \notin \mathscr{S}$ and let

$$X = \sum_{j=1}^{\infty} \xi_j I_{a_j}$$

be the reduced representation of X by indicators. We write

$$X_n = \xi_1 I_{a_1} + \xi_2 I_{a_2} + \ldots + \xi_n I_{a_n}, \qquad n = 1, 2, \ldots;$$

then $X_n \underset{\mathscr{E}}{\overset{o}{\to}} X$, because $X_n = I_{s_n} X$, where $s_n = a_1 \vee a_2 \vee \ldots \vee a_n$, and $I_{s_n} \underset{\mathscr{E}}{\overset{o}{\to}} I_e = 1$; hence, by Theorem 3.4, $I_{s_n} X \underset{\mathscr{E}}{\overset{o}{\to}} 1X = X$.

Theorem 3.6α.

Let $(\mathfrak{A}, p)$ be a pr subalgebra of $(\mathfrak{B}, p)$, which σ-generates $(\mathfrak{B}, p)$, i.e., $\mathfrak{A}^{\sigma\delta} = \mathfrak{B}$; then $\mathscr{S}(\mathfrak{A})$, i.e., the simple stochastic space over $(\mathfrak{A}, p)$, is o-dense in $\mathscr{S}(\mathfrak{B})$.

Proof. Let $X \in \mathscr{S}(\mathfrak{B})$ and let $X = \xi_1 I_{a_1} + \ldots + \xi_n I_{a_n}$ be the reduced representation of X by indicators; then since $\mathfrak{A}$ is o-dense in $\mathfrak{B}$, for every

$i = 1, 2, ..., n$, there exists a sequence $a_{ik} \in \mathfrak{A}$, $k = 1, 2, ...,$ with $a_{ik} \underset{\mathfrak{B}}{\overset{o}{\to}} a_i$. Hence, we have (by the complete regularity of $\mathscr{J}$ as a sublattice of $\mathscr{E}$) $I_{a_{ik}} \underset{\mathscr{E}}{\overset{o}{\to}} I_{a_i}$. We write

$$X_k = \xi_1 I_{a_{1k}} + \xi_2 I_{a_{2k}} + \ldots + \xi_n I_{a_{nk}}, \qquad k = 1, 2, \ldots;$$

then we have (see Theorem 3.4)

$$X_k = \xi_1 I_{a_{1k}} + \xi_2 I_{a_{2k}} + \ldots + \xi_n I_{a_{nk}} \underset{\mathscr{E}}{\overset{o}{\to}} \xi_1 I_{a_1} + \xi_2 I_{a_2} + \ldots + \xi_n I_{a_n} = X.$$

The reader can prove that Theorems 3.6 and 3.6α imply:

Theorem 3.6β.

Any simple stochastic space $\mathscr{S}(\mathfrak{A})$ in which $\mathfrak{A}$ is a Boolean subalgebra of $\mathfrak{B}$ with $\mathfrak{A}^{\sigma\delta} = \mathfrak{B}$ is o-dense in $\mathscr{E} = \mathscr{E}(\mathfrak{B})$.

3.2. For any $X \in \mathscr{E}$ we shall write $s_X(\xi) = [X < \xi]$, or briefly

$$s(\xi) = [X < \xi]$$

without the subscript X, if the same erv X is meant throughout and there is no possibility of confusion. We shall call $s_X(\xi)$ the *lower spectral chain* of X in $\mathfrak{B}$. The *upper spectral chain* of X in $\mathfrak{B}$ is defined as follows:

$$t_X(\xi) = [X \leqslant \xi] = \bigvee_{X(a_j) \leqslant \xi} a_j,$$

if

$$X = \sum_{j \geqslant 1} \xi_j I_{a_j}$$

is the reduced representation of X by indicators, for every ξ with

$$-\infty < \xi < +\infty.$$

We shall also need the complementary chains of s_X and t_X, i.e.,

$$s_X{}^c(\xi) = (s_X(\xi))^c = [X \geqslant \xi] \quad \text{and} \quad t_X{}^c(\xi) = (t_X(\xi))^c = [X > \xi],$$

$$-\infty < \xi < +\infty.$$

Obviously, the smallest Boolean σ-subalgebra of $\mathfrak{B}$ containing one of the four chains contains also all others and is identical with the Boolean σ-subalgebra $\mathfrak{B}_X$ of $\mathfrak{B}$, which is σ-generated by the set $\{a_1, a_2, \ldots\}$ (see Section 2.6). We note that the definition of the functions s_X and t_X is independent of the representation of X by indicators. To define s_X and t_X we may use not only the reduced but also any representation of X by pairwise disjoint indicators. The chains s and t are increasing functions in the real variable ξ, the first of them continuous from the left and the second continuous from the right. Moreover,

$$(s-\infty) = t(-\infty) = \emptyset \quad \text{and} \quad s(+\infty) = t(+\infty) = e.$$

The complement chains s^c and t^c are decreasing functions in the real variable ξ, the first continuous from the left and the second continuous from the right; they satisfy

$$s^c(-\infty) = t^c(-\infty) = e, \qquad s^c(+\infty) = t^c(+\infty) = \emptyset.$$

3.3. The following relations are valid between the four spectral chains, s, t, s^c, t^c:

(a) If $\xi < \eta$, then

$$s(\xi) \leqslant t(\xi) \leqslant s(\eta) \leqslant t(\eta), \qquad s^c(\xi) \geqslant t^c(\xi) \geqslant s^c(\eta) \geqslant t^c(\eta).$$

(b) For every $\xi \in R$, we have

$$t(\xi) = t(\xi+0) = s(\xi+0) \geqslant s(\xi); \quad t^c(\xi) = t^c(\xi+0) = s^c(\xi+0) \leqslant s^c(\xi).$$

$$t(\xi) \geqslant t(\xi-0) = s(\xi-0) = s(\xi); \quad t^c(\xi) \leqslant t^c(\xi-0) = s^c(\xi-0) = s^c(\xi).$$

Remark. $s(\xi+0) = \underset{\substack{\eta\downarrow\xi\\ \eta>\xi}}{o\text{-lim}}\, s(\eta)$, $s(\xi-0) = \underset{\substack{\eta\uparrow\xi\\ \eta<\xi}}{o\text{-lim}}\, s(\eta)$; analogously we define $t(\xi+0)$, $t(\xi-0)$.

(c) Let $\sigma(\xi)$ be an increasing function defined for every $\xi \in R$ with values in $\mathfrak{B}$ which satisfies the condition $s(\xi) \leqslant \sigma(\xi) \leqslant t(\xi)$, for every $\xi \in R$; then we have $\sigma(\xi-0) = s(\xi)$ and $\sigma(\xi+0) = t(\xi)$.

(d) If $X \in \mathscr{E}$, $Y \in \mathscr{E}$, then $X = Y$ if and only if $s_X(\xi) = s_Y(\xi)$, or equivalently $t_X(\xi) = t_Y(\xi)$, for every $\xi \in R$.

(e) If $X \in \mathscr{E}$, $Y \in \mathscr{E}$, then $X \leqslant Y$ if and only if $s_X(\xi) \geqslant s_Y(\xi)$, or dually $s_X{}^c(\xi) \leqslant s_Y{}^c(\xi)$, or equivalently $t_X(\xi) \geqslant t_Y(\xi)$, or dually $t_X{}^c(\xi) \leqslant t_Y{}^c(\xi)$, for every $\xi \in R$.

The reader can prove that the following theorem is true:

Theorem 3.7.

Let $X_i \in \mathscr{E}$, $i \in I$, be a bounded family in $\mathscr{E}$; i.e., let two elements $L \in \mathscr{E}$ and $U \in \mathscr{E}$ exist such that

$$L \leqslant X_i \leqslant U, \qquad i \in I;$$

then:

(1) $\bigwedge_{i\in I} t_{X_i}(\xi) \underset{\text{def}}{=} \tau(\xi) \geqslant t_U(\xi)$, *i.e.*, $\bigvee_{i\in I} t^c_{X_i}(\xi) \leqslant t^c(\xi)$.

(2) *$\tau(\xi)$ is increasing, continuous from the right and $\tau(-\infty) = \varnothing$, $\tau(+\infty) = e$.*

(3) *If $\bigvee_{i\in I} X_i$ exists and is equal to $X \in \mathscr{E}$, then $\tau(\xi) = t_X(\xi)$, $\xi \in R$.*

(4) *$\bigwedge_{i\in I} s_{X_i}(\xi) \underset{\text{def}}{=} \sigma(\xi) \geqslant s_U(\xi)$, for every $\xi \in R$.*

(5) *$\sigma(\xi)$ is increasing and $\sigma(-\infty) = \varnothing$, $\sigma(+\infty) = e$; if $s(\xi) \underset{\text{def}}{=} \sigma(\xi-0)$, then $s(\xi)$ is continuous from the left.*

(6) *If $\bigvee_{i\in I} X_i$ exists and is equal to X in $\mathscr{E}$, then*

$$\sigma(\xi-0) = s(\xi) = s_X(\xi),$$

for every $\xi \in R$.

(7) *$\bigvee_{i\in I} t_{X_i}(\xi) \underset{\text{def}}{=} \lambda(\xi) \leqslant t_L(\xi)$, for every $\xi \in R$.*

(8) *$\lambda(\xi)$ is increasing, and $\lambda(-\infty) = \varnothing$, $\lambda(+\infty) = e$; furthermore, if $t(\xi) \underset{\text{def}}{=} \lambda(\xi+0)$, then $t(\xi)$ is continuous from the right.*

(9) *If $\bigwedge_{i\in I} X_i$ exists and is equal to $Y \in \mathscr{E}$, then*

$$\lambda(\xi+0) = t(\xi) = t_Y(\xi) \quad \textit{for every} \quad \xi \in R.$$

(10) *$\bigvee_{i\in I} s_{X_i}(\xi) \underset{\text{def}}{=} \rho(\xi) \leqslant s_L(\xi)$ for every $\xi \in R$.*

(11) *$\rho(\xi)$ is increasing, continuous from the left and $\rho(-\infty) = \varnothing$, $\rho(+\infty) = e$.*

(12) *If $\bigwedge_{i\in I} X_i$ exists and is equal to $Y \in \mathscr{E}$, then*

$$\rho(\xi) = s_Y(\xi), \quad \textit{for every} \quad \xi \in R.$$

3.4. The spectral chains of a constant variable and of an indicator are given as follows:

(1) For the constant rv 0, we have

$$t_0(\xi) = \begin{cases} \varnothing, & \xi < 0 \\ e, & \xi \geqslant 0 \end{cases}, \qquad t_0{}^c(\xi) = \begin{cases} e, & \xi < 0 \\ \varnothing, & \xi \geqslant 0 \end{cases}$$

$$s_0(\xi) = \begin{cases} \varnothing, & \xi \leqslant 0 \\ e, & \xi > 0 \end{cases}, \qquad s_0{}^c(\xi) = \begin{cases} e, & \xi \leqslant 0 \\ \varnothing, & \xi > 0 \end{cases}$$

(2) For any constant rv $X = \lambda \in R$, we have

$$t_\lambda(\xi) = \begin{cases} \emptyset, & \xi < \lambda \\ e, & \xi \geqslant \lambda \end{cases}, \qquad t_\lambda{}^c(\xi) = \begin{cases} e, & \xi < \lambda \\ \emptyset, & \xi \geqslant \lambda \end{cases}$$

$$s_\lambda(\xi) = \begin{cases} \emptyset, & \xi \leqslant \lambda \\ e, & \xi > \lambda \end{cases}, \qquad s_\lambda{}^c(\xi) = \begin{cases} e, & \xi \leqslant \lambda \\ \emptyset, & \xi > \lambda \end{cases}$$

(3) For any indicator I_a with $a \in \mathfrak{B}$, we have

$$t_{I_a}(\xi) = \begin{cases} \emptyset, & \xi < 0 \\ a^c, & 0 \leqslant \xi < 1 \\ e, & 1 \leqslant \xi \end{cases}, \qquad t_{I_a}^c(\xi) = \begin{cases} e, & \xi < 0 \\ a, & 0 \leqslant \xi < 1 \\ \emptyset, & 1 \leqslant \xi \end{cases}$$

$$s_{I_a}(\xi) = \begin{cases} \emptyset, & \xi \leqslant 0 \\ a^c, & 0 < \xi \leqslant 1 \\ e, & 1 < \xi \end{cases}, \qquad s_{I_a}^c(\xi) = \begin{cases} e, & \xi \leqslant 0 \\ a, & 0 < \xi \leqslant 1 \\ \emptyset, & 1 < \xi \end{cases}$$

3.5. The following theorems are immediate consequences of Theorem 3.7.

Theorem 3.8.

If $X_n \uparrow$ and

$$\bigvee_{n=1}^{\infty} X_n = X \quad \textit{in } \mathscr{E},$$

then

$$\bigvee_{n=1}^{\infty} t_{X_n}^c(\xi) = t_X{}^c(\xi), \qquad \bigwedge_{n=1}^{\infty} t_{X_n}(\xi) = t_X(\xi),$$

for every $\xi \in R$, and dually if $X_n \downarrow$ and

$$\bigwedge_{n=1}^{\infty} X_n = Y \quad \textit{in } \mathscr{E},$$

then

$$\bigwedge_{n=1}^{\infty} s_{X_n}^c(\xi) = s_Y{}^c(\xi) \quad \textit{and} \quad \bigvee_{n=1}^{\infty} s_{X_n}(\xi) = s_Y(\xi),$$

for every $\xi \in R$.

Theorem 3.9.

If $X_n \in \mathscr{E}$, $n = 1, 2, \ldots$, with

$$\bigvee_{n=1}^{\infty} X_n = X \in \mathscr{E},$$

then

$$t_X{}^c(\xi) = \bigvee_{n=1}^{\infty} t^c_{X_n}(\xi), \quad t_X(\xi) = \bigwedge_{n=1}^{\infty} t_{X_n}(\xi), \quad \textit{for every} \quad \xi \in R.$$

If

$$\bigwedge_{n=1}^{\infty} X_n = Y \text{ in } \mathscr{E}$$

and

$$\bigwedge_{n=1}^{\infty} t^c_{X_n}(\xi) \underset{\text{def}}{=} \lambda(\xi),$$

then

$$t_Y{}^c(\xi) = \lambda(\xi+0).$$

Theorem 3.10.

If $X_n \underset{\mathscr{E}}{\overset{o}{\rightarrow}} X$ *and*

$$\bigwedge_{n=1}^{\infty} \bigvee_{j \geqslant n} t^c_{X_j}(\xi) \underset{\text{def}}{=} \lambda(\xi)$$

and

$$\bigvee_{n=1}^{\infty} \bigwedge_{j \geqslant n} t^c_{X_j}(\xi) \underset{\text{def}}{=} \lambda^*(\xi), \quad \xi \in R,$$

then

$$\lambda(\xi+0) = \lambda^*(\xi+0) = t_X{}^c(\xi),$$

for every $\xi \in R$.

4. O-CONVERGENCE IN $\mathscr{E}$ WITH RESPECT TO A VECTOR SUBLATTICE OF $\mathscr{E}$

The o-convergence can be generalized first for nets instead of sequences and moreover to an o-convergence in $\mathscr{E}$ relative to a vector sublattice of $\mathscr{E}$, for example, the vector sublattice R of all constant rv's or the vector sublattice of all srv's. These generalizations will briefly be stated in this section.

4.1. A set I of indices is said to be *directed* if and only if a reflexive and transitive relation " $>$ " is defined in I, having the so-called *Moore–Smith property*:

given $i_1 \in I$, $i_2 \in I$, there is some $i_3 \in I$ such that $i_3 > i_1$ and $i_3 > i_2$. A family $X_i \in \mathscr{E}$, $i \in I$, in which I directed, is called a *directed family* or briefly a *net* $X_i \in \mathscr{E}$, $i \in I$.

The set $\mathbf{N} = \{1, 2, ...\}$ of the natural numbers with the relation " $\geqslant$ " is directed and a sequence $X_n \in \mathscr{E}$, $n \in \mathbf{N}$, can be considered as a net.

The cartesian product $\mathbf{N}\times\mathbf{N}$ can be directed by the so-called cartesian ordering, i.e., $(n_1, n_2) \geqslant (n_1', n_2')$ if and only if $n_1 \geqslant n_2'$ and $n_2 \geqslant n_2'$.

A net $X_i \in \mathscr{E}$, $i \in I$, is decreasing or increasing, denoted by $X_i\downarrow$ or $X_i\uparrow$, if and only if $i > j$ implies $X_i \leqslant X_j$ or $X_i \geqslant X_j$ respectively.

4.2. Let $\mathscr{F}$ be any vector sublattice of $\mathscr{E}$, for example $\mathscr{F} = R$ or $\mathscr{S}$; then we say:

A net $X_i \in \mathscr{E}$, $i \in I$, *$\mathscr{F}$-o-converges* (or *o*-converges with respect to $\mathscr{F}$) to $X \in \mathscr{E}$, denoted

$$\mathscr{F}\text{-}o\text{-}\lim_{i\in I,\, >} X_i = X \quad \text{or} \quad X_i \underset{\mathscr{F}}{\overset{o}{\to}} X \quad \text{in } \mathscr{E},$$

if and only if there is a decreasing net $U_i \in \mathscr{F}$, $i \in I$, such that

$$|X_i - X| \leqslant U_i, \quad i \in I, \quad \text{and} \quad (\mathscr{F}) \bigwedge_{i\in I} U_i = 0.$$

The following theorem is true:

Theorem 4.1.

$\mathscr{F}$-o-limits of nets (resp of sequences) are unique in $\mathscr{E}$ if and only if $\mathscr{F}$ is regular (resp σ-regular) in $\mathscr{E}$.

Remark. $\mathscr{F}$ is *regular* (resp *σ-regular*) in $\mathscr{E}$ if and only if, for every decreasing net $X_i \in \mathscr{F}$, $i \in I$, with

$$(\mathscr{F}) \bigwedge_{i\in I} X_i = 0$$

$$\left(\text{resp for every decreasing sequence } X_n \in \mathscr{F}, n \in \mathbf{N}, \text{ with } (\mathscr{F}) \bigwedge_{n\in\mathbf{N}} X_n = 0\right),$$

we have

$$(\mathscr{E}) \bigwedge_{i\in I} X_i = 0 \qquad \left(\text{resp} \quad (\mathscr{E}) \bigwedge_{n\in\mathbf{N}} X_n = 0\right).$$

Proof. Let the $\mathscr{F}$-*o*-limits of nets be unique and let a decreasing net $X_i \in \mathscr{F}$ with $(\mathscr{F}) \bigwedge_{i\in I} X_i = 0$. Suppose that $\mathscr{F}$ is not regular; then there is an $Y \in \mathscr{E}$ such that $X_i \geqslant Y > 0$, for every $i \in I$, i.e., $(\mathscr{E}) \bigwedge_{i\in I} X_i > 0$ or $\bigwedge_{i\in I} X_i$ does not exist in $\mathscr{E}$; then we have

$$|X_i - Y| = X_i - Y \leqslant X_i, \quad i \in I,$$

with

$$(\mathscr{F}) \bigwedge_{i\in I} X_i = 0,$$

hence

$$\mathscr{F}\text{-}o\text{-}\lim_{i\in I} (X_i - Y) = 0 \quad \text{in } \mathscr{E},$$

i.e.,
$$\mathscr{F}\text{-}o\text{-}\lim_{i\in I} X_i = Y > 0$$
but $(\mathscr{F})\bigwedge_{i\in I} X_i = 0$ and X_i decreasing imply $\mathscr{F}\text{-}o\text{-}\lim_{i\in I} X_i = 0$; in other words in this case the $\mathscr{F}$-o-limit is not unique (contradiction!).

Let now $\mathscr{F}$ be regular in $\mathscr{E}$, and $\mathscr{F}\text{-}o\text{-}\lim_{i\in I} X_i = X$ for the net X_i, $i\in I$. Suppose also $\mathscr{F}\text{-}o\text{-}\lim_{i\in I} X_i = Y$; then there are decreasing nets $U_i \in \mathscr{F}$ and $V_i \in \mathscr{F}$, $i\in I$, with $|X_i - X| \leqslant U_i$ and $|X_i - Y| \leqslant V_i$, $i\in I$, and
$$(\mathscr{F})\bigwedge_{i\in I} U_i = 0 \quad\text{and}\quad (\mathscr{F})\bigwedge_{i\in I} V_i = 0.$$
Then
$$|X-Y| = |X - X_i + X_i - Y| \leqslant |X_i - X| + |X_i - Y| \leqslant U_i + V_i, \quad i\in I;$$
but
$$(\mathscr{F})\bigwedge_{i\in I}(U_i + V_i) = (\mathscr{F})\bigwedge_{i\in I} U_i + (\mathscr{F})\bigwedge_{i\in I} V_i = 0,$$
hence $|X - Y| = 0$, i.e., $X = Y$.

The proof in the case of sequences is analogous.

4.3. We note that in the case $\mathscr{F}$ is regular (resp σ-regular) in $\mathscr{E}$, in which the uniqueness of the $\mathscr{F}$-o-limits is secured, the following statements are true:
$$\mathscr{F}\text{-}o\text{-}\lim_{i\in I,\, >} X_i = X \quad\text{and}\quad \mathscr{F}\text{-}o\text{-}\lim_{j\in J,\, >} Y_j = Y$$
(resp in the case $I = J = \mathbf{N}$), imply
$$\mathscr{F}\text{-}o\text{-}\lim_{(i,j)\in I\times J} (X_i + Y_j) = X + Y,\dagger$$
$$\mathscr{F}\text{-}o\text{-}\lim_{(i,j)\in I\times J} (X_i \vee Y_j) = (X \vee Y) \text{ and dually,}$$
$$\mathscr{F}\text{-}o\text{-}\lim_{i\in I,\, >} (-X_i) = -X, \qquad \mathscr{F}\text{-}o\text{-}\lim_{i\in I,\, >} |X_i| = |X|.$$
If $\lambda_i \in R$, $i\in I$, with $\lim_{i\in I\, >} \lambda_i = \lambda$, then $\mathscr{F}\text{-}o\text{-}\lim_{i\in I,\, >} \lambda_i X_i = \lambda X$.

† As to the product $X_i\, Y_j$, (i,j), one cannot always show $\mathscr{F}\text{-}o\text{-}\lim X_i\, Y_j = XY$, but always $(\mathscr{E})\text{-}o\text{-}\lim X_i\, Y_j = XY$.

If $I = J$, then

$$\mathscr{F}\text{-}o\text{-}\lim_{i\in I,\,>} (X_i + Y_i) = X + Y,$$

$$\mathscr{F}\text{-}o\text{-}\lim_{i\in I,\,>} (X_i \vee Y_i) = X \vee Y \text{ and dually.}$$

4.4. The set R of all real numbers, considered as a vector sublattice of $\mathscr{E}$, is regular in $\mathscr{E}$, hence R-o-limits of nets are unique. It is easy to prove that R-o-convergence in $\mathscr{E}$ is essentially equivalent to the so-called *uniform convergence*, briefly, *u-convergence* in $\mathscr{E}$. We say:

a net $X_i \in \mathscr{E}$, $i \in I$, u-converges to X in $\mathscr{E}$ if and only if, for every $\varepsilon > 0$ there is an index $i_0(\varepsilon) \in I$ such that $|X_i - X| < \varepsilon$, for every $i > i_0(\varepsilon)$, $i \in I$.

We note that $\mathscr{S}$ is also a regular vector sublattice of $\mathscr{E}$. In fact, if $X_i \in \mathscr{S}$, $i \in I$, is a decreasing net with

$$(\mathscr{S}) \bigwedge_{i\in I} X_i = 0,$$

then

$$(\mathscr{E}) \bigwedge_{i\in I} X_i = 0;$$

for, if there is a $Z \in \mathscr{E}$, $Z \notin \mathscr{S}$, i.e.,

$$Z = \sum_{i=1}^{\infty} \zeta_i I_{c_i},$$

with $Z > 0$ and $X_i \geqslant Z > 0$, for every $i \in I$, then there is a $c_i \neq \varnothing$ with $\zeta_i > 0$, hence $\zeta_i I_{c_i} > 0$, and $\zeta_i I_{c_i} \in \mathscr{S}$, with $X_j \geqslant Z > \zeta_i I_{c_i}$, $j \in I$, which is a contradiction to

$$(\mathscr{S}) \bigwedge_{i\in I} X_i = 0.$$

Hence $\mathscr{S}$-o-limits of nets are also unique.

4.5. *$\mathscr{F}$-o-fundamental nets resp sequences.* Let $\mathscr{F}$ be any vector sublattice of $\mathscr{E}$. A net $X_i \in \mathscr{E}$, $i \in I$, is $\mathscr{F}$-o-fundamental if and only if

$$\mathscr{F}\text{-}o\text{-}\lim_{(i,\,j)\in I\times I} (X_i - X_j) = 0 \quad \Big(\text{resp } \mathscr{F}\text{-}o\text{-}\lim_{(n.m)\in N\times N} (X_n - X_m) = 0\Big).$$

Here $I \times I$ (resp $\mathbf{N} \times \mathbf{N}$) is directed by the cartesian (co-ordinatewise) ordering. In view of Theorem 4.1, we assume that $\mathscr{F}$ is regular (resp σ-regular) in $\mathscr{E}$; then it is easy to prove that, if $X_i \in \mathscr{E}$, $i \in I$, and $Y_j \in \mathscr{E}$, $j \in J$, are two $\mathscr{F}$-o-fundamental nets, then $X_i + Y_j$, $X_i \wedge Y_j$, $X_i \vee Y_j$,

$(i, j) \in I \times J$, $-X_i$, $|X_i|$, $i \in I$† are also $\mathscr{F}$-o-fundamental nets. Particularly, that is true if $I = J = \mathbf{N}$, i.e., if the two nets are sequences. Moreover, if $I = J$, then $X_i + Y_i$, $X_i \vee Y_i$, $X_i \wedge Y_i$, $i \in I$, are $\mathscr{F}$-o-fundamental nets.

4.6. We now restrict our study to $\mathscr{F}$-o-convergent and $\mathscr{F}$-o-fundamental sequences, and assume that $\mathscr{F}$ is σ-regular in $\mathscr{E}$. The following propositions are true:

4.6.1. *A sequence $X_n \in \mathscr{E}$, $n = 1, 2, \ldots$, is $\mathscr{F}$-o-fundamental, if and only if there is a decreasing sequence $U_n \in \mathscr{F}$, $n = 1, 2, \ldots$, such that*

$$(\mathscr{F}) \bigwedge_{n=1}^{\infty} U_n = 0 \quad \textit{and} \quad |X_n - X_{n+k}| \leqslant U_n, \quad n = 1, 2, \ldots, \quad k = 1, 2, \ldots .$$

Proof. Let $X_n \in \mathscr{E}$, $n = 1, 2, \ldots$, be $\mathscr{F}$-o-fundamental; then there is a decreasing double sequence $U_{n,k} \in \mathscr{F}$, $(n, k) \in \mathbf{N} \times \mathbf{N}$ with

$$(\mathscr{F}) \bigwedge_{(n,k) \in \mathbf{N} \times \mathbf{N}} U_{n,k} = 0 \quad \text{and} \quad |X_n - X_k| \leqslant U_{n,k}$$

for every $(n, k) \in \mathbf{N} \times \mathbf{N}$. We set $U_n = U_{n,n}$, $n = 1, 2, \ldots$; then

$$|X_n - X_{n+k}| \leqslant U_{n,n+k} \leqslant U_{n,n} = U_n, \quad n = 1, 2, \ldots$$

and obviously

$$(\mathscr{F}) \bigwedge_{n=1}^{\infty} U_n = 0.$$

Conversely, let

$$|X_n - X_{n+k}| \leqslant U_n, \quad n = 1, 2, \ldots, \quad k = 1, 2, \ldots,$$

with $U_n \in \mathscr{F}$, $n = 1, 2, \ldots$, decreasing, and

$$(\mathscr{F}) \bigwedge_{n=1}^{\infty} U_n = 0.$$

We set $U_{n,k} = U_{\min(n,k)}$; then $U_{n,k}$ is decreasing with

$$(\mathscr{F}) \bigwedge_{(n,k) \in \mathbf{N} \times \mathbf{N}} U_{n,k} = 0$$

and
$$|X_n - X_{n+k}| \leqslant U_{n,k}, \quad n = 1, 2, \ldots, \quad k = 1, 2, \ldots .$$

† As to the product $X_i Y_j$, $(i, j) \in I \times J$, one cannot always show that it is $\mathscr{F}$-o-fundamental, but it is always $\mathscr{E}$-o-fundamental.

Hence $X_n \in \mathscr{E}$, $n = 1, 2, \ldots$, is $\mathscr{F}$-o-fundamental.

Now it is obvious that the two following propositions are true:

4.6.2. *$\mathscr{F}$-o-lim $X_n = X$ in $\mathscr{E}$ implies $\mathscr{E}$-o-lim $X_n = X$ in $\mathscr{E}$.*†

4.6.3. *If $X_n \in \mathscr{E}$ is $\mathscr{F}$-o-fundamental, then it is also $\mathscr{E}$-o-fundamental.*

4.6.4. *If $X_n \in \mathscr{E}$, $n = 1, 2, \ldots$, $\mathscr{E}$-o-lim $X_n = X \in \mathscr{E}$, and X_n, $n = 1, 2, \ldots$, is $\mathscr{F}$-o-fundamental, then $\mathscr{F}$-o-lim $X_n = X$.*

Proof. Let $A_n \in \mathscr{E}$, $n = 1, 2, \ldots$, be decreasing with

$$(\mathscr{E}) \bigwedge_{n=1}^{\infty} A_n = 0 \quad \text{and} \quad |X_n - X| \leqslant A_n, \quad n = 1, 2, \ldots,$$

and, moreover, let $U_n \in \mathscr{F}$, $n = 1, 2, \ldots$, be also decreasing with

$$(\mathscr{F}) \bigwedge_{n=1}^{\infty} U_n = 0,$$

and $$|X_n - X_{n+k}| \leqslant U_n, \quad n = 1, 2, \ldots, \quad k = 1, 2, \ldots;$$

then, for every $n = 1, 2, \ldots$,

$$|X_n - X| \leqslant |X_n - X_{n+k}| + |X_{n+k} - X| \leqslant U_n + A_{n+k};$$

hence
$$|X_n - X| \leqslant (\mathscr{E}) \bigwedge_{k=1}^{\infty} (U_n + A_{n+k}) = U_n + (\mathscr{E}) \bigwedge_{k=1}^{\infty} A_{n+k} = U_n$$

i.e., $$|X_n - X| \leqslant U_n, \quad n = 1, 2, \ldots, \quad \text{i.e.,}$$

$$\mathscr{F}\text{-}o\text{-}\lim X_n = X.$$

4.6.5. *If $X_n \in \mathscr{E}$, $n = 1, 2, \ldots$, and $\mathscr{F}$-o-lim $X_n = X$, then X_n, $n = 1, 2, \ldots$, is $\mathscr{F}$-o-fundamental.*

Proof. Let $U_n \in \mathscr{F}$, $n = 1, 2, \ldots$, be decreasing with

$$(\mathscr{F}) \bigwedge_{n=1}^{\infty} U_n = 0 \quad \text{and} \quad |X_n - X| \leqslant U_n, \quad n = 1, 2, \ldots;$$

then

$$|X_n - X_{n+k}| \leqslant |X_n - X| + |X - X_{n+k}| \leqslant U_n + U_{n+k} \leqslant U_n + U_n = 2U_n.$$

But $2U_n$ is decreasing with

$$\bigwedge_{n=1}^{\infty} 2U_n = 0,$$

i.e., X_n, $n = 1, 2, \ldots$, is $\mathscr{F}$-o-fundamental.

† It is obvious that $\mathscr{E}$-o-lim $X_n = X$ in X coincides with $X_n \xrightarrow{o} X$ defined in Section 3.1.

The following fundamental Lemma is true:

Lemma 4.1.

A sequence $X_n \in \mathscr{E}$, $n = 1, 2, \ldots$, *is* $\mathscr{F}$*-o-fundamental, if and only if there are two sequences* $Y_n \in \mathscr{E}$ *and* $Z_n \in \mathscr{E}$, $n = 1, 2, \ldots$, *such that* Y_n, $n = 1, 2, \ldots$, *is increasing,* Z_n, $n = 1, 2, \ldots$, *is decreasing, with* $Y_n \leqslant X_n \leqslant Z_n$, $n = 1, 2, \ldots$, *and* $\mathscr{F}\text{-o-}\lim (Z_n - Y_n) = 0$. *In this case* Y_n *and* Z_n, $n = 1, 2, \ldots$, *are also* $\mathscr{F}$*-o-fundamental, and, if* X_n, $n = 1, 2, \ldots$, *is* $\mathscr{F}$*-o-convergent, then they are also* $\mathscr{F}$*-o-convergent.*

Proof. Suppose there is a decreasing sequence $U_n \in \mathscr{F}$, $n = 1, 2, \ldots$, such that

$$|X_n - X_{n+k}| \leqslant U_n, \quad n = 1, 2, \ldots, \quad k = 1, 2, \ldots,$$

and

$$(\mathscr{F}) \bigwedge_{n=1}^{\infty} U_n = 0;$$

then
$$X_n - U_n \leqslant X_m \leqslant X_n + U_n \quad \text{for all} \quad n \leqslant m,$$

i.e., for $n = 1, 2, \ldots, m$. We set

$$Y_m = \bigvee_{n=1}^{m} (X_n - U_n), \quad Z_m = \bigwedge_{n=1}^{m} (X_n + U_n).$$

Clearly,

$$X_m - U_m \leqslant Y_m \leqslant X_m \leqslant Z_m \leqslant X_m + U_m, \quad m = 1, 2, \ldots,$$

which implies

$$Z_m - Y_m \leqslant (X_m + U_m) - (X_m - U_m) = 2U_m,$$

i.e.,
$$\mathscr{F}\text{-o-}\lim (Z_m - Y_m) = 0.$$

The rest of the proof is obvious.

Thus an $\mathscr{F}$-o-fundamental sequence is of the same nature as an $\mathscr{F}$-o-convergent, except that it may lack a limit.

Lemma 4.2.

If $X_n \in \mathscr{E}$, $n = 1, 2, \ldots$, *is* $\mathscr{F}$*-o-fundamental then so is the sequence* $X_1, X_1 \vee X_2, X_1 \vee X_2 \vee X_3, \ldots$, *and its dual, also.*

Proof. Suppose there is a decreasing sequence $U_n \in \mathscr{F}$ such that

$$|X_n - X_{n+k}| \leqslant U_n, \qquad n = 1, 2, \ldots, \quad \text{and} \quad (\mathscr{F}) \bigwedge_{n=1}^{\infty} U_n = 0.$$

We set
$$Y_n = X_1 \vee X_2 \vee \ldots \vee X_n,$$
then
$$|Y_n - Y_{n+k}| = |(X_1 \vee X_2 \vee \ldots \vee X_n) \vee X_n - (X_1 \vee X_2 \vee \ldots \vee X_n) \vee (X_{n+1} \vee \ldots \vee X_{n+k})|,$$
and according to (13) of Appendix II, Section 1,
$$\leqslant |X_n - (X_{n+1} \vee \ldots \vee X_{n+k})| \leqslant |X_{n+1} - X_n| \vee \ldots \vee |X_{n+k} - X_n| \leqslant U_n$$
i.e.,
$$|Y_n - Y_{n+k}| \leqslant U_n, \qquad n = 1, 2, \ldots, \quad k = 1, 2, \ldots .$$

Theorem 4.2.

Every $\mathscr{F}$-o-fundamental sequence is bounded in $\mathscr{E}$.

Proof. Suppose there is a decreasing sequence $U_n \in \mathscr{F}$, $n = 1, 2, \ldots$, with
$$|X_n - X_{n+k}| \leqslant U_n, \qquad n = 1, 2, \ldots;$$
then
$$|X_1 - X_n| \leqslant U_1, \qquad n = 1, 2, \ldots;$$
hence
$$X_1 - U_1 \leqslant X_n \leqslant X_1 + U_1, \qquad n = 1, 2, \ldots,$$
with $X_1 - U_1 \in \mathscr{E}$, $X_1 + U_1 \in \mathscr{E}$; i.e., X_n, $n = 1, 2, \ldots$, is bounded in $\mathscr{E}$.

Theorem 3.7 and the previous theorems imply the following three theorems:

Theorem 4.3.

If the sequence $X_n \in \mathscr{E}$, $n = 1, 2, \ldots$, is bounded in $\mathscr{E}$, i.e., if L and U exist in $\mathscr{E}$, such that
$$L \leqslant X_n \leqslant U, \qquad n = 1, 2, \ldots,$$
and
$$\bigwedge_{n=1}^{\infty} \bigvee_{j \geqslant n} t^c_{X_j}(\xi) \underset{\text{def}}{=} \lambda(\xi), \quad \bigvee_{n=1}^{\infty} \bigwedge_{j \geqslant n} t^c_{X_j}(\xi) \underset{\text{def}}{=} \lambda^*(\xi), \qquad \xi \in R,$$
then
$$\lambda(\xi+0) \geqslant \lambda^*(\xi+0),$$
for every $\xi \in R$, $\lambda(\xi)$ and $\lambda^*(\xi)$ are decreasing with
$$\lambda(-\infty) = \lambda^*(-\infty) = e, \quad \lambda(+\infty) = \lambda^*(+\infty) = \emptyset,$$
and, moreover,
$$\lambda(\xi+0) = \lambda^*(\xi+0)$$
if and only if $X_n \in \mathscr{E}$, $n = 1, 2, \ldots$, is an $\mathscr{E}$-o-fundamental sequence.

Theorem 4.4.

Let $X_n \in \mathscr{E}$, $n = 1, 2, \ldots$, $X \in \mathscr{E}$, and write

$$\sigma(\xi) = \bigwedge_{n=1}^{\infty} \bigvee_{j \geqslant n} s^c_{|X_j - X|}(\xi), \qquad \xi \in R;$$

then $X_n \underset{\mathscr{E}}{\overset{o}{\rightarrow}} X$ if and only if

$$\sigma(\xi - 0) = \begin{cases} e & \text{for } \xi \leqslant 0 \\ \varnothing & \text{for } \xi > 0 \end{cases},$$

or if and only if

$$\bigwedge_{n=1}^{\infty} \bigvee_{j \geqslant n} s^c_{|X_j - X|}(\xi) = \varnothing$$

for every $\xi > 0$.

Theorem 4.5.

Let $X_n \in \mathscr{E}$, $n = 1, 2, \ldots$, be an $\mathscr{E}$-o-fundamental sequence in $\mathscr{E}$, and write

$$\sigma(\xi) = \bigwedge_{j=1}^{\infty} \bigvee_{n, k \geqslant j} s^c_{|X_n - X_k|}(\xi),$$

for every $\xi \in R$. Then $\sigma(\xi) = \varnothing$, for every $\xi > 0$.

4.7. In this section the $\mathscr{E}$-o-convergence, $\mathscr{S}$-o-convergence and R-o-convergence will be compared. In the following, we will say o-convergence instead of $\mathscr{E}$-o-convergence and uniform convergence briefly u-convergence instead of R-o-convergence, and it will be written $X_n \underset{\mathscr{E}}{\overset{o}{\rightarrow}} X$, or $X_n \underset{\mathscr{E}}{\overset{u}{\rightarrow}} X$, or $X_n \underset{\mathscr{E}}{\overset{\mathscr{F}\text{-}o}{\rightarrow}} X$, if the sequence $X_n \in \mathscr{E}$, $n = 1, 2, \ldots$, o- or u- or $\mathscr{F}$-o-converges to X in $\mathscr{E}$ respectively.

Analogously, we will say o-fundamental, u-fundamental instead of $\mathscr{E}$-o-fundamental, R-o-fundamental respectively. Clearly, u-convergence resp u-fundamentality implies $\mathscr{S}$-o-convergence, resp $\mathscr{S}$-u-fundamentality, and this implies o-convergence, resp o-fundamentality. Conversely, an o-convergent, resp o-fundamental sequence does not always u-converge; for example, if $\mathbf{a} = \{a_1, a_2, \ldots\}$ is a infinite experiment, and we set

$$X_n = \sum_{k=1}^{n} k I_{a_k}, \quad X = \sum_{k=1}^{\infty} k I_{a_k},$$

then we have $X_n \underset{\mathscr{E}}{\overset{o}{\rightarrow}} X$, but X_n, $n = 1, 2, \ldots$, does not u-converge to X.

The following theorem is true:

Theorem 4.6.

If a sequence $X_n \in \mathscr{S}$, $n = 1, 2, \ldots$, is u-fundamental, then there is a positive number δ such that $|X_n| \leqslant \delta$, $n = 1, 2, \ldots$, i.e., any u-fundamental sequence of simple random variables is always bounded by a constant.

Proof. There is a decreasing sequence of positive numbers δ_n, $n = 1, 2, \ldots$, such that

$$|X_n - X_{n+k}| \leqslant \delta_n, \qquad n = 1, 2, \ldots;$$

hence

$$X_1 - \delta_1 \leqslant X_n \leqslant X_1 + \delta_1,$$

for every $n = 1, 2, \ldots$, and

$$|X_n| \leqslant |X_1 - \delta_1| \vee |X_1 + \delta_1| = Y \in \mathscr{S},$$

i.e., there is a finite experiment

$$\mathbf{a} = \{a_1, a_2, \ldots, a_n\}$$

such that

$$Y = \sum_{i=1}^{n} \xi_i I_{a_i}$$

with $\xi_i \geqslant 0$; we set $\delta = \max\{\xi_1, \xi_2, \ldots, \xi_n\} \in R$; then

$$0 \leqslant |X_n| \leqslant Y \leqslant \delta \quad \text{for every} \quad n = 1, 2, \ldots.$$

4.8. Two sequences $X_n \in \mathscr{E}$, $Y_n \in \mathscr{E}$, $n = 1, 2, \ldots$, are said to be o-equivalent or u-equivalent in $\mathscr{E}$ if and only if $X_n - Y_n \underset{\mathscr{E}}{\overset{o}{\rightarrow}} 0$ or $X_n - Y_n \underset{\mathscr{E}}{\overset{u}{\rightarrow}} 0$ respectively; then we shall write

$$\{X_n\} \overset{o}{\approx} \{Y_n\} \quad \text{or} \quad \{X_n\} \overset{u}{\approx} \{Y_n\} \quad \text{respectively.}$$

In general, the previous sequences are said to be $\mathscr{F}$-o-equivalent in $\mathscr{E}$, if and only if $X_n - Y_n \underset{\mathscr{E}}{\overset{\mathscr{F}\text{-}o}{\rightarrow}} 0$; we shall then write

$$\{X_n\} \overset{\mathscr{F}\text{-}o}{\approx} \{Y_n\}.$$

Theorem 4.7.

Let X_n, X_n^*, Y_n, Y_n^*, $n = 1, 2, \ldots$, be any sequence in $\mathscr{E}$; then, if

$$\{X_n\} \overset{\mathscr{F}\text{-}o}{\approx} \{Y_n\} \quad \text{and} \quad \{X_n^*\} \overset{\mathscr{F}\text{-}o}{\approx} \{Y_n^*\},$$

we have

$$\{X_n \pm X_n^*\} \overset{\mathscr{F}\text{-}o}{\approx} \{Y_n \pm Y_n^*\}, \quad \{X_n \vee X_n^*\} \overset{\mathscr{F}\text{-}o}{\approx} \{Y_n \vee Y_n^*\},$$

and dually,

$$\{\lambda X_n\} \overset{\mathscr{F}\text{-}o}{\approx} \{\lambda Y_n\}, \quad \{|X_n|\} \overset{\mathscr{F}\text{-}o}{\approx} \{|Y_n|\};$$

The product $\{X_n X_n^*\}$ is o-equivalent, but not always $\mathscr{F}$-o-equivalent, if $\mathscr{F} \neq \mathscr{E}$.

The reader can prove

Theorem 4.8.

(1) *Let*

$$X_n \in \mathscr{E}, \qquad n = 1, 2, \ldots,$$

and

$$\bigvee_{n=1}^{\infty} \bigwedge_{j \geqslant n} s_{X_j}(\xi) = \lambda(\xi),$$

$$\bigwedge_{n=1}^{\infty} \bigvee_{j \geqslant n} s_{X_j}(\xi) = \rho(\xi), \qquad \xi \in R.$$

Then X_n, $n = 1, 2, \ldots$, is o-fundamental, if and only if $\lambda(\xi - 0) = \rho(\xi - 0)$, for every $\xi \in R$, and moreover, the sequence X_n, $n = 1, 2, \ldots$, is bounded in $\mathscr{E}$, or, equivalently, $\rho(-\infty) = \emptyset$ and $\lambda(+\infty) = e$; in that case we define

$$s_{\{X_n\}}(\xi) = \lambda(\xi - 0) = \rho(\xi - 0).$$

(2) *Two o-fundamental sequences $\{X_n\}$ and $\{Y_n\}$ are o-equivalent if and only if*

$$s_{\{X_n\}}(\xi) = s_{\{Y_n\}}(\xi), \quad \text{for every} \quad \xi \in R.$$

Theorem 4.9.

If $\{X_n\} \overset{\mathscr{F}\text{-}o}{\approx} \{Y_n\}$ and one of the given sequences is $\mathscr{F}$-o-fundamental, then the other is also $\mathscr{F}$-o-fundamental.

Now we shall prove the following:

Fundamental Theorem 4.10.

If $X_n \in \mathscr{E}$ is an o-fundamental sequence in $\mathscr{E}$, and $\mathscr{F}$ is a σ-regular vector sublattice of $\mathscr{E}$ such that $R \subseteq \mathscr{F}$, then there exists an $\mathscr{F}$-o-fundamental sequence $Y_n \in \mathscr{E}$, $n = 1, 2, \ldots$, such that

$$\{X_n\} \overset{o}{\approx} \{Y_n\} \text{ in } \mathscr{E}.$$

Proof. It is sufficient to prove the theorem if $\mathscr{F} = R$, i.e., for a u-fundamental sequence; then every u-fundamental sequence is also a $\mathscr{F}$-o-fundamental sequence. We may assume $X_n \uparrow$ and $X_n \geqslant 0$ without any loss of generality, for Lemma 4.1 implies the existence in $\mathscr{E}$ of an o-fundamental increasing sequence, which is o-equivalent to the given o-fundamental sequence X_n, $n = 1, 2, \ldots$,. We write

$$\sigma(\xi) = \bigwedge_{n=1}^{\infty} s_{X_n}(\xi) \quad \text{and} \quad s(\xi) = \sigma(\xi - 0), \qquad \xi \in R;$$

then $s(\xi)$ is increasing and continuous from the left, with $s(-\infty) = \emptyset$, $s(+\infty) = e$. Now we define

$$a_{mn} = s\left(\frac{m+1}{2^n}\right) \wedge s^c\left(\frac{m}{2^n}\right), \qquad \begin{cases} m = 0, 1, 2, \ldots \\ n = 0, 1, 2, \ldots \end{cases},$$

and we have
$$e = a_{0n} \vee a_{1n} \vee \ldots$$
with
$$a_{in} \wedge a_{jn} = \emptyset, \qquad i \neq j.$$
Hence
$$\mathbf{a}_n = \{a_{0n}, a_{1n}, \ldots\}$$
is, for every $n = 0, 1, 2, \ldots$, an experiment. We define the following erv's:

$$Y_n\colon\ \mathbf{a}_n \ni a_{mn} \Rightarrow \frac{m}{2^n} = Y_n(a_{mn}) \in R,$$

$$\tilde{Y}_n\colon\ \mathbf{a}_n \ni a_{mn} \Rightarrow \frac{m+1}{2^n} = \tilde{Y}_n(a_{mn}) \in R.$$

Obviously, we have

$$\tilde{Y}_n - Y_n = \frac{1}{2^n}, \qquad n = 1, 2, \ldots,$$

i.e., $\{\tilde{Y}_n\}$ and $\{Y_n\}$ are u-equivalent. We shall prove that $Y_n\uparrow$, $\tilde{Y}_n\downarrow$, and both are u-fundamental. We have

$$Y_{n+1} - Y_n\colon\ \mathbf{a}_n \wedge \mathbf{a}_{n+1} \ni a_{m,n} \wedge a_{k,n+1} \Rightarrow \frac{k}{2^{n+1}} - \frac{m}{2^n},$$

for all

$$a_{m,\,n} \wedge a_{k,\,n+1} \neq \varnothing.$$

But if

$$\frac{k}{2^{n+1}} - \frac{m}{2^n} < 0,$$

i.e., $k < 2m$, then

$$a_{mn} \wedge a_{k,\,n+1} = \varnothing.$$

In fact, since $k < 2m$, we have

$$\frac{k+1}{2^{n+1}} \leqslant \frac{m}{2^n},$$

i.e.,

$$\frac{k}{2^{n+1}} < \frac{k+1}{2^{n+1}} \leqslant \frac{m}{2^n} < \frac{m+1}{2^n};$$

hence, since s is increasing, we have

$$s\left(\frac{k}{2^{n+1}}\right) \leqslant s\left(\frac{k+1}{2^{n+1}}\right) \leqslant s\left(\frac{m}{2^n}\right) \leqslant s\left(\frac{m+1}{2^n}\right),$$

and

$$a_{mn} = s^c\left(\frac{m}{2^n}\right) \wedge s\left(\frac{m+1}{2^n}\right) = s\left(\frac{m+1}{2^n}\right) - s\left(\frac{m}{2^n}\right)$$

$$a_{k,\,n+1} = s^c\left(\frac{k}{2^{n+1}}\right) \wedge s\left(\frac{k+1}{2^{n+1}}\right) = s\left(\frac{k+1}{2^{n+1}}\right) - s\left(\frac{k}{2^{n+1}}\right).$$

Thus, a_{mn} and $a_{k,\,n+1}$ are disjoint. If

$$a_{mn} \wedge a_{k,\,n+1} \neq \varnothing,$$

we always have

$$\frac{k}{2^{n+1}} - \frac{m}{2^n} \geqslant 0;$$

hence $Y_{n+1} - Y_n \geqslant 0$, i.e., $Y_n\uparrow$. Similarly it can be proved $\tilde{Y}_n\downarrow$. Now, $Y_n\uparrow$, $\tilde{Y}_n\downarrow$, and

$$\tilde{Y}_n - Y_n = \frac{1}{2^n}$$

implies that both sequences Y_n, $\tilde{Y}_n$ are u-fundamental.

We shall prove now that both sequences Y_n, $\tilde{Y}_n$, $n = 1, 2, \ldots$, are o-equivalent to the sequence X_n, $n = 1, 2, \ldots$. In fact, $X_n\uparrow$, $Y_n\uparrow$; hence $s_{X_n}(\xi)\downarrow$, $s_{Y_n}(\xi)\downarrow$. Moreover,

$$\bigwedge_{n=1}^{\infty} s_{X_n}(\xi) = \sigma(\xi)$$

and $$\bigwedge_{n=1}^{\infty} s_{Y_n}(\xi) = s(\xi) = \sigma(\xi - 0);$$

hence, by Theorem 4.8
$$\{X_n\} \overset{o}{\approx} \{Y_n\} \overset{u}{\approx} \{\tilde{Y}_n\}.$$

The following theorem is also true:

Theorem 4.11.

If $X_n \in \mathscr{E}$, $n = 1, 2, \ldots$, is an o-fundamental sequence in $\mathscr{E}$, then there exists an o-fundamental sequence $Y_n \in \mathscr{S}$, $n = 1, 2, \ldots$, such that $\{X_n\} \overset{o}{\approx} \{Y_n\}$.

5. EXTENSION OF THE ELEMENTARY STOCHASTIC SPACE

5.1. The vector lattice $\mathscr{E}$ of all erv's over a pr σ-algebra $(\mathfrak{B}, p)$, considered as a lattice, is conditionally complete (with respect to the lattice operations), hence closed with respect to the o-convergence and u-convergence, if and only if the Boolean σ-algebra $\mathfrak{B}$ is atomic. If the Boolean σ-algebra $\mathfrak{B}$ is not atomic, then $\mathscr{E}$ is not conditionally complete. We shall call a vector lattice $\mathscr{A}$ a *σ-extension* of $\mathscr{E}$, if and only if (1) $\mathscr{A}$ is conditionally σ-complete, considered as a lattice, and (2) the vector lattice $\mathscr{E}$ can be embedded σ-regularly (σ-invariantly) in $\mathscr{A}$. Let $\mathscr{A}$ be a σ-extension of $\mathscr{E}$ and $\mathscr{E}^*$ an isomorphic and σ-regular (σ-invariant) image of $\mathscr{E}$ into $\mathscr{A}$; then the σ-extension $\mathscr{A}$ is said to be a *minimal* σ-extension, if and only if the smallest vector σ-sublattice of $\mathscr{A}$ containing $\mathscr{E}^*$ is $\mathscr{A}$ itself. We shall prove that a σ-extension of $\mathscr{E}$ can always be defined, which, considered as a lattice, is even conditionally complete, hence closed with respect to the o-convergence and u-convergence, and moreover closed with respect to the convergence in probability and almost uniform convergence, which will be defined in Section 7.

5.2. Let $\mathscr{F}$, $\mathscr{K}$, and $\mathscr{N}$ be the set of all o-fundamental, all o-convergent, and all o-null sequences in $\mathscr{F}$ respectively. Then we have
$$\mathscr{F} \supseteq \mathscr{K} \supseteq \mathscr{N}.$$

We define in $\mathscr{F}$ the following operations:

Let $\{X_n\} \in \mathscr{F}$ and $\{Y_n\} \in \mathscr{F}$; then we define
$$\{X_n\} + \{Y_n\} = \{X_n + Y_n\}, \quad \{X_n\}\{Y_n\} = \{X_n Y_n\}$$
$$\lambda\{X_n\} = \{\lambda X_n\}$$
$$\{X_n\} \vee \{Y_n\} = \{X_n \vee Y_n\}$$
$$\{X_n\} \wedge \{Y_n\} = \{X_n \wedge Y_n\}.$$

It is easy to prove that $\mathscr{F}$ is a vector lattice with respect to these operations, $\mathscr{K}$ is a vector sublattice of $\mathscr{F}$, and $\mathscr{N}$ is an l-ideal in $\mathscr{K}$. The map

$$\mathscr{E} \ni X \Rightarrow \{X_n\} \in \mathscr{F} \quad \text{with} \quad X_n = X, \qquad n = 1, 2, \ldots, \tag{1}$$

embeds the vector lattice $\mathscr{E}$ isomorphically with respect to all operations and completely regularly (invariantly) with respect to the lattice operations in $\mathscr{F}$, i.e., if $X = (\mathscr{E}) \bigvee_{i \in I} X_i$, where I any set of indices then

$$\{X_n\} = (\mathscr{F}) \bigvee_{i \in I} X_{in}$$

and dually, where $X_n = X$, and $X_{in} = X_i$, $n = 1, 2, \ldots, i \in I$.

It is easy to prove this statement, if one notices that $\{Y_n\} \leqslant \{Z_n\}$ in $\mathscr{F}$ if and only if $Y_n \leqslant Z_n$ in $\mathscr{E}$, $n = 1, 2, \ldots$. The sets $\mathscr{F}/\mathscr{N}$ and $\mathscr{K}/\mathscr{N}$ ($\mathscr{F}$ mod $\mathscr{N}$ and $\mathscr{K}$ mod $\mathscr{N}$) of all classes $\{X_n\}/\mathscr{N}$, $\{X_n\} \in \mathscr{F}$, and $\{Y_n\}/\mathscr{N}$, $\{Y_n\} \in \mathscr{K}$, constitute respectively a vector lattice $\mathscr{V} = \mathscr{F}/\mathscr{N}$ and a vector sublattice $\mathscr{E}^* = \mathscr{K}/\mathscr{N}$ of $\mathscr{V}$. Note that a multiplication of two elements of $\mathscr{V}$ if defined in $\mathscr{V}$, and $\mathscr{V}$ is, with respect to addition and this multiplication, an algebra with unit. The map

$$\mathscr{E} \ni X \Rightarrow \{Y_n\}/\mathscr{N}, \quad \text{where} \quad Y_n = X, \qquad n = 1, 2, \ldots, \tag{2}$$

embeds the vector lattice $\mathscr{E}$ isomorphically and completely regularly (invariantly) in $\mathscr{V}$. The vector sublattice $\mathscr{E}^*$ is the image of $\mathscr{E}$ under this map. Hence, the vector lattice $\mathscr{V}$ may be considered as an extension of the vector lattice $\mathscr{E}$. We shall prove that $\mathscr{V}$ is a minimal σ-extension of $\mathscr{E}$. The image of $X \in \mathscr{E}$ in $\mathscr{E}^*$ will be denoted by $\tilde{X}$. In general, a class $\{X_n\}/\mathscr{N} \in \mathscr{V}$ will be denoted by $[\{X_n\}]$. We shall prove the following:

Theorem 5.1.

If $\{X_n\} \in \mathscr{F}$, then o-$\lim \tilde{X}_n = [\{X_n\}]$ in $\mathscr{V}$; i.e., the vector lattice $\mathscr{E}$, identified with its image $\mathscr{E}^*$ and in this way regarded as a vector sublattice of $\mathscr{V}$, is o-dense in $\mathscr{V}$.

Proof. Lemma 4.1 implies the existence of two o-fundamental monotone sequences $A_n \uparrow$, $B_n \downarrow$ in $\mathscr{E}$ such that $A_n \leqslant X_n \leqslant B_n$, $n = 1, 2, \ldots$, and $B_n - A_n \underset{\mathscr{E}}{\overset{o}{\to}} 0$; then $\tilde{A}_n \uparrow$, $\tilde{B}_n \downarrow$, $\tilde{A}_n \leqslant \tilde{X}_n \leqslant \tilde{B}_n$, $n = 1, 2, \ldots$, and

$$\tilde{B}_n - \tilde{A}_n \overset{o}{\to} 0$$

in $\mathscr{V}$. Obviously $[\{X_n\}] \geqslant \tilde{A}_k$, $k = 1, 2, \ldots$. Let $[\{Y_n\}] \geqslant \tilde{A}_k$, $k = 1, 2, \ldots$.

We can suppose that $Y_n \downarrow$; then $Y_k \geqslant A_k$, $k = 1, 2, \dots$; i.e.,

$$[\{Y_k\}] \geqslant [\{A_k\}] = [\{X_k\}].$$

Hence we have

$$[\{X_n\}] = \bigvee_{k=1}^{\infty} \tilde{A}_k = \bigwedge_{k=1}^{\infty} \tilde{B}_k \quad \text{in} \quad \mathscr{V},$$

and

$$o\text{-}\lim_{k\to\infty} \tilde{X}_k = [\{X_n\}] \quad \text{in} \quad \mathscr{V}.$$

The following theorem is true:

Theorem 5.2.

The vector lattice $\mathscr{V}$ is closed with respect to the o-convergence, i.e., every o-fundamental sequence in $\mathscr{V}$ o-converges in $\mathscr{V}$.

Proof.† In order to prove this theorem we can prove, as in Theorem 4.10, that to every o-fundamental sequence in $\mathscr{V}$ there corresponds a u-fundamental sequence in $\mathscr{V}$, which is o-equivalent to it. Hence, it suffices to prove that $\mathscr{V}$ is closed with respect to the u-convergence.

Let $[\{X_{nk}\}] \in \mathscr{V}$, $k = 1, 2, \dots$, be a u-fundamental sequence, where the sequence X_{nk}, $n = 1, 2, \dots$, is a u-fundamental sequence in $\mathscr{E}$, for every $k = 1, 2, \dots$; then we have

$$u\text{-}\lim_{n\to\infty} \tilde{X}_{nk} = [\{X_{nk}\}], \quad k = 1, 2, \dots .$$

Let ε_k, $k = 1, 2, \dots$, be a monotone null sequence of positive numbers. Then, for every k, there exists an index n_k such that

$$|\tilde{X}_{n_k, k} - [\{X_{nk}\}]| < \varepsilon_k, \quad k = 1, 2, \dots .$$

$\tilde{X}_{n_k, k}$, $k = 1, 2, \dots$, is a u-fundamental sequence in $\mathscr{E}^*$; hence $X_{n_k, k}$, $k = 1, 2, \dots$, is a u-fundamental sequence in $\mathscr{E}$, which defines a class $[\{X_{n_k, k}\}] \in \mathscr{V}$. This class is the u-limit of the sequence $\tilde{X}_{n_k, k}$, $k = 1, 2, \dots$, in $\mathscr{V}$. But the sequence $[\{X_{n, k}\}]$, $k = 1, 2, \dots$, is u-equivalent to $\tilde{X}_{n_k, k}$, $k = 1, 2, \dots$. Hence, we have

$$u\text{-}\lim_{k\to\infty} [\{X_{n, k}\}] = [\{X_{n_k, k}\}] \quad \text{in} \quad \mathscr{V}.$$

† A different proof, dealing with the general case of an abelian lattice-group, is given in Papangelou [2], Corollary 4.5. See also Theorem 3.2, Chapter VII.

The Theorems 4.1 and 4.2 and the property 4.6.4 formulated for $\mathscr{V}$ instead of $\mathscr{E}$ imply:

Corollary 5.1.

If $\mathscr{B}$ is any vector sublattice of $\mathscr{V}$ and regular in $\mathscr{V}$ (for example, R, $\mathscr{S}$, $\mathscr{E}$, considered as vector sublattices of $\mathscr{V}$ are regular in $\mathscr{V}$), then $\mathscr{V}$ is closed with respect to the $\mathscr{B}$-o-convergence, i.e., every $\mathscr{B}$-o-fundamental sequence in $\mathscr{V}$ is $\mathscr{B}$-o-convergent in $\mathscr{V}$, hence also $\mathscr{V}$-o-convergent = o-convergent in $\mathscr{V}$. The vector lattice $\mathscr{E}$ considered as a vector sublattice of $\mathscr{V}$ is $\mathscr{B}$-o-dense in $\mathscr{V}$.

5.3. We shall call the vector lattice $\mathscr{V}$ the *stochastic space* over $(\mathfrak{B}, p)$ and shall regard the elementary and simple stochastic spaces $\mathscr{E}$ and $\mathscr{S}$ respectively as vector sublattices of $\mathscr{V}$, i.e., $\mathscr{S} \subseteq \mathscr{E} \subseteq \mathscr{V}$. We will denote the elements of $\mathscr{V}$ by capital letters $X, Y, Z, \ldots$ and will call them *random variables*, briefly rv's. We proved in the preceding section that $\mathscr{E}$ is o-dense and u-dense in $\mathscr{V}$. Since $\mathscr{S}$ is o-dense, exactly $\mathscr{E}$-o-dense, in $\mathscr{E}$, we can prove that $\mathscr{S}$ is $\mathscr{E}$-o-dense in $\mathscr{V}$ and therefore o-dense = $\mathscr{V}$-o-dense. Obviously $\mathscr{S}$ is not u-dense in $\mathscr{V}$; because every u-fundamental sequence in $\mathscr{S}$ is always bounded by a constant (see Theorem 4.6); hence it u-converges to an element in $\mathscr{V}$, which is bounded by a constant, whereas there exist elements in $\mathscr{V}$, which are not bounded by a constant. For every rv $X \in \mathscr{V}$, there exist sequences $X_n \in \mathscr{E}$ and $Y_n \in \mathscr{S}$ respectively, such that

$$u\text{-}\lim_{n\to\infty} X_n = X \quad \text{and} \quad o\text{-}\lim Y_n = X \quad \text{in} \quad \mathscr{V}.$$

Then we can prove that

$$o\text{-}\lim\sup s_{X_n}(\xi) = \bar{\sigma}(\xi)$$

and

$$o\text{-}\lim\inf s_{X_n}(\xi) = \underline{\sigma}(\xi)$$

exist and we have

$$\bar{\sigma}(\xi-0) = \underline{\sigma}(\xi-0) = \sigma(\xi);$$

the function $\sigma(\xi)$ is increasing, continuous from the left, and such that $\sigma(-\infty) = \emptyset$, $\sigma(+\infty) = e$; it is independent of the choice of the sequence X_n with $u\text{-}\lim_{n\to\infty} X_n = X$. Moreover, every sequence $Y_n \in \mathscr{E}$ or $\mathscr{S}$ with $o\text{-}\lim Y_n = X$ defines by the foregoing the same function $\sigma(\xi)$. Thus we write

$$[X < \xi] \underset{\text{def}}{=} s_X(\xi) \underset{\text{def}}{=} \sigma(\xi) = \bar{\sigma}(\xi-0) = \underline{\sigma}(\xi-0), \quad -\infty < \xi < +\infty.$$

There also exist

$$o\text{-}\lim\sup t_{X_n}(\xi) = \bar{\tau}(\xi) \quad \text{and} \quad o\text{-}\lim\inf t_{X_n}(\xi) = \underline{\tau}(\xi),$$

and we have:

$$\tau(\xi) \underset{\text{def}}{=} \bar{\tau}(\xi+0) = \underline{\tau}(\xi+0), \text{ for every } \xi \in R,$$

independently of the choice of the sequence $X_n \in \mathscr{E}$ with u-lim $X_n = X$ or o-lim $X_n = X$ in $\mathscr{V}$. Hence we write

$$[X \leqslant \xi] \underset{\text{def}}{=} t_X(\xi) \underset{\text{def}}{=} \tau(\xi) = \bar{\tau}(\xi+0) = \underline{\tau}(\xi+0), \text{ for every } \xi \in R.$$

The function $t_X(\xi)$ is monotone, increasing, continuous from the right, and such that $t_X(-\infty) = \emptyset$, $t_X(+\infty) = e$. Now, we define

$$[X \geqslant \xi] = s_X{}^c(\xi) \text{ and } [X > \xi] = t_X{}^c(\xi), \quad -\infty < \xi < +\infty.$$

The chain $[X < \xi] = s_X(\xi)$ or $[X \leqslant \xi] = t_X(\xi)$ characterizes uniquely the rv X, i.e., $s_X(\xi) = s_Y(\xi)$, for every $\xi \in R$, if and only if $X = Y$ in $\mathscr{V}$. Conversely, if $s(\xi)$, $-\infty < \xi < +\infty$, is any monotone, increasing, continuous from the left function in ξ with $s(-\infty) = \emptyset$, $s(+\infty) = e$, then there exists a rv $X \in \mathscr{V}$ such that $s(\xi) = s_X(\xi)$, for every $\xi \in R$.

5.4. Let $X_i \in \mathscr{V}$, $i \in I$, be a bounded family in $\mathscr{V}$; then

$$\bigwedge_{i \in I} s_{X_i}(\xi) = \sigma(\xi)$$

exists for every $\xi \in R$. We write $\sigma(\xi-0) = s(\xi)$, $\xi \in R$; $s(\xi)$ is increasing, continuous from the left, with $s(-\infty) = \emptyset$, $s(+\infty) = e$. Hence, there exists a rv $X \in \mathscr{V}$ such that $s_X(\xi) = s(\xi)$. It is easy to prove that $X = \bigvee_{i \in I} X_i$. Moreover, $\bigvee_{i \in I} s_{X_i}(\xi) = \rho(\xi)$ exists for every $\xi \in R$ and is increasing, continuous from the left, with $\rho(-\infty) = \emptyset$, $\rho(+\infty) = e$. Hence there exists a rv $Y \in \mathscr{V}$ such that $\rho(\xi) = s_Y(\xi)$ for every $\xi \in R$. It is easy to prove that $Y = \bigwedge_{i \in I} X_i$. Thus the following theorem is true:

Theorem 5.3.

The stochastic space $\mathscr{V}$, considered as a lattice, is conditionally complete, i.e., for any bounded family X_i, $i \in I$, in $\mathscr{V}$, there exist $\bigvee_{i \in I} X_i$ and $\bigwedge_{i \in I} X_i$ respectively in $\mathscr{V}$.

5.5. Let $\mathbf{B}$ be the σ-field of all Borel subsets of the real line R; then the set $\mathfrak{D}$ of all intervals $(-\infty, \xi) = i_\xi$, $\xi \in R$, is a chain in $\mathbf{B}$, which σ-generates $\mathbf{B}$, i.e., $\mathbf{B}$ is the smallest σ-field of subsets of R containing $\mathfrak{D}$.

The map

$$\mathfrak{D} \in i_\xi \Rightarrow h_X(i_\xi) \underset{\text{def}}{=} [X < \xi] \in \mathfrak{B}, \quad \xi \in R,$$

can be extended from $\mathfrak{D}$ to $\mathbf{B}$ so that the extension $\hat{h}_X$ so obtained is a σ-homomorphism of the σ-field $\mathbf{B}$ into the Boolean σ-algebra $\mathfrak{B}$. We have $\hat{h}_X(A) = \hat{h}_Y(A)$, for every $A \in \mathbf{B}$, if and only if $X = Y$. Moreover, if ϕ is any σ-homomorphism of $\mathbf{B}$ into $\mathfrak{B}$, then the restriction $\phi(i_\xi)$, $i_\xi \in \mathfrak{D}$, defines a chain: $s(\xi) = \phi(i_\xi) \in \mathfrak{B}$, $\xi \in R$, in $\mathfrak{B}$, which is increasing, continuous from the left, with $s(-\infty) = \emptyset$, $s(+\infty) = e$. Hence there exists a rv $X \in \mathscr{V}$ such that $s_X(\xi) = s(\xi)$ for every $\xi \in R$. In this way, a one-to-one correspondence between the set of all σ-homomorphisms of $\mathbf{B}$ into $\mathfrak{B}$ and the set $\mathscr{V}$ of all rv's is defined.

Now, since the function:

$$P_X(A) = p(\hat{h}_X(A)), \quad \text{for every} \quad A \in \mathbf{B},$$

is a quasi-probability on $\mathfrak{B}$, we can assign to X a uniquely defined pr space, the so-called "*distribution pr space*" or the "*sample pr space*" of X: $(R, \mathbf{B}, P_X)$. The chain $s_X(\xi)$, $-\infty < \xi < +\infty$, σ-generates a Boolean σ-subalgebra $\mathfrak{B}_X$ of the Boolean σ-algebra $\mathfrak{B}$; then $(\mathfrak{B}_X, p)$ is the *distribution pr σ-algebra* of X. Thus to every rv $X \in \mathscr{V}$, there correspond uniquely a pr space $(R, \mathbf{B}, P_X)$ and a pr σ-algebra $(\mathfrak{B}_X, p)$. Let $\mathfrak{N}_X$ be the σ-ideal of all $K \in \mathbf{B}$ with $P_X(K) = 0$; then $\mathbf{B}/\mathfrak{N}_X$ is a Boolean σ-algebra isomorphic to $\mathfrak{B}_X$, and, if we define

$$\pi(K/\mathfrak{N}_X) = P_X(K) = p(\hat{h}_X(K)),$$

for every class $K/\mathfrak{N}_X \in \mathbf{B}/\mathfrak{N}_X$, then $(\mathbf{B}/\mathfrak{N}_X, \pi)$ is a pr σ-algebra isometric to the pr σ-algebra $(\mathfrak{B}_X, p)$.

5.6. Let $(\Omega, \mathfrak{F}, P)$ be a pr space, which represents set-theoretically the pr σ-algebra $(\mathfrak{B}, p)$. Let, moreover, $\mathfrak{N}$ be the σ-ideal of all $A \in \mathfrak{F}$ with $P(A) = 0$; then there exists an isomorphism ϕ of the quotient Boolean σ-algebra $\mathfrak{F}/\mathfrak{N}$ onto the Boolean σ-algebra $\mathfrak{B}$. A real-valued point function f defined on Ω is said to be *measurable* $(\mathfrak{F})$ if and only if $f^{-1}(K) \in \mathfrak{F}$, for every $K \in \mathbf{B}$; f^{-1} is, obviously, a Boolean σ-homomorphism of the σ-field $\mathbf{B}$ of all Borel subsets of the real line R into the σ-field $\mathfrak{F}$, i.e., $\psi(K) = f^{-1}(K)/\mathfrak{N}$ is a Boolean σ-homomorphism from $\mathbf{B}$ into the quotient Boolean σ-algebra $\mathfrak{F}/\mathfrak{N}$. We write

$$h(K) = \phi(\psi(K)) = \phi(f^{-1}(K)/\mathfrak{N}) \in \mathfrak{B}, \quad \text{for every} \quad K \in \mathbf{B};$$

then h is a Boolean σ-homomorphism of $\mathbf{B}$ into $\mathfrak{B}$ induced by the measurable ($\mathfrak{F}$) function f; obviously, if another measurable ($\mathfrak{F}$) function g is almost everywhere equal to f, then g induces, by the foregoing, the same Boolean σ-homomorphism h of $\mathbf{B}$ into $\mathfrak{B}$. A class of almost everywhere equal measurable ($\mathfrak{F}$) functions defines uniquely a Boolean σ-homomorphism h of $\mathbf{B}$ into $\mathfrak{B}$. Now a rv $X \in \mathscr{V}$ corresponds uniquely to this σ-homomorphism h. Conversely, R. Sikorski [1] has proved that for each Boolean σ-homomorphism h of $\mathbf{B}$ into $\mathfrak{B}$, there exists a measurable ($\mathfrak{F}$) function f defined on Ω, such that f induces, in the same manner as above, the given Boolean σ-homomorphism of $\mathbf{B}$ into $\mathfrak{B}$. Now, if X is a rv in $\mathscr{V}$, then there exists a Boolean σ-homomorphism h_X of $\mathbf{B}$ into $\mathfrak{B}$, such that $h_X(i_\xi) = [X < \xi]$, $i_\xi \in \mathfrak{D}$. To this Boolean σ-homomorphism corresponds a measurable ($\mathfrak{F}$) function f defined on Ω such that f induces h_X. The function f belongs to a class of almost everywhere equal measurable ($\mathfrak{F}$) functions and each of them induces the given Boolean σ-homomorphism. In this way, we have a one-to-one correspondence between the rv's of the stochastic space $\mathscr{V}$ and the classes of almost everywhere equal measurable ($\mathfrak{F}$) functions defined on Ω. The following theorem is true:

Theorem 5.4.

Let $(\mathfrak{B}, p)$ be a pr σ-algebra and $(\Omega, \mathfrak{F}, P)$ be any pr space, which represents set-theoretically the pr σ-algebra $(\mathfrak{B}, p)$. Let $\mathscr{V}$ be the stochastic space over $(\mathfrak{B}, p)$ and $\mathscr{M}$ the set of all measurable ($\mathfrak{F}$) real-valued functions defined on Ω. Then every rv $X \in \mathscr{V}$ is uniquely characterized by a class of almost everywhere equal elements of $\mathscr{M}$, or by a Boolean σ-homomorphism h_X of $\mathbf{B}$ (= the σ-field of all Borel subsets of the real line R) into $\mathfrak{B}$, or by an increasing, continuous from the left, chain $s_X(\xi)$, $-\infty < \xi + \infty$, with $s_X(-\infty) = \emptyset$ and $s_X(+\infty) = e$. To every $X \in \mathscr{V}$ there corresponds a real-valued function $\phi_X(\xi) = p([X < \xi])$, $-\infty < \xi + \infty$, the so-called distribution function of X, which is increasing, continuous from the left, with $\phi_X(-\infty) = 0$, $\phi_X(+\infty) = 1$.

5.7. Note that is is possible for different rv's X and Y to have equal distribution functions ϕ_X and ϕ_Y. Two rv's X and Y are said to be *equimeasurable* if and only if $\phi_X(\xi) = \phi_Y(\xi)$ for every $\xi \in R$. The rv's of a family $X_i \in \mathscr{V}$, $i \in I$, are said to be *p-independent*, if and only if the Boolean σ-algebras $\mathfrak{B}_{X_i}$, $i \in I$, are p-independent (see this concept, Section 5, Chapter III).

Let now X_i, $i \in I$, be a family of not constant rv's in the stochastic space $\mathscr{V}(\mathfrak{B}, p)$ over any pr σ-algebra $(\mathfrak{B}, p)$; then the following problem can be set: Is there in $\mathscr{V}(\mathfrak{B}, p)$ a family Y_i, $i \in I$, of p-independent rv's, such that every rv X_i is equimeasurable to Y_i, i.e., $\phi_{X_i}(\xi) = \phi_{Y_i}(\xi)$, $i \in I$? Section 5 Chapter III implies an answer to this question: there is in $\mathscr{V}(\mathfrak{B}, p)$ a family Y_i, $i \in I$, of p-independent not constant rv's if and only if there exists a pr σ-subalgebra $(\mathfrak{A}, p)$ of $(\mathfrak{B}, p)$ isometric to the product σ-algebra $(\tilde{\mathfrak{F}}, \tilde{\pi}) = \tilde{\mathbf{P}}_{i \in I} (\mathfrak{B}_{X_i}, p_i)$, where $p_i = p$, $i \in I$.

6. THE STOCHASTIC SPACE OF ALL BOUNDED RANDOM VARIABLES [rv's]

6.1. A rv $X \in \mathscr{V}$ is said to be *bounded* if and only if there exists a constant η such that $|X| \leqslant \eta$. Now let $\mathscr{M}$ be the set of all bounded rv's in $\mathscr{V}$; then $\mathscr{M}$ is obviously, a vector sublattice of $\mathscr{V}$, the so-called stochastic space of all bounded rv's. Obviously, the stochastic space $\mathscr{S}$ of all simple rv's is contained in $\mathscr{M}$ and, moreover, is u-dense in $\mathscr{M}$, i.e., for every $X \in \mathscr{M}$ there exists a sequence $X_n \in \mathscr{S}$, $n = 1, 2, \ldots$, such that $X_n \underset{\mathscr{M}}{\overset{u}{\to}} X$. In fact, there exists two constants α and β such that $\alpha < X < \beta$. Let $\varepsilon_n \downarrow 0$, $\varepsilon_n \in R$; then there exists a finite set of real numbers $\alpha = \xi_{n0} < \xi_{n1} < \ldots \xi_{n, k_n} = \beta$ such that $\xi_{n, i+1} - \xi_{n, i} < \varepsilon_n$ for every $i = 0, 1, 2, \ldots, k_n - 1$. We obviously have $[X < \xi_{n, 0}] = \emptyset$ and

$$[X < \xi_{n, k_n}] = e, \qquad n = 1, 2, \ldots .$$

Now, if we write

$$a_{n, i} = s_X(\xi_{n, i+1}) - s_X(\xi_{n, i}) \qquad i = 0, 1, 2, \ldots, k_n - 1,$$

then $e = a_{n, 0} \vee a_{n, 1} \vee \ldots \vee a_{n, k_n - 1}$. We consider the simple rv's:

$$X_n = \sum_{i=0}^{k_n - 1} \xi_{n, i} I_{a_{n, i}}, \qquad n = 1, 2, \ldots;$$

then $|X_n - X| < \varepsilon_n$, $n = 1, 2, \ldots$, i.e., $X_n \underset{\mathscr{M}}{\overset{u}{\to}} X$.

6.2. It is easy to prove that $\mathscr{M}$ is closed with respect to the u-convergence. In fact, if $X_n \in \mathscr{M}$, $n = 1, 2, \ldots$, is a u-fundamental sequence, then $X_n \overset{u}{\to} X \in \mathscr{V}$ and X is bounded, i.e., $X \in \mathscr{M}$. The stochastic space $\mathscr{M}$ of all bounded rv's can be endowed with a norm, namely: we define for every $X \in \mathscr{M}$, $X \geqslant 0$: $\|X\| = \inf \eta$, for all $\eta \in R$ with $X \leqslant \eta$. Then we have

(1) $\|0\| = 0$;

(2) If $X > 0$, then $\|X\| > 0$;

(3) If $0 \leqslant X \leqslant Y$, then $0 \leqslant \|X\| \leqslant \|Y\|$;

(4) If $X \geqslant 0$ and $\xi \geqslant 0$, then $\|\xi X\| = \xi\|X\|$;

(5) If $X \geqslant 0$ and $Y \geqslant 0$, then $\|X+Y\| \leqslant \|X\|+\|Y\|$.

For arbitrary $X \in \mathscr{M}$ we define $\|X\| = \|\,|X|\,\|$; then $\|X\|$ is a norm on $\mathscr{M}$. We can easily prove that the convergence which is defined in $\mathscr{M}$ by this norm, i.e.; $X_n \xrightarrow{\|\ \|} X$ if and only if $\|X_n - X\| \to 0$, is equivalent to the u-convergence. Hence $\mathscr{M}$ is closed with respect to the $\|\ \|$-convergence, i.e., $\mathscr{M}$ is actually a Banach space (Banach lattice) with respect to this norm.

7. CONVERGENCE IN PROBABILITY AND ALMOST UNIFORM CONVERGENCE

7.1. (P_1). A sequence $X_n \in \mathscr{V}$, $n = 1, 2, \ldots$, is said to be *convergent in probability* (briefly *p-convergent*) to $X \in \mathscr{V}$, if and only if for every $\varepsilon > 0$, we have

$$p([|X_n - X| \geqslant \varepsilon]) \to 0, \quad \text{or equivalently} \quad p([|X_n - X| < \varepsilon]) \to 1.$$

We shall then write $X_n \xrightarrow{p} X$ in $\mathscr{V}$.

(P_2) A sequence $X_n \in \mathscr{V}$, $n = 1, 2, \ldots$, is said to be *fundamental in probability* (briefly *p-fundamental*), if and only if, for every $\varepsilon > 0$, we have $p([|X_n - X_k| \geqslant \varepsilon]) \to 0$ or equivalently $p([|X_n - X_k| < \varepsilon]) \to 1$.

(U_1) A sequence $X_n \in \mathscr{V}$, $n = 1, 2, \ldots$, is said to be *almost uniformly convergent* (briefly *au-convergent*) to $X \in \mathscr{V}$, if and only if, for every $\varepsilon > 0$, there exists an element $a_\varepsilon \in \mathfrak{B}$ such that $p(a_\varepsilon^c) < \varepsilon$, where $a_\varepsilon^c = e - a_\varepsilon$, and $I_{a_\varepsilon} X_n \xrightarrow{u} I_{a_\varepsilon} X$ in $\mathscr{V}$.

(U_2) A sequence $X_n \in \mathscr{V}$, $n = 1, 2, \ldots$, is said to be *almost uniformly fundamental* (briefly *au-fundamental*) if and only if for every $\varepsilon > 0$ there exists an element $a_\varepsilon \in \mathfrak{B}$ such that $p(a_\varepsilon^c) < \varepsilon$ and $I_{a_\varepsilon} X_n \in \mathscr{V}$, $n = 1, 2, \ldots$, is a u-fundamental sequence.

Theorem 7.1.

$X_n \xrightarrow{o} X$ in $\mathscr{V}$ if and only if $X_n \xrightarrow{au} X$ in $\mathscr{V}$.

Proof. (a) If $X_n \xrightarrow{au} X$ in $\mathscr{V}$, then $X_n \xrightarrow{o} X$ in $\mathscr{V}$. Since $X_n \xrightarrow{au} X$ in $\mathscr{V}$, for every $1/k > 0$, $k = 1, 2, \ldots$, there exists an element $b_k \in \mathfrak{B}$ with

$p(b_k{}^c) < 1/k$ and $I_{b_k} X_n \xrightarrow{u} I_{b_k} X$. Setting

$$x = \bigvee_{k=1}^{\infty} b_k$$

we have

$$x^c = \bigwedge_{k=1}^{\infty} b_k{}^c \leqslant b_n{}^c, \qquad n = 1, 2, \ldots,$$

i.e.,

$$p(x^c) \leqslant p(b_n{}^c) < \frac{1}{k}, \qquad k = 1, 2, \ldots;$$

this implies $p(x^c) = 0$ and $x^c = \emptyset$, i.e., $x = e$. Hence, we have $e = \bigvee_{k=1}^{\infty} b_k$. We shall prove: there exists $\bigvee_{n=1}^{\infty} |X_n - X|$ in $\mathscr{V}$. In fact, $\bigvee_{n=1}^{\infty} I_{b_k}|X_n - X|$ exists in $\mathscr{V}$, for every $k = 1, 2, \ldots$, since $I_{b_k}|X_n - X| \xrightarrow{u} 0$, $k = 1, 2, \ldots$, i.e., the sequence $I_{b_k}|X_n - X|$, $n = 1, 2, \ldots$, is bounded and, moreover, bounded by a real number η_k, for $k = 1, 2, \ldots$. We write

$$a_1 = b_1, \ldots, a_k = b_k - (b_1 \vee b_2 \vee \ldots \vee b_{k-1}), \qquad k = 2, 3, \ldots;$$

then $a_1, a_2, \ldots$ are pairwise disjoint with $e = \bigvee_{k=1}^{\infty} a_k$. We consider the erv

$$Y\colon\ a_k \Rightarrow \eta_1 + \eta_2 + \ldots + \eta_k;$$

then we have

$$I_{b_k}|X_n - X| \leqslant Y, \qquad k = 1, 2, \ldots;$$

but

$$\bigvee_{k=1}^{\infty} I_{b_k} = I_e = 1;$$

hence,

$$|X_n - X| \leqslant Y, \qquad n = 1, 2, \ldots,$$

i.e., bounded, and $\bigvee_{n=1}^{\infty} |X_n - X|$ exists in $\mathscr{V}$. Now $I_{b_k} X_n \xrightarrow{u} I_{b_k} X$, i.e., for $\varepsilon > 0$, there exists an index $\rho(k)$ such that

$$I_{b_k}|X_n - X| < \frac{\varepsilon}{2}, \qquad n \geqslant \rho(k);$$

hence, we have

$$\bigvee_{n \geqslant \rho(k)} I_{b_k}|X_n - X| < \varepsilon \quad \text{and} \quad \left[\bigvee_{n \geqslant \rho(k)} I_{b_k}|X_n - X| < \varepsilon\right] = e \geqslant b_k.$$

Since the indicator I_{b_k} is equal to 1 on b_k, we also have

$$\left[\bigvee_{n \geqslant \rho(k)} |X_n - X| < \varepsilon\right] \geqslant b_k$$

and

$$\left[\bigwedge_{\rho=1}^{\infty} \bigvee_{n \geqslant \rho} |X_n - X| < \varepsilon\right] \geqslant \left[\bigvee_{n \geqslant \rho(k)} |X_n - X| < \varepsilon\right] \geqslant b_k, \qquad k = 1, 2, \ldots,$$

i.e.,

$$\left[\bigwedge_{\rho=1}^{\infty} \bigvee_{n \geqslant \rho} |X_n - X| < \varepsilon\right] \geqslant \bigvee_{k=1}^{\infty} b_k = e, \quad \text{for every} \quad \varepsilon > 0.$$

We write

$$Z = \bigwedge_{\rho=1}^{\infty} \bigvee_{n \geqslant \rho} |X_n - X|,$$

then $Z \geqslant 0$ and, moreover,

$$s_Z(\varepsilon) = [Z < \varepsilon] = e, \quad \text{for every} \quad \varepsilon > 0,$$

i.e.,

$$s_Z(\varepsilon) = s_0(\varepsilon), \qquad \varepsilon > 0.$$

Hence, $Z = 0$. Thus

$$o\text{-}\limsup_{n \to \infty} |X_n - X| = 0,$$

i.e.,

$$|X_n - X| \xrightarrow{o} 0 \quad \text{and} \quad X_n \xrightarrow{o} X.$$

(b) If $X_n \xrightarrow{o} X$, then $X_n \xrightarrow{au} X$ in $\mathscr{V}$.

Since $X_n \xrightarrow{o} X$, i.e., $|X_n - X| \xrightarrow{o} 0$, we have

$$\bigwedge_{\rho=1}^{\infty} \bigvee_{n \geqslant \rho} |X_n - X| = 0, \quad \text{and} \quad \bigvee_{n \geqslant \rho} |X_n - X| = Y_\rho \overset{o}{\downarrow} 0.$$

We write

$$a_{\rho k} = \left[\bigvee_{n=\rho}^{\infty} |X_n - X| < \frac{1}{k}\right],$$

then $a_{\rho k} \uparrow e$, for every $k = 1, 2, \ldots$, i.e., $p(a_{\rho k}) \uparrow 1$; hence, given $\varepsilon > 0$, there exists for every $k = 1, 2, \ldots$, and index ρ_k such that

$$p(a^c_{\rho_k, k}) < \frac{\varepsilon}{2^k}\,.$$

We write

$$b = \bigwedge_{k=1}^{\infty} a_{\rho_k, k},$$

i.e.,

$$b^c = \bigvee_{k=1}^{\infty} a^c_{\rho_k, k};$$

then

$$p(b^c) \leqslant \sum_{k=1}^{\infty} \frac{\varepsilon}{2^k} = \varepsilon.$$

We shall prove: $I_b X_n \overset{u}{\to} I_b X$. In fact, we have

$$b = \bigwedge_{k=1}^{\infty} a_{\rho_k, k} \leqslant a_{\rho_k, k}, \qquad k = 1, 2, \ldots, \quad \text{i.e.,}$$

$$b \leqslant \left[\bigvee_{n \geqslant \rho_k} |X_n - X| < \frac{1}{k}\right] \leqslant \left[\bigvee_{n \geqslant \rho_k} I_b |X_n - X| < \frac{1}{k}\right],$$

since the indicator I_b is equal to 1 on b; thus,

$$b \leqslant \left[\bigvee_{n \geqslant \rho_k} |I_b X_n - I_b X| < \frac{1}{k}\right].$$

Now, since $I_b X_n$ equals 0 on b^c, we also have

$$b^c \leqslant \left[\bigvee_{n \geqslant \rho_k} |I_b X_n - I_b X| < \frac{1}{k}\right];$$

hence,

$$e = b \vee b^c \leqslant \left[\bigvee_{n \geqslant \rho_k} |I_b X_n - I_b X| < \frac{1}{k}\right],$$

i.e.,

$$|I_b X_n - I_b X| < \frac{1}{k}, \qquad n \geqslant \rho_k, \quad k = 1, 2, \ldots,$$

and

$$I_b X_n \overset{u}{\to} I_b X.$$

Hence, we have proved:

$$X_n \overset{au}{\to} X \quad \text{in} \quad \mathscr{V}.$$

The following theorem is true:

Theorem 7.2.

(a) *If* $X_n \overset{u}{\to} X$ *in* $\mathscr{V}$, *then* $X_n \overset{o}{\to} X$ *in* $\mathscr{V}$.

(b) *If* $X_n \overset{u}{\to} X$ *in* $\mathscr{V}$, *then* $X_n \overset{au}{\to} X$ *in* $\mathscr{V}$.

(c) *If* $X_n \overset{o}{\to} X$ *in* $\mathscr{V}$, *then* $X_n \overset{p}{\to} X$ *in* $\mathscr{V}$.

(d) *If* $X_n \overset{p}{\to} X$ *in* $\mathscr{V}$, *then there exists a subsequence* X_{k_n}, $n = 1, 2, \ldots$, *such that* $X_{k_n} \overset{o}{\to} X$ *in* $\mathscr{V}$.

(e) $X_n \overset{au}{\to} X$ *in* $\mathscr{V}$ *if and only if* $X_n \overset{o}{\to} X$ *in* $\mathscr{V}$.

(f) *If* $X_n \xrightarrow{p} X$, *then there exists a subsequence* X_{k_n}, $n = 1, 2, \ldots$, *such that* $X_{k_n} \xrightarrow{au} X$ *in* $\mathscr{V}$.

Proof. Obviously (a) and (b) are true; (e) is proved by Theorem 7.1; (f) is implied by (d) and (e). Hence we shall prove (c) and (d).

Proof of (c). Since $X_n \xrightarrow{o} X$ in $\mathscr{V}$, i.e.,

$$|X_n - X| \xrightarrow{o} 0 \quad \text{in} \quad \mathscr{V},$$

we have

$$o\text{-}\lim_{n\to\infty} S_{|X_n - X|}(\varepsilon) = o\text{-}\lim_{n\to\infty} [|X_n - X| < \varepsilon] = e, \quad \text{for every} \quad \varepsilon > 0;$$

i.e.,

$$\lim_{n\to\infty} p([|X_n - X| < \varepsilon]) = 1$$

or equivalently

$$\lim_{n\to\infty} p([|X_n - X| \geqslant \varepsilon]) = 0,$$

i.e.,

$$X_n \xrightarrow{p} X \quad \text{in } \mathscr{V}.$$

Proof of (d). Since $X_n \xrightarrow{p} X$ in $\mathscr{V}$, the sequence X_n, $n = 1, 2, \ldots$, is p-fundamental. Hence, for any index k, an index $n^*(k)$ exists, such that if $n \geqslant n^*(k)$ and $m \geqslant n^*(k)$; then

$$p\left(\left[|X_n - X_m| \geqslant \frac{1}{2^k}\right]\right) < \frac{1}{2^k}.$$

We write

$$\begin{aligned} n_1 &= n^*(1), \\ n_2 &= \max\{n^*(2), n_1 + 1\}, \ldots, \\ n_k &= \max\{n^*(k), n_{k-1} + 1\}, \qquad k = 1, 2, \ldots; \end{aligned}$$

then $n_1 < n_2 < n_3 < \ldots$, and X_{n_k}, $k = 1, 2, \ldots$, is a subsequence of X_n, $n = 1, 2, \ldots$.

Now we write

$$a_k = \left[|X_{n_k} - X_{n_{k+1}}| \geqslant \frac{1}{2^k}\right] \quad \text{and} \quad b_k{}^c = a_k \vee a_{k+1} \vee \ldots,$$

i.e.,

$$b_k = a_k{}^c \wedge a_{k+1}^c \wedge \ldots.$$

If $k \leqslant i \leqslant j$, then

$$I_{b_k} |X_{n_i} - X_{n_j}| \leqslant \sum_{m=i}^{\infty} I_{b_k} |X_{n_m} - X_{n_{m+1}}| < \frac{1}{2^i} + \frac{1}{2^{i+1}} + \ldots = \frac{1}{2^{i-1}},$$

i.e.,

$$I_{b_k}|X_{n_i} - X_{n_j}| < \frac{1}{2^{i-1}}, \quad \text{for} \quad k \leqslant i \leqslant j.$$

Hence, for every $k = 1, 2, \ldots$, the sequence $I_{b_k} X_{n_j}$, $j = 1, 2, \ldots$, is u-fundamental with

$$p(b_k{}^c) < \frac{1}{2^k} + \frac{1}{2^{k+1}} + \ldots = \frac{1}{2^{k-1}}.$$

Thus, the sequence X_{n_j}, $j = 1, 2, \ldots$ is *au*-fundamental and, by (e), o-fundamental. Hence, an element $Y \in \mathscr{V}$ exists such that $o\text{-}\lim\limits_{j\to\infty} X_{n_j} = Y$. By (c), we then have $p\text{-}\lim\limits_{j\to\infty} X_{n_j} = Y$. Since $p\text{-}\lim\limits_{n\to\infty} X_n = X$, the element Y must be equal to X, hence $X_{n_j} \overset{o}{\to} X$ in $\mathscr{V}$.

The following theorem is true:

Theorem 7.3.

If a sequence X_n, $n = 1, 2, \ldots$, is p-fundamental, then there exists an element X in $\mathscr{V}$ such that $X_n \overset{p}{\to} X$ in $\mathscr{V}$.

Proof. Just as in the proof of Theorem 7.2 (d), one can prove the existence of an o-fundamental subsequence X_{j_n}, $n = 1, 2, \ldots$, of X_n, $n = 1, 2, \ldots$; then there exists an element X in $\mathscr{V}$ such that $X_{j_n} \overset{o}{\to} X$ or equivalently $x_{j_n} \overset{au}{\to} X$ in $\mathscr{V}$. We remark that, for every $\varepsilon > 0$,

$$[|X_n - X| \geqslant \varepsilon] \leqslant \left[|X_n - X_{j_n}| \geqslant \frac{\varepsilon}{2}\right] \vee \left[|X_{j_n} - X| \geqslant \frac{\varepsilon}{2}\right].$$

Since $\{X_n\}$ is p-fundamental and $j_n \geqslant n$,

$$\lim_{n\to\infty} p\left(\left[|X_n - X_{j_n}| \geqslant \frac{\varepsilon}{2}\right]\right) = 0.$$

Since $o\text{-}\lim\limits_{n\to\infty} |X_{j_n} - X| = 0$, $p\text{-}\lim\limits_{n\to\infty} |X_{j_n} - X| = 0$, i.e.,

$$\lim_{n\to\infty} p\left(\left[|X_{j_n} - X| \geqslant \frac{\varepsilon}{2}\right]\right) = 0;$$

then we have

$$\lim_{n\to\infty} p([|X_n - X| \geqslant \varepsilon]) = 0,$$

i.e.,

$$X_n \overset{p}{\to} X \quad \text{in} \quad \mathscr{V}.$$

The previous theorems imply:

Theorem 7.4.

The stochastic space $\mathscr{V}$ is closed for o-convergence, u-convergence, au-convergence and p-convergence.

8. GENERATORS OF THE STOCHASTIC SPACE

8.1. Let $(\mathfrak{B}, p)$ be a pr σ-algebra and $\mathfrak{A}$ a Boolean subalgebra of $\mathfrak{B}$, which σ-genrates $\mathfrak{B}$, i.e., $\mathfrak{A}^{\sigma\delta} = \mathfrak{A}^{\delta\sigma} = \mathfrak{B}$. Then we have

$$\mathscr{J}(\mathfrak{A}) \subseteq \mathscr{J}(\mathfrak{B}), \quad \mathscr{S}(\mathfrak{A}) \subseteq \mathscr{S}(\mathfrak{B}) \subseteq \mathscr{V}(\mathfrak{B}).$$

For every element $X \in \mathscr{S}(\mathfrak{A})$ there exists a reduced representation

$$X = \xi_1 I_{x_1} + \xi_2 I_{x_2} + \ldots + \xi_k I_{x_k},$$

where $x_j \in \mathfrak{A}$, $j = 1, 2, \ldots, k$. We write $\mathscr{S}_0(\mathfrak{A}) = \{X \in \mathscr{S}(\mathfrak{A})$ with $X = \xi_1 I_{x_1} + \xi_2 I_{x_2} + \ldots + \xi_k I_{x_k}$, where $\xi_1, \xi_2, \ldots, \xi_k$ are rational numbers$\}$.
where $\xi_1, \xi_2, \ldots, \xi_k$ are rational numbers$\}$.

The following theorem is true:

Theorem 8.1.

The space $\mathscr{S}_0(\mathfrak{A})$ is o-dense in $\mathscr{V}(\mathfrak{B})$, i.e., for every $X \in \mathscr{V}(\mathfrak{B})$, there exists a sequence $X_n \in \mathscr{S}_0(\mathfrak{A})$, $n = 1, 2, \ldots$, such that $X_n \overset{o}{\to} X$.

We shall first prove the following fundamental lemma:

Lemma (Fréchet) 8.1.

If $X_n \in \mathscr{V}$ and $X_{n,k} \in \mathscr{V}$, $n = 1, 2, \ldots$, $k = 1, 2, \ldots$, such that $X_n \overset{au}{\to} X$ and $X_{n,k} \underset{k\to\infty}{\overset{au}{\to}} X_n$, $n = 1, 2, \ldots$, i.e., if

$$(au)\text{-}\lim_{n\to\infty} ((au)\text{-}\lim_{k\to\infty} X_{n,k}) = X,$$

then there exists a simple subsequence X_{n_ν, k_ν}, $\nu = 1, 2, \ldots$ of the double sequence $X_{n,k}$, $(n, k) \in \mathbf{N} \times \mathbf{N}$, such that $X_{n_\nu, k_\nu} \overset{au}{\to} X$.

Remark. Since *au*-convergence is equivalent to the *o*-convergence, this lemma is also true, if the *au*-convergence is replaced by the *o*-convergence.

Proof of the lemma. Let ε_ν and δ_ν be positive real numbers, $\nu = 1, 2, \ldots$, such that $\varepsilon_\nu \to 0$ and

$$\sum_{\nu=1}^{\infty} \delta_\nu < +\infty.$$

Since $X_n \overset{au}{\to} X$, for every δ_ν there exists an element $b_\nu \in \mathfrak{B}$, such that

$$p(b_\nu{}^c) < \frac{\delta_\nu}{2} \quad \text{and} \quad I_{b_\nu} X_n \underset{n\to\infty}{\overset{u}{\to}} I_{b_\nu} X;$$

thus there exists an index n_ν with

$$|I_{b_\nu} X_{n_\nu} - I_{b_\nu} X| < \frac{\varepsilon_\nu}{2} \quad \text{and} \quad p(b_\nu{}^c) < \frac{\delta_\nu}{2}. \tag{1}$$

Since, obviously, we have

$$I_{b_\nu} X_{n_\nu, k} \underset{k\to\infty}{\overset{au}{\to}} I_{b_\nu} X_{n_\nu},$$

for every $\delta_\nu > 0$ there exists an element $b_\nu{}^* \leqslant b_\nu$, $b_\nu{}^* \in \mathfrak{B}$, such that

$$p(b_\nu - b_\nu{}^*) < \frac{\delta_\nu}{2} \quad \text{and} \quad I^*_{b_\nu} X_{n_\nu, k} \underset{k\to\infty}{\overset{u}{\to}} I^*_{b_\nu} X_{n_\nu}, \quad \nu = 1, 2, \ldots;$$

thus there exists an index k_ν with

$$|I^*_{b_\nu} X_{n_\nu, k_\nu} - I^*_{b_\nu} X_{n_\nu}| < \frac{\varepsilon_\nu}{2} \quad \text{and} \quad p(b_\nu - b_\nu{}^*) < \frac{\delta_\nu}{2}. \tag{2}$$

Since $b_\nu{}^* \leqslant b_\nu$ and $b_\nu{}^c \vee (b_\nu - b_\nu{}^*) = b_\nu{}^{*c}$, the relations (1) and (2) imply

$$|I^*_{b_\nu} X_{n_\nu} - I^*_{b_\nu} X| < \frac{\varepsilon_\nu}{2} \quad \text{and} \quad |I^*_{b_\nu} X_{n_\nu, k_\nu} - I^*_{b_\nu} X_{n_\nu}| < \frac{\varepsilon_\nu}{2}.$$

This implies

$$|I^*_{b_\nu} X_{n_\nu, k_\nu} - I^*_{b_\nu} X| < \varepsilon_\nu, \quad \text{with} \quad p(b_\nu{}^{*c}) < \delta_\nu.$$

Now consider any $\varepsilon > 0$; then there exists an index m such that

$$\varepsilon > \delta_m + \delta_{m+1} + \ldots.$$

We write

$$a_\varepsilon = \bigwedge_{j=m}^{\infty} b_j{}^*;$$

then, for every $j \geqslant m$ we have

$$|I_{a_\varepsilon} X_{n_j, k_j} - I_{a_\varepsilon} X| < \varepsilon_j$$

and since

$$a_\varepsilon^c = \bigvee_{j=m}^{\infty} b_j{}^{*c},$$

we have

$$p(a_\varepsilon^c) < \delta_m + \delta_{m+1} + \ldots, \quad \text{i.e.,} \quad p(a_\varepsilon^c) < \varepsilon.$$

This implies: for any $\varepsilon > 0$ there exists an element a_ε with $p(a_\varepsilon^c) < \varepsilon$ and

$$I_{a_\varepsilon} X_{n_\nu, k_\nu} \underset{\nu \to \infty}{\overset{u}{\to}} I_{a_\varepsilon} X, \quad \text{i.e.,} \quad X_{n_\nu, k_\nu} \underset{\nu \to \infty}{\overset{au}{\to}} X.$$

Proof of Theorem 8.1. (A) By Theorem 7.2(e), it suffices to prove that $\mathscr{S}_0(\mathfrak{A})$ is *au*-dense in $\mathscr{V}(\mathfrak{B})$. We write briefly:

$$\mathscr{S}_0(\mathfrak{A}) = \mathscr{S}_0, \quad \mathscr{S}(\mathfrak{A}) = \mathscr{S}_1, \quad \mathscr{S}(\mathfrak{B}) = \mathscr{S}_2, \quad \mathscr{E}(\mathfrak{B}) = \mathscr{S}_3, \quad \mathscr{V}(\mathfrak{B}) = \mathscr{S}_4,$$

and we shall prove first that if $X \in \mathscr{S}_j$, then there exists a sequence $X_n \in \mathscr{S}_{j-1}$, $n = 1, 2, \ldots$ such that $X_n \overset{au}{\to} X$ for every $j = 1, 2, 3, 4$.

(1) For $j = 1$, (A) is obvious, because if $X \in \mathscr{S}_1$, then

$$X = \xi_1 I_{x_1} + \ldots + \xi_k I_{x_k} \quad \text{with} \quad x_i \in \mathfrak{A}, \quad \xi_i \in R.$$

Now let

$$\lim_{n \to \infty} \rho_{in} = \xi_i$$

with ρ_{in} rational numbers, $n = 1, 2, \ldots$; then, if we write

$$X_n = \rho_{1n} I_{x_1} + \ldots + \rho_{kn} I_{x_k}, \quad n = 1, 2, \ldots,$$

we have

$$X_n \overset{u}{\to} X, \quad \text{i.e.,} \quad X_n \overset{au}{\to} X \quad \text{with} \quad X_n \in \mathscr{S}_0, \quad n = 1, 2, \ldots.$$

(2) For $j = 2$: Let $X \in \mathscr{S}_2$. Then

$$X = \xi_1 I_{x_1} + \xi_2 I_{x_2} + \ldots + \xi_k I_{x_k}$$

with $x_i \in \mathfrak{B}$, $\xi_i \in R$, $i = 1, 2, \ldots, k$. Since $\mathfrak{A}$ is o-dense in $\mathfrak{B}$, for every $i = 1, 2, \ldots k$, there exists a sequence

$$x_{in} \underset{\mathfrak{B}}{\overset{o}{\to}} x_i \quad \text{with} \quad x_{in} \in \mathfrak{A}, \quad n = 1, 2, \ldots.$$

We write

$$X_n = \xi_1 I_{x_{1n}} + \xi_2 I_{x_{2n}} + \ldots + \xi_k I_{x_{kn}};$$

then $X_n \overset{o}{\to} X$ or equivalently $X_n \overset{au}{\to} X$ with $X_n \in \mathscr{S}_2 = \mathscr{S}(\mathfrak{A})$.

(3) For $j = 3$: Let $X \in \mathscr{S}_3$. Then

$$X = \sum_{j \geqslant 1} \xi_j I_{x_j} \quad \text{with} \quad x_j \in \mathfrak{B}, \quad j \geqslant 1.$$

We write

$$X_n = \sum_{j=1}^{n} \xi_j I_{x_j};$$

then $X_n \overset{o}{\to} X$, or equivalently $X_n \overset{au}{\to} X$ with $X_n \in \mathscr{S}_2 = \mathscr{S}(\mathfrak{B})$.

(4) For $j = 4$: Let $X \in \mathscr{S}_4$. Then, by Section 5, $\mathscr{S}_3$ is o-dense or equivalently au-dense in $\mathscr{S}_4$; i.e., there exists a sequence $X_n \in \mathscr{S}_3$, $n = 1, 2, \ldots$, such that $X_n \overset{au}{\to} X$ in $\mathscr{V}$.

Now, the Fréchet Lemma and (A) imply the truth of Theorem 8.1.

8.2. Section 8.1 implies: If the Boolean σ-algebra $\mathfrak{B}$ of a pr σ-algebra $(\mathfrak{B}, p)$ is separable, then $\mathscr{V}$ is, with respect to o-convergence, au-convergence, and p-convergence, separable. In general, if $\mathfrak{B}$ possesses a σ-generating basis $\mathfrak{A}$ of cardinality $|\mathfrak{A}| = \aleph_\beta$, $\beta \geqslant 0$, then $\mathscr{V}(\mathfrak{B})$ also possesses a stochastic subspace of the same cardinality, which is o-dense, au-dense and p-dense in $\mathscr{V}(\mathfrak{B})$. Hence, if we define the character of $\mathscr{V}(\mathfrak{B})$ as the min $|\mathscr{A}|$ for all stochastic subspaces $\mathscr{A}$ of $\mathscr{V}(\mathfrak{B})$ which are o-dense in $\mathscr{V}(\mathfrak{B})$, then the character of $\mathscr{V}(\mathfrak{B})$ is equal to the character of $\mathfrak{B}$, if $|\mathfrak{B}| \geqslant \aleph_0$.

9. OTHER CONVERGENCES IN THE STOCHASTIC SPACE

9.1. Let $(\mathfrak{B}, p)$ be a pr σ-algebra and $\mathscr{V}(\mathfrak{B}, p)$, briefly $\mathscr{V}$, be the stochastic space over $(\mathfrak{B}, p)$. In Section 7, the convergence in probability and the almost uniform convergence are defined and compared with the order and uniform convergence. In the present section, it will be said on other kinds of convergences, which can be defined in the stochastic space $\mathscr{V}$. It is well known that a fundamental property of a sequential convergence with respect to a topology is the following: a sequence of points in a topological space converges to a point of this space if and only if every subsequence of the given sequence has a subsequence that converges to the same point. This property of convergence is the so-called *star property*. Unfortunately, order convergence does not always possess this property. On account of that, we introduce the following definition:

A sequence $X_n \in \mathscr{V}$, $n = 1, 2, \ldots$, *order star converges* (briefly o^*-converges) to $X \in \mathscr{V}$ if and only if every subsequence of X_n, $n = 1, 2, \ldots$, contains a subsequence that order converges to X.

This kind of convergence will be denoted by

$$X_n \overset{o*}{\to} X \quad \text{or} \quad o^*\text{-lim}\, X_n = X \quad \text{in } \mathscr{V}.$$

It is clear that $X_n \xrightarrow{o} X$ implies $X_n \xrightarrow{o*} X$ in $\mathscr{V}$. In Section 6 it has been proved that $X_n \xrightarrow{o} X$ implies $X_n \xrightarrow{p} X$ in $\mathscr{V}$; conversely if $X_n \xrightarrow{p} X$, then we do not always have $X_n \xrightarrow{o} X$, but there always exists a subsequence X_{k_n}, $n = 1, 2, \ldots$, such that $X_{k_n} \xrightarrow{o} X$. On the other hand it is easy to verify that the p-convergence is equivalent to the o^*-convergence, i.e., p-convergence has the star property. Since p-convergence does not imply o-convergence, we conclude that o-convergence does not have the star property. The following example shows that p-convergence does not imply o-convergence:

Example. Let $(\mathfrak{A}, m)$ be the interval algebra of Section 4.5, Chapter I, and $(\mathfrak{B}, p) \equiv (\widetilde{\mathfrak{A}}, \tilde{m}) \equiv (\mathfrak{J}, \mu)$, i.e., the so-called linear Lebesgue pr σ-algebra (Section 3, Chapter II). Consider $\mathfrak{A}$ embedded in the Boolean σ-algebra $\mathfrak{J}$, and the events (intervals):

$$a_{n,\,i} \equiv \left[\frac{i-1}{n}, \frac{i}{n}\right), \quad i = 1, 2, \ldots, n, \quad n = 1, 2, \ldots,$$

as elements of $\mathfrak{J}$. Let now $X_{n,\,i}$ be the indicator of the element $a_{n,\,i}$ i.e.,

$$X_{n,\,i} = I_{a_{n,\,i}} \in \mathscr{V}(\mathfrak{J});$$

then the sequence

$$X_{1,\,1},\ X_{2,\,1},\ X_{2,\,2},\ X_{3,\,1},\ X_{3,\,2},\ X_{3,\,3}, \ldots, X_{n,\,1},\ X_{n,\,2}, \ldots, X_{n,\,n}, \ldots \qquad n = 1, 2, \ldots$$

converges in probability $m = p$ to $X = 0$ in $\mathscr{V}(\mathfrak{J})$, but fails to o-converge to an element of $\mathscr{V}(\mathfrak{J})$ (*cf.* Halmos [2], p. 94).

9.2. Another type of convergence is also determined by the ordering in $\mathscr{V}$, the so-called *relative uniform convergence* (see G. Birkhoff [1] p. 243).

We shall say that a sequence $X_n \in \mathscr{V}$, $n = 1, 2, \ldots$, *converges relatively uniformly* to an element $X \in \mathscr{V}$, briefly $X_n \xrightarrow{ru} X$ in $\mathscr{V}$ or ru-lim $X_n = X$ in $\mathscr{V}$, if and only if, for some $U \in \mathscr{V}$ and λ_n, $n = 1, 2, \ldots$, a decreasing sequence of real numbers with $\lambda_n \to 0$ we have

$$|X_n - X| \leqslant \lambda_n U.$$

The *relative uniform star convergence*, denoted by

$$X_n \xrightarrow{ru^*} X \quad \text{or} \quad ru^*\text{-lim}\, X_n = X \quad \text{in } \mathscr{V},$$

can also be defined, i.e., $X_n \overset{ru*}{\to} X$ if and only if every subsequence of X_n, $n = 1, 2, \ldots$, contains a subsequence that is ru-convergent to X. Clearly

(A) $$X_n \overset{ru}{\to} X \quad \text{implies} \quad X_n \overset{o}{\to} X \quad \text{in } \mathscr{V},$$

for $\lambda_n U = U_n \in \mathscr{V}$ is a decreasing sequence such that

$$\bigwedge_{n=1}^{\infty} U_n = o\text{-}\lim_{n\to\infty} \lambda_n U = 0.U = 0.$$

We shall prove that conversely:

(B) $$X_n \overset{o}{\to} X \quad \text{implies} \quad X_n \overset{ru}{\to} X \quad \text{in } \mathscr{V}.$$

Now (A) and (B) imply:

Theorem 9.1.

Relative uniform convergence and o-convergence are equivalent in $\mathscr{V}$.

We first notice that $\mathscr{V}$ possesses the following property:

(C) If $U_n \in \mathscr{V}$, $n = 1, 2, \ldots$, is a decreasing sequence such that $U_n \overset{o}{\to} 0$, then there exists a subsequence U_{n_ν}, $\nu = 1, 2, \ldots$, of U_n, $n = 1, 2, \ldots$, such that

$$\nu . U_{n_\nu} \underset{\nu\to\infty}{\overset{o}{\to}} 0.$$

Proof of (C). Let us set

$$X_{n,k} \equiv kU_n, \quad (k, n) \in \mathbf{N}\times\mathbf{N}; \tag{1}$$

then obviously,

$$X_{n,k} \underset{n\to\infty}{\overset{o}{\to}} X_k = 0 \quad \text{for every} \quad k = 1, 2, \ldots \quad \text{and} \quad X_k \underset{k\to 0}{\overset{o}{\to}} 0;$$

since Fréchet Lemma of Section 8 is true also for o-convergence, there exists a subsequence

$$X_{n_\nu, k_\nu} = k_\nu U_{n_\nu}, \quad \nu = 1, 2, \ldots,$$

such that

$$k_\nu U_{n_\nu} \underset{\nu\to\infty}{\overset{o}{\to}} 0,$$

and since n_ν, $\nu = 1, 2, \ldots$, can be chosen strictly increasing, when, obviously, $\nu \leqslant k_\nu$, $\nu = 1, 2, \ldots$, we have

$$\nu U_{n_\nu} \underset{\nu\to\infty}{\overset{o}{\to}} 0.$$

Proof of (B). Let $U_n \in \mathscr{V}$, $n = 1, 2, \ldots$, be a decreasing sequence such that

$$U_n \overset{o}{\to} 0 \quad \text{and} \quad |X_n - X| \leqslant U_n;$$

then by (C) there exists a subsequence U_{n_ν}, $\nu = 1, 2, \ldots$, with $\nu U_{n_\nu} \to 0$; for every $n = 1, 2, \ldots$, there exists an index ν such that $n_\nu \leqslant n < n_{\nu+1}$; we set $\rho_n = \nu$; then we have, obviously,

$$\rho_n U_n \overset{o}{\to} 0 \quad \text{with} \quad \rho_n \to \infty.$$

Let now U be an upper bound of the sequence $\rho_n U_n$, then we have

$$\rho_n |X_n - X| \leqslant \rho_n U_n \leqslant U$$

and

$$|X_n - X| \leqslant \lambda_n U \quad \text{where} \quad \lambda_n = \frac{1}{\rho_n} \to 0, \quad n = 1, 2, \ldots,$$

i.e.,

$$X_n \overset{ru}{\to} X.$$

Corollary 9.1.

The concepts: o-convergence, au-convergence and ru-convergence are equivalent in $\mathscr{V}$. Moreover, each of the concepts o-convergence, au*-convergence and ru*-convergence is equivalent to p-convergence.*

10. CLOSURE OPERATOR IN THE STOCHASTIC SPACE

10.1. Let $\mathfrak{P}(\mathscr{V})$ be the set of all subsets of $\mathscr{V}$; then the mapping

$$\mathfrak{P}(\mathscr{V}) \ni \mathscr{B} \Rightarrow \overline{\mathscr{B}}$$

defined by

(1) $\overline{\mathscr{B}} \equiv \{X \in \mathscr{V}$: there is a sequence $X_n \in \mathscr{B}$ such that $X = o\text{-}\lim_{n\to\infty} X_n\}$ is a closure operator on $\mathscr{V}$; because, obviously, we have

(a) $\overline{\emptyset} = \emptyset$.

(b) $\mathscr{B} \subseteq \overline{\mathscr{B}}$, for every $\mathscr{B} \in \mathfrak{P}(\mathscr{V})$.

Furthermore, we can prove

(c) $\overline{\overline{\mathscr{B}}} = \overline{\mathscr{B}}$.

Proof. Let $X \in \overline{\overline{\mathscr{B}}}$; then there is a sequence $X_n \in \overline{\mathscr{B}}$, $n = 1, 2, \ldots$, such that $o\text{-}\lim_{n\to\infty} X_n = X$; since now $X_n \in \overline{\mathscr{B}}$, there is a sequence X_{nk}, $k = 1, 2, \ldots$, such that $o\text{-}\lim_{k\to\infty} X_{nk} = X_n$; Fréchet Lemma 8.1 implies now the existence of a sequence X_{n_ν, k_ν}, $\nu = 1, 2, \ldots$, such that $o\text{-}\lim_{\nu\to\infty} X_{n_\nu, k_\nu} = X$, i.e., $X \in \bar{\mathscr{B}}$; hence $\bar{\bar{\mathscr{B}}} \subseteq \bar{\mathscr{B}}$; however (b) implies $\bar{\mathscr{B}} = \bar{\bar{\mathscr{B}}}$, i.e., $\bar{\bar{\mathscr{B}}} = \bar{\mathscr{B}}$.

The following property is also true

(d) $\overline{\mathscr{B}_1 \cup \mathscr{B}_2} = \bar{\mathscr{B}}_1 \cup \bar{\mathscr{B}}_2$.

The properties (a), (b), (c) and (d) imply that the mapping (1) is a closure operator.

The following theorem is true:

Theorem 10.1.

Let $\mathfrak{T}$ be the unique topology on $\mathscr{V}$ determined by this closure operator; then a sequence $X_n \in \mathscr{V}$ converges for $\mathfrak{T}$ to X in $\mathscr{V}$ if and only if

$$o^*\text{-}\lim X_n = X \quad in \quad \mathscr{V}.$$

Proof. Suppose $X_n \in \mathscr{V}$, $n = 1, 2, \ldots$, is $\mathfrak{T}$-convergent to X, without being o^*-convergent to X; then there is a subsequence X_{n_ν}, $\nu = 1, 2, \ldots$ such that no subsequence of X_{n_ν}, $\nu = 1, 2, \ldots$, o-converges to X, i.e., $X \notin \bar{\mathscr{B}}$, where $\mathscr{B} = \{Y \in \mathscr{V}\colon$ there is n_ν with $Y = X_{n_\nu}\}$. Therefore, the $\bar{\mathscr{B}}^c$ is a neighbourhood of X and for every index k, we have $X_{n_k} \notin \bar{\mathscr{B}}^c$. Hence, X_n, $n = 1, 2, \ldots$, does not $\mathfrak{T}$-converge to X (contradiction).

Conversely, if X_n, $n = 1, 2, \ldots$, is o^*-convergent to X without being $\mathfrak{T}$-convergent to X, then there is an open neighbourhood $\mathscr{U}$ of X and a subsequence X_{n_ν}, $\nu = 1, 2, \ldots$, of X_n, $n = 1, 2, \ldots$, such that for all n_ν, $X_{n_\nu} \notin \mathscr{U}$. The limit of any subsequence of X_{n_ν}, $\nu = 1, 2, \ldots$, that o-converges must be in the complement $\mathscr{U}^c$ by definition of the closure operator determining $\mathfrak{T}$; hence no subsequence of X_{n_ν}, $\nu = 1, 2, \ldots$, o-converges to X. Thus, o^*-convergence coincides with $\mathfrak{T}$-convergence in $\mathscr{V}$.

11. SERIES CONVERGENCE

11.1. Let $\sum_{i=1}^{\infty} X_i$ be a series with $X_i \in \mathscr{V}$, $i = 1, 2, \ldots$. We shall say the series $\sum_{i=1}^{\infty} X_i$ is *order convergent* to $X \in \mathscr{V}$ and write

$$\text{(I)} \qquad (o)\text{-} \sum_{i=1}^{\infty} X_i = X \quad \text{in} \quad \mathscr{V},$$

if and only if $o\text{-}\lim (X_1 + X_2 + \ldots + X_n) = X$ in $\mathscr{V}$.

Analogously, one can define:

$$\text{(II)}\quad \begin{cases} (u)\text{-}\sum_{i=1}^{\infty} X_i = X, & (au)\text{-}\sum_{i=1}^{\infty} X_i = X \\ & & \text{in } \mathscr{V}. \\ (p)\text{-}\sum_{i=1}^{\infty} X_i = X, & (\text{ru})\text{-}\sum_{i=1}^{\infty} X_i = X \end{cases}$$

One can easily prove:

(1) $(o)\text{-}\sum_{i=1}^{\infty} X_i = X$ in $\mathscr{V}$ implies $X_i \xrightarrow{o} 0$.

(2) $(o)\text{-}\sum_{i=1}^{\infty} X_i = X$ and $(o)\text{-}\sum_{i=1}^{\infty} Y_i = Y$ in $\mathscr{V}$ imply

$$(o)\text{-}\sum_{i=1}^{\infty} (\alpha X_i + \beta Y_i) = \alpha X + \beta Y \quad \text{in} \quad \mathscr{V}.$$

(1) and (2) are also true for the other kinds (II) of convergence.

Theorem 11.1.

Let the series $\sum_{i=1}^{\infty} |X_i|$ *with* $X_i \in \mathscr{V}$ *be o-convergent and* Y_i, $i = 1, 2, \ldots,$ *be a sequence in* $\mathscr{V}$, *such that* $|Y_i| \leqslant |X_i|$, $i = 1, 2, \ldots$; *then the series* $\sum_{i=1}^{\infty} Y_i$ *o-converges and we have*

$$\text{(III)} \qquad \left|(o)\text{-}\sum_{i=1}^{\infty} Y_i\right| \leqslant (o)\text{-}\sum_{i=1}^{\infty} |X_i|.$$

Proof. Let

$U_n = Y_1 + Y_2 + \ldots + Y_n$, $\quad V_n = |X_1| + |X_2| + \ldots + |X_n|$, $\quad n = 1, 2, \ldots$;

then we have

$$\begin{aligned} |U_{n+k} - U_n| &= |Y_{n+1} + \ldots + Y_{n+k}| \\ &\leqslant |X_{n+1}| + \ldots + |X_{n+k}| \\ &= V_{n+k} - V_n \\ &\equiv A_{n,k}, \qquad k = 1, 2, \ldots . \end{aligned}$$

But $A_{n,k} \xrightarrow{o} 0$; hence

$$o\text{-}\lim_{n,\, k \to \infty} |U_{n+k} - U_n| = 0,$$

i.e., the series $\sum_{i=1}^{\infty} Y_i$ o-converges.

Now

$$|Y_1 + Y_2 + \ldots + Y_n| \leqslant |X_1| + |X_2| + \ldots + |X_n| \quad \text{implies (III).}$$

Theorem 11.1 implies:

If the series $\sum_{i=1}^{\infty} |X_i|$ o-converges, then the series $\sum_{i=1}^{\infty} X_i$ o-converges. Hence we define: A series $\sum_{i=1}^{\infty} X_i$ is *absolutely o-convergent* if and only if the series $\sum_{i=1}^{\infty} |X_i|$ is o-convergent. It is easy to prove that if the series $\sum_{i=1}^{\infty} X_i$ is absolutely o-convergent, then the series

$$\sum_{i=1}^{\infty} X_i^+ \quad \text{and} \quad \sum_{i=1}^{\infty} X_i^-$$

respectively are o-convergent and we have

$$(o)\text{-}\sum_{i=1}^{\infty} X_i = (o)\text{-}\sum_{i=1}^{\infty} X_i^+ - (o)\text{-}\sum_{i=1}^{\infty} X_i^-.$$

Theorem 11.2.

Let X_n, $n = 1, 2, \ldots$, be a sequence in $\mathscr{V}$ such that $|X_j| \wedge |X_i| = 0$, for every $(j, i) \in \mathbf{N} \times \mathbf{N}$ with $j \neq i$, i.e., the terms X_n, $n = 1, 2, \ldots$, are pairwise orthogonal; then the series $\sum_{n=1}^{\infty} X_n$ is absolutely o-convergent.

Proof. Since

$$|X_j| \wedge |X_i| = 0, \qquad (j, i) \in \mathbf{N} \times \mathbf{N},\ i \neq j,$$

there exists a sequence a_n, $n = 1, 2, \ldots$, of pairwise disjoint elements of $\mathfrak{B}$ such that $\bigvee_{n=1}^{\infty} a_n = e$ and $X_n = I_{a_n} X_n$, for every $n = 1, 2, \ldots$. But, for every $n = 1, 2, \ldots$, there exists an erv $Y_n \in \mathscr{E} \subseteq \mathscr{V}$ such that $|X_n| \leqslant Y_n$; hence, we have

$$|X_n| = I_{a_n} |X_n| \leqslant I_{a_n} Y_n;$$

and, since a_n, $n = 1, 2, \ldots$, are pairwise disjoint, there exists

$$(o)\text{-}\sum_{n=1}^{\infty} I_{a_n} Y_n \equiv Y,$$

and is an erv. In fact, let:

$$Y_n\colon\ \mathbf{b}_n \ni b_{nj} \to \eta_{nj}, \qquad j \geqslant 1,\ n = 1, 2, \ldots;$$

then

$$I_{a_n} Y_n\colon \begin{array}{l} a_n \wedge b_{nj} \to \eta_{nj} \\ \\ a_n^c \wedge b_{nj} \to 0 \end{array}, \qquad j \geqslant 1,\ n = 1, 2, \ldots.$$

Obviously, all elements:

(E) $$a_n \wedge b_{nj} \neq \varnothing, \qquad j \geqslant 1,\ n = 1, 2, \ldots,$$

are pairwise disjoint, and the union of all of them is equal to e, i.e., the elements (E) define an experiment in $\mathfrak{B}$. Hence, we can define:

$$Y:\quad a_n \wedge b_{nj} \to \eta_{nj}, \quad \text{for all} \quad a_n \wedge b_{nj} \neq \emptyset, \qquad j \geqslant 1,\ n = 1, 2, \ldots .$$

According to Section 2.5, we have:

$$Y = (o)\text{-} \sum_{j,\, n \geqslant 1} \eta_{nj} I_{a_n \wedge b_{nj}}$$

and

$$I_{a_n} Y_n = (o)\text{-} \sum_{j \geqslant 1} \eta_{nj} I_{a_n \wedge b_{nj}}, \qquad n = 1, 2, \ldots .$$

It is easy now to prove that:

$$Y = (o)\text{-} \sum_{n=1}^{\infty} I_{a_n} Y_n.$$

It follows from the above that:

$$|X_n| \leqslant I_{a_n} Y_n \leqslant Y, \quad \text{for every} \quad n = 1, 2, \ldots .$$

But, then, since $|X_n|$, $n = 1, 2, \ldots$, are pairwise orthogonal, we have

$$|X_1| + |X_2| + \ldots + |X_n| = |X_1| \vee |X_2| \vee \ldots \vee |X_n| \leqslant Y.$$

Hence, there exists:

$$\bigvee_{n=1}^{\infty} \left(\sum_{i=1}^{n} |X_i| \right) = o\text{-}\lim_{n \to \infty} \sum_{i=1}^{n} |X_i| \equiv (o)\text{-} \sum_{i=1}^{\infty} |X_i| \quad \text{in } \mathscr{V}.$$

Remark. Theorem 11.2 may be proved without using Section 2.5, if the assumption is made that the sequence X_j is bounded. In fact, if $Y \geqslant |X_j|$, $j = 1, 2, \ldots$, then

$$\sum_{j=1}^{n} |X_j| = \bigvee_{j=1}^{n} |X_j| \leqslant Y;$$

hence there exists

$$o\text{-}\lim_{n \to \infty} \sum_{j=1}^{n} |X_j| \quad \text{in } \mathscr{V}.$$

Theorem 11.3.

Let $\sum_{i=1}^{\infty} X_i$ *be an o-convergent series in* $\mathscr{V}$ *such that the terms* X_i *are pairwise orthogonal*; *then*

$$(o)\text{-} \sum_{i=1}^{\infty} X_i = \bigvee_{i=1}^{\infty} X_i^{+} - \bigvee_{i=1}^{\infty} X_i^{-}.$$

Proof. According to (I) we have $o\text{-}\lim_{n \to \infty} X_n = 0$, i.e., the sequence X_i,

$i = 1, 2, \ldots$, and therefore the sequences X_i^+ and X_i^-, $i = 1, 2, \ldots$, are bounded in $\mathscr{V}$. Hence there exist

$$\bigvee_{i=1}^{\infty} X_i^+ \quad \text{and} \quad \bigvee_{i=1}^{\infty} X_i^- \text{ in } \mathscr{V}.$$

Now we have

$$\begin{aligned}(o)\text{-} \sum_{i=1}^{\infty} X_i &= o\text{-}\lim_{n\to\infty} \sum_{i=1}^{n} X_i \\ &= o\text{-}\lim_{n\to\infty} \left(\sum_{i=1}^{n} X_i^+ - \sum_{i=1}^{n} X_i^- \right) \\ &= o\text{-}\lim_{n\to\infty} \left(\bigvee_{i=1}^{n} X_i^+ - \bigvee_{i=1}^{n} X_i^- \right) \\ &= \bigvee_{i=1}^{\infty} X_i^+ - \bigvee_{i=1}^{\infty} X_i^-. \end{aligned}$$

Corollary 11.1.

Let X be an elementary random variable (see Section 2) defined by

$$X: \quad \mathbf{a} \ni a_j \Rightarrow X(a_j) = \xi_j \in R, \qquad j = 1, 2, \ldots,$$

where $\mathbf{a} = \{a_j, \ j \geqslant 1\}$ is an experiment; then

$$X = (o)\text{-} \sum_{j=1}^{\infty} \xi_j I_{a_j} \quad \text{in} \quad \mathscr{V}.$$

Proof.† In fact, we have $|\xi_j I_{a_j}| \leqslant |X|$, $j = 1, 2, \ldots$, and the terms $\xi_j I_{a_j}$ are pairwise orthogonal; furthermore, if the real numbers $\xi_i \geqslant 0$ for every $i \in I \subseteq \mathbf{N}$ and $\xi_j \leqslant 0$ for every $j \in J = \mathbf{N} - I$, then there exist

$$\bigvee_{i\in I} \xi_i I_{a_i} \quad \text{and} \quad \bigvee_{j\in J} (-\xi_j I_{a_j}) \quad \text{in } \mathscr{V},$$

and we have

$$X^+ = \bigvee_{i\in I} \xi_i I_{a_i} \quad \text{and} \quad X^- = \bigvee_{j\in J} (-\xi_j I_{a_j}) \quad \text{in } \mathscr{V}.$$

Hence (according to Theorem 11.3):

$$X = X^+ - X^- = \bigvee_{i\in I} \xi_i I_{a_i} - \bigvee_{j\in J} (-\xi_j I_{a_j}) = (o)\text{-} \sum_{j=1}^{\infty} \xi_j I_{a_j}.$$

11.2. Let now correspond to every integer ν a rv $X_\nu \in \mathscr{V}$ and suppose that the two series

$$\sum_{\nu=1}^{\infty} X_\nu \quad \text{and} \quad \sum_{\nu=0}^{\infty} X_{-\nu}$$

---

† Compare this proof with that of Section 2.5, Chapter 4.

o-converge in $\mathscr{V}$; then we define

$$(o)\text{-}\sum_{-\infty}^{+\infty} X_\nu \equiv (o)\text{-}\sum_{\nu=1}^{\infty} X_\nu + (o)\text{-}\sum_{\nu=0}^{\infty} X_{-\nu} \quad \text{in } \mathscr{V}.$$

It is easy to verify that

$$(o)\text{-}\sum_{-\infty}^{+\infty} X_\nu = (o)\text{-}\sum_{\nu=k}^{\infty} X_\nu + (o)\text{-}\sum_{\nu=1}^{\infty} X_{k-\nu} \quad \text{in } \mathscr{V}$$

for every integer k.

11.3. In Section 5.3 we defined a chain

$$[X < \xi] = s_X(\xi), \qquad -\infty < \xi < +\infty$$

of events $s_X(\xi) \in \mathfrak{B}$, which characterizes uniquely a rv $X \in \mathscr{V}$; $s_X(\xi)$, as a function of $\xi \in R$ with values in $\mathfrak{B}$, is monotone, increasing, and continuous from the left with

$$o\text{-}\lim_{\xi \to +\infty} s_X(\xi) = e \quad \text{and} \quad o\text{-}\lim_{\xi \to -\infty} s_X(\xi) = \emptyset.$$

Conversely, if $s(\xi)$, $-\infty < \xi < +\infty$ is any chain in $\mathfrak{B}$ which is monotone, increasing, and continuous from the left with

$$o\text{-}\lim_{\xi \to +\infty} s(\xi) = e \quad \text{and} \quad o\text{-}\lim_{\xi \to -\infty} s(\xi) = \emptyset,$$

then there exists exactly a rv X in $\mathscr{V}$ such that $s(\xi) = s_X(\xi)$, for every $\xi \in R$.

Every such chain $s(\xi)$, $-\infty < \xi < +\infty$ determines a chain of indicators $I_{s(\xi)}$, $\xi \in R$ in $\mathscr{V}$, which is also monotone, increasing, and continuous from the left with

$$o\text{-}\lim_{\xi \to +\infty} I_{s(\xi)} = 1 \quad \text{and} \quad o\text{-}\lim_{\xi \to -\infty} I_{s(\xi)} = 0,$$

i.e., a so-called *spectrum* in $\mathscr{V}$.†

11.4. A *partition* δ of the real line R is defined as a family $\xi_\nu \in R$, $\nu = 0, \pm 1, \pm 2, \ldots$, with $\xi_{\nu-1} < \xi_\nu$, $\nu = 0, \pm 1, \pm 2, \ldots$, and $\sup \xi_\nu = +\infty$, $\inf \xi_\nu = -\infty$, and denoted by

(P) $$\delta = \{\ldots < \xi_{-2} < \xi_{-1} < \xi_0 < \xi_1 < \xi_2 < \ldots\}.$$

Let Δ be the set of all partitions δ of the real line R; then Δ can be directed with respect to the norm

$$N(\delta) = \sup\{\xi_\nu - \xi_{\nu-1}, \quad \nu = 0, \pm 1, \pm 2, \ldots\}$$

† Or a *resolution* of the weak unit 1 (see Birkhoff [1], p. 251).

or with respect to the inclusion relation, i.e., $\delta > \delta'$ if and only if $N(\delta) \leqslant N(\delta')$ or $\delta \supseteq \delta'$ if and only if δ as a set of points contains the set δ'.

11.5. Let X be any rv in $\mathscr{V}$, then

$$I^X(\xi) \equiv I_{s_X(\xi)}, \qquad \xi \in R,$$

is said to be the spectrum of X in $\mathscr{V}$. To every partition δ of the real line, there corresponds the chain

$$\ldots \subseteq I_{\xi_{-2}} \subseteq I_{\xi_{-1}} \subseteq I_{\xi_0} \subseteq I_{\xi_1} \subseteq I_{\xi_2} \subseteq \ldots \dagger.$$

Obviously, the terms (differences)

$$I_{\xi_\nu} - I_{\xi_{\nu-1}} \in \mathscr{V}, \qquad \nu = 0, \pm 1, \pm 2, \ldots,$$

are pairwise orthogonal. Now pick out any family

$$\Gamma\colon\ \gamma_\nu \in R, \qquad \nu = 0, \pm 1, \pm 2, \ldots,$$

with

$$\xi_{\nu-1} \leqslant \gamma_\nu \leqslant \xi_\nu, \qquad \nu = 0, \pm 1, \pm 2, \ldots;$$

then the series

$$(\Delta) \qquad \sum_{\nu=-\infty}^{+\infty} \gamma_\nu (I_{\xi_\nu} - I_{\xi_{\nu-1}})$$

o-converges. In order to prove it, we notice that the union of all differences

$$(\mathrm{E}) \qquad d_\nu = s_X(\xi_\nu) - s_X(\xi_{\nu-1}) \in \mathfrak{B}, \qquad \nu = 0, +1, -1, \ldots$$

is equal to e in $\mathfrak{B}$ and the non-zero elements of (E) are pairwise disjoint, hence the sequence (E) defines an experiment in $\mathfrak{B}$. The erv which corresponds to the experiment (E) by the mapping

$$(\mathrm{m}) \qquad d_\nu \Rightarrow \gamma_\nu \in R, \qquad \nu = 0, +1, -1, +2, -2, \ldots$$

will be denoted by $X_{\delta, \gamma}$. One can easily prove, as in Section 2.5, that the series (Δ) o-converges to this erv $X_{\delta, \gamma}$ in $\mathscr{E}$. However $\mathscr{E}$ is a regular vector sublattice $\mathscr{V}$, hence the o-convergence of the series (Δ) to $X_{\delta, \gamma}$ holds also in $\mathscr{V}$; we have also

$$(\mathrm{Z}) \qquad (o)\text{-}\sum_{\nu=-\infty}^{+\infty} \gamma_\nu (I_{\xi_\nu} - I_{\xi_{\nu-1}}) = X_{\delta, \gamma} \quad \text{in } \mathscr{V}.$$

---

† We write in this section instead of $I^X(\xi)$ also I_ξ as far as X remains the same.

Without using Section 2.5, we can prove directly:

Theorem 11.4.

The series (Δ) *o-conveges to an elementary random variable* $X_{\delta,\gamma}$ *in* $\mathscr{V}$ *for every partition* $\delta \in \Delta$ *and every*

$$\gamma = (\gamma_\nu, \quad \nu = 0, \pm 1, \pm 2, \ldots)$$

with
$$\xi_{\nu-1} \leqslant \gamma_\nu \leqslant \xi_\nu, \qquad \nu = 0, \pm 1, \pm 2, \ldots .$$

The family $X_{\delta,\gamma} \in \mathscr{V}$, $\delta \in \Delta$, *directed with respect to* $>$ *or* $\supseteq$ *u-converges (i.e., also o-converges) to the rv* X *in* $\mathscr{V}$.

Proof. In order to prove this theorem, we notice that

$$(o)\text{-}\sum_{\nu=-\infty}^{+\infty} (I_{\xi_\nu} - I_{\xi_{\nu-1}}) = 1 \quad \text{in} \quad \mathscr{V}.$$

In fact, the terms (differences) $I(\xi_\nu) - I(\xi_{\nu-1})$ of this series are pairwise orthogonal and bounded by 1, hence the series o-converges. Now we have

$$\sum_{\nu=-k}^{+n} (I_{\xi_\nu} - I_{\xi_{\nu-1}}) = I_{\xi_n} - I_{\xi_{-k-1}}.$$

This implies

$$\begin{aligned}
(o)\text{-}\sum_{\nu=-\infty}^{+\infty} (I_{\xi_\nu} - I_{\xi_{\nu-1}}) &= \underset{n\to+\infty}{o\text{-lim}} \left(\underset{k\to+\infty}{o\text{-lim}} \sum_{\nu=-k}^{+n} (I_{\xi_\nu} - I_{\xi_{\nu-1}}) \right) \\
&= \underset{k\to+\infty}{o\text{-lim}} \left(\underset{n\to+\infty}{o\text{-lim}} (I_{\xi_n} - I_{\xi_{-k-1}}) \right) \\
&= \underset{n\to+\infty}{o\text{-lim}}\, I_{\xi_n} - \underset{k\to+\infty}{o\text{-lim}}\, I_{\xi_{-k-1}} \\
&= 1 - 0 = 1.
\end{aligned}$$

Obviously, we have now

$$X = (o)\text{-}\sum_{\nu=-\infty}^{+\infty} (I_{\xi_\nu} - I_{\xi_{\nu-1}})\, X.$$

The relation $\xi_{\nu-1} \leqslant \gamma_\nu \leqslant \xi_\nu$, implies

$$\xi_\nu - \gamma_\nu \leqslant \xi_\nu - \xi_{\nu-1} \leqslant \sup\{\xi_\nu - \xi_{\nu-1}, \quad \nu = 0, \pm 1, \pm 2, \ldots\} = N(\delta),$$

and furthermore

$$\xi_{\nu-1}(I_{\xi_\nu} - I_{\xi_{\nu-1}}) \leqslant (I_{\xi_\nu} - I_{\xi_{\nu-1}})\, X \leqslant \xi_\nu(I_{\xi_\nu} - I_{\xi_{\nu-1}})$$

and
$$|(I_{\xi_\nu} - I_{\xi_{\nu-1}})\, X - \gamma_\nu(I_{\xi_\nu} - I_{\xi_{\nu-1}})| \leqslant N(\delta).1;$$

all these imply

$$\left|\sum_{\nu=-k}^{n}(I_{\xi_\nu}-I_{\xi_{\nu-1}})\,X-\sum_{\nu=-k}^{n}\gamma_\nu(I_{\xi_\nu}-I_{\xi_{\nu-1}})\right|\leqslant N(\delta).1$$

for every $n>0$ and $k\geqslant 0$.

However, there exists

$$o\text{-}\sum_{\nu=-\infty}^{+\infty}\gamma_\nu(I_{\xi_\nu}-I_{\xi_{\nu-1}}).$$

In fact, let us consider the experiment (E) and the erv:

$$Y\colon\ \mathfrak{B}\ni d_\nu\Rightarrow|\gamma_\nu|,\qquad \nu=0,\pm1,\pm2,\ldots.$$

Then we have:

$$|\gamma_\nu(I_{\xi_\nu}-I_{\xi_{\nu-1}})|\leqslant Y.$$

That is, the terms of the above series are absolutely bounded by Y. Moreover, they are pairwise orthogonal. Hence, according to the Remark, Section 11.1, the series converges absolutely.

Let us denote

$$(o)\text{-}\sum_{\nu=-\infty}^{+\infty}\gamma_\nu(I_{\xi_\nu}-I_{\xi_{\nu-1}})\equiv X_{\delta,\gamma};\dagger$$

then, we have

$$\text{(R)}\qquad |X-X_{\delta,\gamma}|\leqslant N(\delta).1=N(\delta),$$

independently of the choice of the γ_ν, with

$$\xi_{\nu-1}\leqslant\gamma_\nu\leqslant\xi_\nu,\qquad \nu=0,\pm1,\pm2,\ldots.$$

Since the family $N(\delta)\in R$, $\delta\in\Delta$, directed by $>$ or $\supseteq$ converges to zero, (R) implies that the family $X_{\delta,\gamma}$, $\delta\in\Delta$, directed by $>$ or $\supseteq$ u-converges to X, hence it also o-converges to X, independently of the choice of the γ_ν with

$$\xi_{\nu-1}\leqslant\gamma_\nu\leqslant\xi_\nu,\qquad \nu=0,\pm1,\pm2,\ldots.$$

Hence Theorem 11.4 is proved.

Remark. Especially, we have for $\gamma_\nu=\xi_{\nu-1}$ or $\gamma_\nu=\xi_\nu$, $\nu=0,\pm1,\pm2,\ldots$

$$(o)\text{-}\sum_{\nu=-\infty}^{+\infty}\xi_{\nu-1}(I_{\xi_\nu}-I_{\xi_{\nu-1}})\equiv \underset{\sim}{X}_\delta,$$

or

$$(o)\text{-}\sum_{\nu=-\infty}^{+\infty}\xi_\nu(I_{\xi_\nu}-I_{\xi_{\nu-1}})\equiv \tilde{X}_\delta,$$

---

† It can be easily proved that $X_{\delta,\gamma}$ is an elementary random variable.

respectively, and then:

$$(\rho) \qquad \underset{\sim}{X}_\delta \leqslant X_{\delta,\gamma} \leqslant \tilde{X}_\delta \quad \text{and} \quad \underset{\sim}{X}_\delta \leqslant X \leqslant \tilde{X}_\delta.$$

Obviously, we furthermore have

$$(\mathrm{s}) \qquad X = \bigvee_{\delta \in \Delta} \underset{\sim}{X}_\delta = \bigwedge_{\delta \in \Delta} \tilde{X}_\delta = o\text{-}\lim_{\delta \in \Delta} X_{\delta,\gamma}.$$

We define the limit on the right side to be signified by the symbol

$$\int_{-\infty}^{+\infty} \xi dI^X(\xi),$$

i.e., for every $X \in \mathscr{V}$ we have, using its spectrum $I^X(\xi)$, $-\infty < \xi < +\infty$, the so-called *integral* (*spectral*) *representation*:

$$X = \int_{-\infty}^{+\infty} \xi dI^X(\xi).$$

11.6. Let $s(\xi)$, $-\infty < \xi < +\infty$, be a chain in $\mathfrak{B}$, which is monotone, increasing, continuous from the left, with

$$o\text{-}\lim_{\xi \to +\infty} s(\xi) = e, \quad o\text{-}\lim_{\xi \to -\infty} s(\xi) = \emptyset,$$

and let us consider the corresponding spectrum

$$I(\xi) = I_{s(\xi)}, \quad -\infty < \xi < +\infty \quad \text{in} \quad \mathscr{V}.$$

We shall say there exists the (spectral) integral

$$\int_{-\infty}^{+\infty} \xi dI(\xi)$$

if and only if there is a rv $X \in \mathscr{V}$ such that $I^X(\xi) = I(\xi)$, for every $\xi \in R$; we then shall say, the spectrum $I(\xi)$, $-\infty < \xi < +\infty$, defines the rv X. In order to symbolize briefly this statement, we shall write

$$(\Sigma) \qquad X = \int_{-\infty}^{+\infty} \xi dI(\xi).$$

In Section 11.3 (respectively in 5.3) we asserted, but we did not prove, that there exists exactly a rv $X \in \mathscr{V}$, such that $s(\xi) = s_X(\xi)$ and accordingly

$$I^X(\xi) = I(\xi) \quad \text{for every} \quad \xi \in R,$$

i.e., every spectrum $I(\xi)$, $-\infty < \xi < +\infty$, defines exactly a rv $X \in \mathscr{V}$. Now it is not difficult, following the steps used to prove Theorem 11.4, to prove also the existence of a rv X such that (Σ) is true. Namely, the properties generally of a spectrum in $\mathscr{V}$, and not especially, the spectrum of a rv $X \in \mathscr{V}$ imply, that the series (Δ) o-converges to an elementary random variable defined on the experiment (E) by the mapping (compare Section 11.5) denoted by $X_{\delta,\gamma}$. Just so the existence of the erv's $\underset{\sim}{X_\delta}$ and $\tilde{X}_\delta$ is secured; the first relation of (ρ), namely

$$\underset{\sim}{X_\delta} \leqslant X_{\delta,\gamma} \leqslant \tilde{X}_\delta$$

is also obvious, as, further, the relation

$$\tilde{X}_\delta - \underset{\sim}{X_\delta} \leqslant N(\delta) . 1;$$

this relation secures the existence of a rv X such that the relation (s) (of Section 11.5) is true and (s) implies

$$X = \int_{-\infty}^{+\infty} \xi dI(\xi).$$

12. COMPOSITION OF RANDOM VARIABLES

Let $\phi(\xi_1, \xi_2, ..., \xi_n)$ be a real-valued function of n real variables which is defined for every point $(\xi_1, \xi_2, ..., \xi_n)$ of the n-dimensional space R^n. Furthermore, let us consider the erv's

$$X_1, X_2, ..., X_n, \quad \text{in} \quad \mathscr{E} \subseteq \mathscr{V};$$

then there exist at least an experiment $\mathbf{a} \equiv \{a_1, a_2, ...\}$ such that every X_j is defined on $\mathbf{a}$, i.e.,

$$X_j : \mathbf{a} \ni a_i \Rightarrow X_j(a_i) = \xi_{ji} \in R, \quad i = 1, 2, ...,$$

for every $j = 1, 2, ..., n$. Now, we can define the following erv $Y \equiv \phi(X_1, X_2, ..., X_n)$:

$$\mathbf{a} \ni a_i \Rightarrow \phi(X_1(a_i), X_2(a_i), ..., X_n(a_i)) \equiv \phi(\xi_{1i}, \xi_{2i}, ..., \xi_{ni}) \equiv \eta_i \in R.$$

The erv $Y = \phi(X_1, X_2, ..., X_n)$ is said to be the *composition* of the erv's $X_1, X_2, ..., X_n$ with respect to the function ϕ.

Let now

$$X_1, X_2, ..., X_n$$

be any rv's in $\mathscr{V}$; then for every X_j, there exists a sequence $X_{j\nu} \in \mathscr{E}$, $\nu = 1, 2, ...,$ such that

$$X_{j\nu} \underset{\nu \to \infty}{\overset{o}{\to}} X_j, \quad j = 1, 2, ...;$$

then a sequence

$$Y_\nu \equiv \phi(X_{1\nu}, X_{2\nu}, \ldots, X_{n\nu}), \qquad \nu = 1, 2, \ldots$$

is also defined in $\mathscr{E}$. We shall say the composition $\phi(X_1, X_2, \ldots, X_n)$ can be defined, where $X_1, X_2, \ldots, X_n$ are any rv's in $\mathscr{V}$, if and only if the sequence

$$Y_\nu \equiv \phi(X_{1\nu}, X_{2\nu}, \ldots, X_{n\nu}), \qquad \nu = 1, 2, \ldots,$$

is o-fundamental and, furthermore, the o-limit

$$Y = o\text{-}\lim_{\nu \to \infty} \phi(X_{1\nu}, X_{2\nu}, \ldots, X_{n\nu}) \quad \text{in} \quad \mathscr{V}$$

is the same for every choice of sequences $X_{j\nu} \in \mathscr{E}$, $\nu = 1, 2, \ldots$, with $X_{j\nu} \overset{o}{\to} X_j$ in $\mathscr{E}$, $j = 1, 2, \ldots, n$.

We then define:

$$\phi(X_1, X_2, \ldots, X_n) = o\text{-}\lim_{\nu \to \infty} \phi(X_{1\nu}, X_{2\nu}, \ldots, X_{n\nu}).$$

The reader can prove that, if the function ϕ is continuous throughout the space R^n, then there exists the composition $\phi(X_1, X_2, \ldots, X_n)$ for any $X_j \in \mathscr{V}$, $j = 1, 2, \ldots, n$. (see a proof of this statement in a more general case, in C. Carathéodory [6], Chapter IV, Section 113).

In the special case $\phi(\xi) = e^\xi$, $\phi(\xi) = \xi^\rho$, $\xi \geqslant 0$, $\rho > 0$, and generally $\phi(\xi)$ continuous in R, the proof is easier.

V

EXPECTATION OF RANDOM VARIABLES

1. EXPECTATION OF ELEMENTARY RANDOM VARIABLES

1.1. Let $(\mathfrak{B}, p)$ be a pr σ-algebra. We shall briefly denote by $\mathscr{V}$ the complete stochastic space over $(\mathfrak{B}, p)$, by $\mathscr{E}$ and $\mathscr{S}$ the elementary and simple stochastic space over $(\mathfrak{B}, p)$, respectively. We shall consider $\mathscr{E}$ and $\mathscr{S}$ as stochastic subspaces of $\mathscr{V}$, i.e., as identical with the isomorphic images $\mathscr{E}^*$ and $\mathscr{S}^*$ of them in $\mathscr{V}$. Now let $X \in \mathscr{E}$ and let

$$X = \sum_{j \geqslant 1} \xi_j I_{a_j}$$

be the reduced representation of X by indicators. We shall say: the erv X possesses an *expectation* $E(X)$ if and only if the series $\sum_{j \geqslant 1} \xi_j p(a_j)$ is absolutely convergent; i.e.,

$$\sum_{j \geqslant 1} |\xi_j| \, p(a_j) < +\infty.$$

We then define

$$E(X) = \sum_{j \geqslant 1} \xi_j p(a_j).$$

It is easy to prove: If

$$X = \sum_{j \geqslant 1} \lambda_j I_{b_j}$$

is another representation of X by indicators, where $b_1, b_2, \ldots,$ are pairwise disjoint, then

$$E(X) = \sum_{j \geqslant 1} \lambda_j p(b_j),$$

i.e. the expectation of X is independent of the particular representation of X by indicators. We shall denote by $\mathscr{K} = \mathscr{K}(\mathfrak{B}, p)$ the set of all $X \in \mathscr{E}$, which possess an expectation $E(X)$. Obviously we have $\mathscr{S} \subseteq \mathscr{K}$. The following theorems are true:

Theorem 1.1.

Let $X \in \mathscr{E}$, with $X \notin \mathscr{S}$, and

$$X = \sum_{j=1}^{\infty} \xi_j I_{a_j}$$

be the reduced representation of X by indicators; further, let Δ be the set of all finite subsets of the set $\mathbf{N} = \{1, 2, \ldots\}$ and define

$$\sigma_\delta = \sum_{j \in \delta} \xi_j p(a_j), \text{ for every } \delta \in \Delta.$$

Then σ_δ, $\delta \in \Delta$, is a directed family with respect to the inclusion relation $\subseteq$ in Δ and the following statement holds:

$X \in \mathscr{K}$ if and only if $\lim_{\delta \in \Delta} \sigma_\delta$ exists in R and

$$E(X) = \lim_{\delta \in \Delta} \sigma_\delta.$$

Proof. Obvious, for $\lim_{\delta \in \Delta} \sigma_\delta$ exists in R if and only if

$$\sum_{j=1}^{\infty} |\xi_j| \, p(a_j) < +\infty \quad \text{and} \quad \sum_{j=1}^{\infty} \xi_j p(a_j) = \lim_{\delta \in \Delta} \sigma_\delta.$$

Theorem 1.2.

The set $\mathscr{K}$ forms a vector sublattice of the vector lattice $\mathscr{E}$ and the expectation $E(X)$ is linear and strictly monotone.

Proof. Let X and Y be elements in $\mathscr{K}$ with the reduced representations by indicators

$$X = \sum_{i \geqslant 1} \xi_i I_{a_i} \quad \text{and} \quad Y = \sum_{j \geqslant 1} \eta_j I_{b_j}$$

respectively.

Then

(1) $\sum(\xi_i+\eta_j)\,p(a_i \wedge b_j)$ converges absolutely and we have

$$\begin{aligned} E(X+Y) &= \sum_{i,\,j\geqslant 1}(\xi_i+\eta_j)\,p(a_i \wedge b_j) \\ &= \sum_{i,\,j\geqslant 1}\xi_i\,p(a_i \wedge b_j)+\sum_{i,\,j\geqslant 1}\eta_j\,p(a_i \wedge b_j) \\ &= \sum_{i\geqslant 1}\xi_i\sum_{j\geqslant 1}p(a_i \wedge b_j)+\sum_{j\geqslant 1}\eta_j\sum_{i\geqslant 1}p(a_i \wedge b_j) \\ &= \sum_{i\geqslant 1}\xi_i\,p(a_i)+\sum_{j\geqslant 1}\eta_j\,p(b_j) = E(X)+E(Y). \end{aligned}$$

(2) $\sum_{i\geqslant 1}\lambda\,\xi_i\,p(a_i)$, where λ is a real number, converges absolutely and we have

$$E(\lambda X) = \sum_{i\geqslant 1}\lambda\xi_i\,p(a_i) = \lambda\sum_{i\geqslant 1}\xi_i\,p(a_i) = \lambda E(X).$$

(3) Obviously,

$$\sum_{i,\,j\geqslant 1}(\xi_i \wedge \eta_j)\,p(a_i \wedge b_j) \quad\text{and}\quad \sum_{i,\,j\geqslant 1}(\xi_i \vee \eta_j)\,p(a_i \wedge b_j)$$

converge absolutely; i.e., $X \wedge Y$ and $X \vee Y$ belong to $\mathscr{K}$; hence $\mathscr{K}$ is a vector sublattice of $\mathscr{E}$ and E is linear.

(4) $E(X)$ is strictly monotone. In fact, if $X \geqslant Y$ then

$$E(X)-E(Y) = E(X-Y) = \sum_{i,\,j\geqslant 1}(\xi_i-\eta_j)\,p(a_i \wedge b_j) \geqslant 0,$$

because $\xi_i-\eta_j \geqslant 0$ for all $a_i \wedge b_j \neq \varnothing$, hence $E(X) \geqslant E(Y)$; if $X > Y$, then there exists, at least, a pair (i, j) such that $\xi_i-\eta_j > 0$ and $a_i \wedge b_j \neq \varnothing$; thus $E(X)-E(Y) > 0$; hence $E(X) > E(Y)$.

Theorem 1.3.

If $X \in \mathscr{E}$ and X is absolutely bounded by a constant, i.e., $|X| \leqslant \eta$, then $X \in \mathscr{K}$ and $|E(X)| \leqslant \eta$, i.e., $\mathscr{E} \cap \mathscr{M} \subseteq \mathscr{K}$.

Proof. Let

$$X = \sum_{i\geqslant 1}\xi_i I_{a_i}$$

be the reduced representation of X by indicators; then

$$\sum_{i\geqslant 1}|\xi_i|\,p(a_i) \leqslant \eta\sum_{i\geqslant 1}p(a_i) = \eta,$$

i.e., $\sum_{i\geqslant 1}|\xi_i|\,p(a_i)$ converges and

$$|E(X)| \leqslant \sum_{i\geqslant 1}|\xi_i|\,p(a_i) \leqslant \eta.$$

1.2. We define $\|X\|_1 = E(|X|)$ for every $X \in \mathscr{K}$. Then $\| \ \|_1$ is a norm on $\mathscr{K}$. In fact, we have

(α) $\|0\|_1 = 0$ and $\|X\|_1 > 0$, for any $X \neq 0$, $X \in \mathscr{K}$.

(β) $\|X+Y\|_1 \leqslant \|X\|_1 + \|Y\|_1$.

(γ) $\|\lambda X\|_1 = |\lambda| . \|X\|_1$ for any $X \in \mathscr{K}$ and any real number λ.

Now, we define the distance

$$\delta(X, Y) = \|X-Y\|_1, \quad X \in \mathscr{K}, \quad Y \in \mathscr{K}.$$

This distance induces in $\mathscr{K}$ a metric convergence, which we shall call *norm*$_1$*-convergence* (briefly N_1*-convergence*) and denote by

$$X_n \underset{\mathscr{K}}{\overset{N_1}{\rightarrow}} X;$$

i.e., we define

$$X_n \underset{\mathscr{K}}{\overset{N_1}{\rightarrow}} X \text{ if and only if } \|X_n - X\|_1 \rightarrow 0.$$

A sequence $X_n \in \mathscr{K}$, $n = 1, 2, \ldots$, is said to be N_1*-fundamental* if and only if the double sequence $\|X_n - X_k\|_1$ converges to zero.

1.3. The normed vector lattice $\mathscr{K}$ so defined, is not N_1-complete, if the Boolean σ-algebra is not atomic. We shall extend $\mathscr{K}$, in Section 2, to an N_1-complete vector sublattice of $\mathscr{V}$. We shall first prove some theorems:

Theorem 1.4.

If $X \in \mathscr{K}$, $Y \in \mathscr{K}$ *with* $X \leqslant Y$ *and any* $Z \in \mathscr{E}$ *satisfies the condition* $X \leqslant Z \leqslant Y$, *then* $Z \in \mathscr{K}$.

Proof. Obvious.

Theorem 1.5.

For any $X \in \mathscr{K}$, $Y \in \mathscr{K}$, *we have*

$$|E(X) - E(Y)| \leqslant E(|X-Y|) = \|X-Y\|_1 \leqslant \|X\|_1 + \|Y\|_1.$$

Proof. Obvious.

Theorem 1.6.

If $X_n \in \mathscr{K}$, $n = 1, 2, \ldots$, *and* $X_n \underset{\mathscr{E}}{\overset{u}{\rightarrow}} X \in \mathscr{E}$, *then* $X \in \mathscr{K}$ *and*

$$E(X_n) \rightarrow E(X).$$

Proof. We have, for $\varepsilon = 1$, an index n_0 such that $X_{n_0}-1 < X < X_{n_0}+1$. But $X_{n_0}-1$ and $X_{n_0}+1$ belong to $\mathscr{K}$, hence, by Theorem 1.4, $X \in \mathscr{K}$. Now, for any $\varepsilon > 0$, there exists an index $n(\varepsilon)$ such that $|X_n - X| < \varepsilon$ for all $n \geqslant n(\varepsilon)$; then

$$|E(X_n)-E(X)| \leqslant E(|X_n - X|) < E(\varepsilon) = \varepsilon, \quad n \geqslant n(\varepsilon),$$

i.e.,

$$E(X_n) \to E(X).$$

Theorem 1.7.

If $X_n \in \mathscr{K}$, $n = 1, 2, \ldots$, is a u-fundamental sequence in $\mathscr{E}$, then $E(X_n)$ is a fundamental sequence of real numbers, i.e., $E(X_n)$ converges to a real number.

Proof. For any $\varepsilon > 0$, an index $n(\varepsilon)$ exists such that $|X_n - X_k| < \varepsilon$, for all $n \geqslant n(\varepsilon)$, $k \geqslant n(\varepsilon)$; then

$$|E(X_n)-E(X_k)| \leqslant E(|X_n - X_k|) < E(\varepsilon) = \varepsilon;$$

i.e., the $\lim_{n\to\infty} E(X_n)$ exists in R.

Theorem 1.8.

Let $X_n \in \mathscr{K}$ and $Y_n \in \mathscr{K}$, $n = 1, 2, \ldots$, be two u-equivalent u-fundamental sequences in $\mathscr{E}$, then

$$\lim_{n\to\infty} E(X_n) = \lim_{n\to\infty} E(Y_n).$$

Proof. Since $X_n - Y_n \overset{u}{\to} 0$, we have by Theorem 1.6,

$$E(X_n)-E(Y_n) = E(X_n - Y_n) \to 0,$$

i.e.,

$$\lim_{n\to\infty} E(X_n) = \lim_{n\to\infty} E(Y_n).$$

Theorem 1.9.

If $X_n \in \mathscr{K}$, $n = 1, 2, \ldots$, and $X_n \geqslant X_{n+1}$, $n = 1, 2, \ldots$, with $\bigwedge_{n=1}^{\infty} X_n = 0$ in $\mathscr{E}$, i.e., $X_n \underset{\mathscr{E}}{\overset{o}{\downarrow}} 0$, then $E(X_n) \downarrow 0$.

Proof. (A). We shall first prove the theorem with

$$X_n \in \mathscr{S} \subseteq \mathscr{K}, \quad n = 1, 2, \ldots.$$

Let

$$X_n = \sum_{i=1}^{\lambda_n} \xi_{ni} I_{a_{ni}}$$

be the reduced representation of X_n by indicators, $n = 1, 2, \ldots$, in which the indices i are enumerated in such a way that

$$\xi_{ni} \geqslant \xi_{n,\, i+1}, \quad i = 1, 2, \ldots, \lambda_n - 1.$$

Now, let ε be > 0; then an index $k_n \leqslant \lambda_n$ exists, such that $\xi_{ni} > \varepsilon$ for all $i \leqslant k_n$ and $\xi_{ni} \leqslant \varepsilon$ for all $i > k_n$, $n = 1, 2, \ldots$. We set

$$\eta = \operatorname{Max} \{\xi_{11}, \xi_{12}, \ldots, \xi_{1,\, \lambda_1}\};$$

then
$$0 \leqslant X_n \leqslant \eta,$$
hence
$$0 \leqslant E(X_n) \leqslant \eta.$$

We write

$$b_n = a_{n1} \vee a_{n2} \vee \ldots \vee a_{nk_n};$$

then
$$b_n \geqslant b_{n+1}, \quad n = 1, 2, \ldots,$$
and
$$\bigwedge_{n=1}^{\infty} b_n = \emptyset,$$

hence $p(b_n) \downarrow 0$, i.e., there exists an index $n(\varepsilon)$ such that $p(b_n) < \varepsilon/\eta$, for all $n \geqslant n(\varepsilon)$. Now, for $n \geqslant n(\varepsilon)$, we have

$$\begin{aligned} E(X_n) &= \sum_{i=1}^{k_n} \xi_{ni}\, p(a_{ni}) + \sum_{i=k_n+1}^{\lambda_n} \xi_{ni}\, p(a_{ni}) \\ &\leqslant \eta \sum_{i=1}^{k_n} p(a_{ni}) + \varepsilon \sum_{i=k_n+1}^{\lambda_n} p(a_{ni}) \\ &\leqslant \eta p(b_n) + \varepsilon < \varepsilon + \varepsilon \\ &= 2\varepsilon. \end{aligned}$$

i.e., $0 \leqslant E(X_n) \leqslant 2\varepsilon$, $n \geqslant n(\varepsilon)$, hence $E(X_n) \to 0$.

Another proof. Let $X_n \in \mathscr{S}$ and $X_n \underset{\mathscr{E}}{\overset{o}{\downarrow}} 0$; then there exists a constant η such that $0 \leqslant X_n \leqslant \eta$, for every $n = 1, 2, \ldots$; but o-convergence is equivalent to au-convergence, i.e., $X_n \xrightarrow{au} 0$. Hence for any $\varepsilon > 0$, an element b_ε exists in $\mathfrak{B}$ such that $p(b_\varepsilon{}^c) < \varepsilon/\eta$ and $I_{b_\varepsilon} X_n \xrightarrow{u} 0$; then by Theorem 1.6, we have $E(I_{b_\varepsilon} X_n) \to 0$. Now, since

$$X_n = I_{b_\varepsilon} X_n + I_{b_\varepsilon^c} X_n,$$

i.e.
$$\begin{aligned} 0 &\leqslant E(X_n) \\ &= E(I_{b_\varepsilon} X_n) + E(I_{b_\varepsilon^c} X_n) < E(I_{b_\varepsilon} X_n) + \frac{\varepsilon}{\eta} \cdot \eta \\ &= E(I_{b_\varepsilon} X_n) + \varepsilon, \end{aligned}$$

we have $$0 \leqslant \limsup E(X_n) \leqslant \varepsilon;$$

hence $$\lim_{n\to\infty} E(X_n) = 0.$$

(B) Now, let X_n be any element in $\mathscr{K}$ and let

$$X_n = \sum_{j\geqslant 1} \xi_{nj} I_{a_{nj}}$$

be the reduced representation of X_n by indicators, $n = 1, 2, \ldots$. Then

$$E(X_n) = \sum_{j\geqslant 1} \xi_{nj}\, p(a_{nj}).$$

Hence, for any $\varepsilon > 0$ and $\varepsilon/2^n > 0$, there exists an index $k_n = k_n(\varepsilon)$ such that

$$0 \leqslant \sum_{j\geqslant k_n} \xi_{nj}\, p(a_{nj}) < \frac{\varepsilon}{2^n}, \quad n = 1, 2, \ldots.$$

We write

$$Z_n = \sum_{j=1}^{k_n} \xi_{nj} I_{a_{nj}}$$

and $$R_n = X_n - Z_n = \sum_{j\geqslant k_n+1} \xi_{nj} I_{a_{nj}};$$

then $$Z_n \in \mathscr{S} \subseteq \mathscr{K}, \quad R_n \in \mathscr{K},$$

and

$$0 \leqslant E(X_n) = E(Z_n) + E(R_n) < E(Z_n) + \frac{\varepsilon}{2^n},$$

i.e.,

$$\text{(0)} \qquad 0 \leqslant E(X_n) < E(Z_n) + \frac{\varepsilon}{2^n}, \quad n = 1, 2, \ldots.$$

Now, we have $Z_n \in \mathscr{S}$, $0 \leqslant Z_n \leqslant X_n$, hence $Z_n \overset{o}{\to} 0$, and

$$\text{(I)} \qquad E(X_n) - \frac{\varepsilon}{2^n} \leqslant E(Z_n) \leqslant E(X_n), \quad n = 1, 2, \ldots.$$

Note that Z_n, $n = 1, 2, \ldots$, is not in general decreasing. We take instead

$$Z_1,\ Z_1 \wedge Z_2,\ Z_1 \wedge Z_2 \wedge Z_3,\ \ldots.$$

Since $Z_1 \wedge Z_2 \wedge \ldots \wedge Z_n = U_n \in \mathscr{S}$ and $U_n \overset{o}{\underset{\mathscr{E}}{\downarrow}} 0$, we have, according to (A), $E(U_n) \downarrow 0$. We now prove, by induction

$$\text{(II)} \qquad E(Z_n) \leqslant E(U_n) + \left(\frac{1}{2} + \frac{1}{2^2} + \ldots + \frac{1}{2^{n-1}}\right)\varepsilon < E(U_n) + \varepsilon.$$

Obviously (II) is true for $n = 1$. Now we have†

$$(Z_1 \wedge Z_2 \wedge \ldots \wedge Z_n)+Z_{n+1} = Z_1 \wedge Z_2 \wedge \ldots \wedge Z_{n+1}+(Z_1 \wedge Z_2 \wedge \ldots \wedge Z_n) \vee Z_{n+1},$$

i.e.

$$U_n+Z_{n+1} = U_{n+1}+U_n \vee Z_{n+1}.$$

Now
$$E(Z_{n+1}) = E(U_{n+1})+E(U_n \vee Z_{n+1})-E(U_n)$$
and
$$U_n \vee Z_{n+1} \leqslant Z_n \vee Z_{n+1} \leqslant X_n \vee X_{n+1} = X_n;$$
from this, (I), and (II), the following relation follows:

$$E(Z_{n+1}) \leqslant E(U_{n+1})+E(X_n)-E(Z_n)+\left(\frac{1}{2}+\frac{1}{2^2}+\ldots+\frac{1}{2^{n-1}}\right)\varepsilon,$$

i.e.,
$$E(Z_{n+1}) \leqslant E(U_{n+1})+\left(\frac{1}{2}+\frac{1}{2^2}+\ldots+\frac{1}{2^{n-1}}+\frac{1}{2^n}\right)\varepsilon,$$

i.e.,
$$E(Z_{n+1}) < E(U_{n+1})+\varepsilon.$$

Hence the relation (II) is true for every $n = 1, 2, \ldots$. The relations (II) and (0) imply

$$0 \leqslant E(X_n) < E(Z_n)+\frac{\varepsilon}{2^n} < E(U_n)+\frac{\varepsilon}{2^n}+\varepsilon, \quad n = 1, 2, \ldots,$$

hence

$$0 \leqslant \limsup_{n\to\infty} E(X_n) \leqslant \lim_{n\to\infty} E(U_n)+\varepsilon \leqslant \varepsilon, \text{ i.e., } \lim_{n\to\infty} E(X_n) = 0.$$

1.4. Note the following properties of the expectation, which are consequences of the previous theorems:

(a) $E(X+Y) = E(X)+E(Y)$ for any $X \in \mathscr{K}$, $Y \in \mathscr{K}$.

(b) $E(\lambda X) = \lambda E(X)$ for any $X \in \mathscr{K}$ and $\lambda \in R$.

(c) If $X \geqslant 0$, then $E(X) \geqslant 0$ and, moreover, $E(X) > 0$ if $X > 0$, $X \in \mathscr{K}$.

(c*) If $X \geqslant Y$, then $E(X) \geqslant E(Y)$ and, moreover, if $X > Y$ then $E(X) > E(Y)$, $X \in \mathscr{K}$, $Y \in \mathscr{K}$.

(d) If $X_n \in \mathscr{K}$, $n = 1, 2, \ldots$, and $X_n \overset{o}{\underset{\mathscr{E}}{\downarrow}} 0$, then $E(X_n) \downarrow 0$, i.e., the expectation is a linear, strictly monotone, and order-continuous functional on $\mathscr{K}$.

† We apply the relation $A+B = A \wedge B+A \vee B$, which is true in every vector lattice.

Now, let $X \in \mathscr{K}$; then $I_b X \in \mathscr{K}$ for every $b \in \mathfrak{B}$. Hence we may define

$$\psi_X(b) = E(I_b X) \text{ for every } b \in \mathfrak{B}.$$

Then

(1) ψ_X is an additive real-valued function defined on $\mathfrak{B}$. In fact, if a and b are disjoint elements of B, then

$$\psi_X(a \vee b) = E(I_{a \vee b} X) = E(I_a X) + E(I_b X) = \psi_X(a) + \psi_X(b).$$

(2) ψ_X is o-continuous, i.e., if $a_n \in \mathfrak{B}$, $n = 1, 2, \ldots$, with $a_n \overset{o}{\downarrow} \varnothing$, then $\psi_X(a_n) \to 0$. In fact, if $X \geqslant 0$, then $I_{a_n} X \overset{o}{\downarrow} I_\varnothing X = 0$, and, according to (d), $\psi_X(a_n) = E(I_{a_n} X) \downarrow 0$.

Let X be any element of $\mathscr{K}$; then

$$X = X^+ - X^-, \ X^+ \geqslant 0, \ X^- \geqslant 0.$$

But,

$$I_{a_n} X = I_{a_n} X^+ - I_{a_n} X^-,$$

i.e.,

$$\psi_X(a_n) = \psi_{X^+}(a_n) - \psi_{X^-}(a_n),$$

where $\psi_{X^+}(a_n) \downarrow 0$ and $\psi_{X^-}(a_n) \downarrow 0$, hence $\psi_X(a_n) \to 0$.

(3) ψ_X is σ-additive. In fact, if $X > 0$, then $\psi_X(a)/\psi_X(e)$, $a \in \mathfrak{B}$, is non-negative, additive and continuous, i.e., a continuous quasi-probability on $\mathfrak{B}$, hence σ-additive. Thus ψ_X is σ-additive. For any $X \in \mathscr{K}$, the σ-additivity follows from the σ-additivity of ψ_{X^+} and ψ_{X^-}.

We shall denote by $|\psi_X|$ the sum of ψ_{X^+} and ψ_{X^-}, i.e.,

$$|\psi_X|(a) \underset{\text{def}}{=} \psi_{X^+}(a) + \psi_{X^-}(a)$$

for every $a \in \mathfrak{B}$. Note that by $|\psi_X(a)|$ is denoted the absolute value of $\psi_X(a)$, for every $a \in \mathfrak{B}$, i.e., $|\psi_X|(a)$ is in general different from $|\psi_X(a)|$.

The properties (1), (2), (or (1), (3)) characterize ψ_X as a signed measure on $\mathfrak{B}$ (see Section 3.1). ψ_{X^+}, ψ_{X^-}, and $|\psi_X|$ are (non-negative) finite measures on $\mathfrak{B}$. We will also call $\psi_X(b)$, $b \in \mathfrak{B}$, the *indefinite integral* of X.

2. THE SPACE $\mathscr{L}_1$ OF ALL RV'S WITH EXPECTATION

2.1. We shall say that an element $X \in \mathscr{V}$ possesses an expectation $E(X) \in R$ if and only if a sequence $X_n \in \mathscr{K}$, $n = 1, 2, \ldots$, exists such that $X_n \overset{u}{\to} X$. We then define

$$E(X) = \lim_{n \to \infty} E(X_n).$$

Note that by Theorems 1.7 and 1.8, $\lim_{n\to\infty} E(X_n)$ always exists in R and $E(X)$ is independent of the choice of the sequence $X_n \in \mathscr{K}$, $n = 1, 2, \ldots$, which u-converges to X, i.e., $E(X)$ is, by the previous definition, uniquely determined.

We shall denote by $\mathscr{L}_1 = \mathscr{L}_1(\mathfrak{B}, p)$ the set of all $X \in \mathscr{V}$ which possess an expectation $E(X) \in R$; then $\mathscr{L}_1$ is a vector sublattice of $\mathscr{V}$ containing $\mathscr{K}$ (proof easy).

The following theorems are true:

Theorem 2.1.

$$\mathscr{M} \subseteq \mathscr{L}_1.$$

Proof. $\mathscr{S}$ is u-dense in $\mathscr{M}$ (see Section 6, Chapter 4) and $\mathscr{S} \subseteq \mathscr{K}$, hence by the previous definition every $X \in \mathscr{M}$ belongs to $\mathscr{L}_1$.

Theorem 2.2.

If $X \in \mathscr{V}$, then $X \in \mathscr{L}_1$ if and only if there exist two elements Y and Z in $\mathscr{K}$ such that $Y \leqslant X \leqslant Z$.

Proof. If $X \in \mathscr{L}_1$, then there exists a sequence $X_n \in \mathscr{K}$, $n = 1, 2, \ldots$, with $X_n \overset{u}{\to} X$; hence there exists an index n_0 such that

$$X_{n_0} - 1 < X_n < X_{n_0} + 1.$$

We write $Y = X_{n_0} - 1$ and $Z = X_{n_0} + 1$; then the condition of the theorem is satisfied, for $Y \in \mathscr{K}$ and $Z \in \mathscr{K}$ and $Y \leqslant X \leqslant Z$.

Now let $X \in \mathscr{V}$ with $Y \leqslant X \leqslant Z$, where $Y \in \mathscr{K}$ and $Z \in \mathscr{K}$. Since $\mathscr{E}$ is u-dense in $\mathscr{V}$, there exists a sequence $X_n \in \mathscr{E}$, $n = 1, 2, \ldots$, such that $X_n \overset{u}{\to} X$ and this sequence can be so determined that $Y - 1 < X_n < Z + 1$. Since $Y - 1 \in \mathscr{K}$ and $Z + 1 \in \mathscr{K}$, the elements X_n must (according to Theorem 1.4) belong to $\mathscr{K}$, $n = 1, 2, \ldots$, i.e., $X \in \mathscr{L}_1$.

Theorem 2.3.

The vector lattice $\mathscr{L}_1$ is closed for the u-convergence, i.e., every u-fundamental sequence $X_n \in \mathscr{L}_1$, $n = 1, 2, \ldots$, u-converges to an element $X \in \mathscr{L}_1$.

Proof. For every $n = 1, 2, \ldots$, let us determine an element $Y_n \in \mathscr{K}$ such that $|X_n - Y_n| < \varepsilon_n$, where $\varepsilon_n \downarrow 0$; then Y_n, $n = 1, 2, \ldots$, u-converges to an element $X \in \mathscr{V}$ and obviously this element X belongs to $\mathscr{L}_1$. But X_n, $n = 1, 2, \ldots$, is u-equivalent to Y_n, $n = 1, 2, \ldots$, hence $X_n \overset{u}{\to} X$.

Theorem 2.4.

The vector lattice $\mathscr{L}_1$ is (conditionally) complete with respect to the lattice operations.

Proof. If $X_i \in \mathscr{L}_1$, $i \in I$, is any family of elements in $\mathscr{L}_1$, bounded in $\mathscr{L}_1$, i.e., $Y \leqslant X_i \leqslant Z$, $i \in I$, with $Y \in \mathscr{L}_1$, $Z \in \mathscr{L}_1$, then $U = \bigvee_{i \in I} X_i$ and $D = \bigwedge_{i \in I} X_i$ exist in $\mathscr{V}$ and $Y \leqslant D \leqslant U \leqslant Z$; hence $D \in \mathscr{L}_1$ and $U \in \mathscr{L}_1$.

2.2. The reader can easily prove:

Theorem 2.5.

The expectation $E(X)$, $X \in \mathscr{L}_1$, is a finite, linear, strictly monotone, real-valued function on $\mathscr{L}_1$ with $E(1) = 1$, $E(0) = 0$. Furthermore, if we define $\|X\|_1 = E(|X|)$ for every $X \in \mathscr{L}_1$, then the function $\| \ \|_1$ is a norm on $\mathscr{L}_1$, i.e., is a normed vector lattice with respect to $\| \ \|_1$.

We shall prove that the expectation E is o-continuous, i.e.,

Theorem 2.6.

If $X_n \in \mathscr{L}_1$, $n = 1, 2, \ldots$, is a monotone decreasing sequence with $X_n \overset{o}{\downarrow} 0$, then $E(X_n) \downarrow 0$.

Proof. Since $\mathscr{K}$ is u-dense in $\mathscr{L}_1$, there exists a sequence Y_{nk}, $k = 1, 2, \ldots$, in $\mathscr{K}$ with $Y_{nk} \overset{u}{\to} X_n$, $n = 1, 2, \ldots$. The sequence $Y_{nk} \in \mathscr{K}$, $k = 1, 2, \ldots$, may be assumed decreasing, i.e.,

$$Y_{nk} \geqslant Y_{n,\,k+1}, \quad k = 1, 2, \ldots .$$

Now, if we write

$$Z_n = Y_{1n} \wedge Y_{2n} \wedge \ldots \wedge Y_{nn}, \quad n = 1, 2, \ldots,$$

then obviously

$$Z_n \geqslant X_n, \ Z_n \geqslant Z_{n+1}, \quad n = 1, 2, \ldots;$$

hence

(a) $$o\text{-}\lim Z_n \text{ exists and is } \geqslant o\text{-}\lim X_n = 0.$$

If m is any fixed index, then $Y_{mn} \geqslant Z_n$ for all $n \geqslant m$; therefore

$$X_m = u\text{-}\lim_{n\to\infty} Y_{mn} = o\text{-}\lim_{n\to\infty} Y_{mn} \geqslant o\text{-}\lim_{n\to\infty} Z_n \geqslant 0$$

for any fixed m; thus

$$0 = o\text{-}\lim_{n\to\infty} X_m \geqslant o\text{-}\lim_{n\to\infty} Z_n \geqslant 0, \text{ i.e., } o\text{-}\lim_{n\to\infty} Z_n = 0.$$

From this, with $Z_n \downarrow$ and $Z_n \in \mathscr{K}$, it follows that $E(Z_n) \downarrow 0$ and, since $Z_n \geqslant X_n \geqslant 0$, we also have $E(X_n) \downarrow 0$. Hence the continuity of E on $\mathscr{L}_1$ is proved.

Theorem 2.7. (Lebesgue's theorem on monotone sequences).

If $X_n \in \mathscr{L}_1$ *with*

$$X_n \leqslant X_{n+1}, \quad n = 1, 2, \ldots,$$

and

$$\bigvee_{n=1}^{\infty} X_n = X \in \mathscr{L}_1,$$

or with

$$X_n \geqslant X_{n+1}, \quad n = 1, 2, \ldots,$$

and

$$\bigwedge_{n=1}^{\infty} X_n = X \in \mathscr{L}_1,$$

then $E(X_n) \uparrow E(X)$ *or* $E(X_n) \downarrow E(X)$ *respectively.*

Proof of the case $X_n \uparrow$: Since $X - X_n \in \mathscr{L}_1$ and $X - X_n \overset{o}{\downarrow} 0$, we have

$$E(X) - E(X_n) = E(X - X_n) \downarrow 0,$$

i.e., $E(X_n) \uparrow E(X)$.

The proof of the case $X_n \downarrow$ is analogous.

Lemma 2.1.

If

$$X_n \in \mathscr{L}_1, \quad Y_n \in \mathscr{L}_1, \quad n = 1, 2, \ldots,$$

with

$$0 \leqslant X_n \uparrow, \; 0 \leqslant Y_n \uparrow, \; \bigvee_{n=1}^{\infty} X_n, \; \bigvee_{n=1}^{\infty} Y_n \text{ in } \mathscr{V}$$

and

$$\bigvee_{n=1}^{\infty} X_n \leqslant \bigvee_{n=1}^{\infty} Y_n,$$

then

$$\lim_{n\to\infty} E(X_n) \leqslant \lim_{n\to\infty} E(Y_n),$$

where $\lim E(X_n)$ *and* $\lim E(Y_n)$ *may be finite or* $+\infty$.

Proof. Since

$$o\text{-}\lim_{n\to\infty} X_k \wedge Y_n = X_k \wedge \bigvee_{n=1}^{\infty} Y_n = X_k \in \mathscr{L}_1,$$

for every $k = 1, 2, \ldots$, and $X_k \wedge Y_n \uparrow$, we have (see Theorem 2.7)

$$\lim_{n\to\infty} E(X_k \wedge Y_n) = E(X_k), \quad k = 1, 2, \ldots .$$

Now, since

$$Y_n \geqslant X_k \wedge Y_n, \quad n = 1, 2, \ldots,$$

we have

$$\lim_{n\to\infty} E(Y_n) \geqslant \lim_{n\to\infty} E(X_k \wedge Y_n) = E(X_k), \quad k = 1, 2, \ldots .$$

Hence

$$\lim_{n\to\infty} E(Y_n) \geqslant \lim_{n\to\infty} E(X_n).$$

Lemma 2.2 (Fatou).

If $X_n \in \mathscr{L}_1$, $n = 1, 2, \ldots$, *with* $0 \leqslant X_n \leqslant Y$, $Y \in \mathscr{L}_1$, $n = 1, 2, \ldots$, *then* $o\text{-}\liminf_{n\to\infty} X_n$ *exists in* $\mathscr{L}_1$ *and we have*

$$E(o\text{-}\liminf_{n\to\infty} X_n) \leqslant \liminf_{n\to\infty} E(X_n).$$

Proof. Since $D_n = \bigwedge_{k \geqslant n} X_k$ exists in $\mathscr{L}_1$ for every $n = 1, 2, \ldots$, with $D_n \uparrow$ and is bounded by $Y \in \mathscr{L}_1$, we have

$$o\text{-}\liminf_{n\to\infty} X_n = \bigvee_{n=1}^{\infty} D_n \in \mathscr{L}_1,$$

hence

$$E(o\text{-}\liminf_{n\to\infty} X_n) = \lim_{n\to\infty} E(D_n). \tag{1}$$

Now, since $D_n \leqslant X_n$, we have

$$E(D_n) \leqslant E(X_n), \quad n = 1, 2, \ldots,$$

so that

$$\lim_{n\to\infty} E(D_n) \leqslant \liminf_{n\to\infty} E(X_n).$$

From this and (1), it follows:

$$E(o\text{-}\liminf_{n\to\infty} X_n) \leqslant \liminf_{n\to\infty} E(X_n).$$

Theorem 2.8 (Lebesgue's theorem on term by term integration).

If $X_n \in \mathscr{L}_1$, $n = 1, 2, \ldots$, *and bounded in* $\mathscr{L}_1$, *i.e.*, $|X_n| \leqslant Y$ *with* $Y \in \mathscr{L}_1$, $n = 1, 2, \ldots$, *then* $o\text{-}\liminf_{n\to\infty} X_n$ *and* $o\text{-}\limsup_{n\to\infty} X_n$ *exist in* $\mathscr{L}_1$ *and we have*

$$\text{(R)} \qquad \begin{cases} \liminf_{n\to\infty} E(X_n) \geqslant E(o\text{-}\liminf_{n\to\infty} X_n) \text{ and} \\ \limsup_{n\to\infty} E(X_n) \leqslant E(o\text{-}\limsup_{n\to\infty} X_n). \end{cases}$$

If, further, $X_n \xrightarrow{o} X \in \mathscr{V}$, *then* $X \in \mathscr{L}_1$ *and*

$$\text{(S)} \qquad \lim_{n\to\infty} E(X_n) = E(X).$$

Proof. Let

$$\underline{X} = o\text{-}\liminf_{n\to\infty} X_n \text{ and } \overline{X} = o\text{-}\limsup_{n\to\infty} X_n;$$

then $|\underline{X}| \leqslant Y$ and $|\overline{X}| \leqslant Y$, hence $\underline{X} \in \mathscr{L}_1$ and $\overline{X} \in \mathscr{L}_1$. Since $Y \pm X_n \geqslant 0$ and $Y \pm X_n \in \mathscr{L}_1$, $n = 1, 2, \ldots$, and moreover $Y \pm X_n$, $n = 1, 2, \ldots$, is bounded in $\mathscr{L}_1$, we have, from Fatou's lemma,

$$\liminf_{n\to\infty} E(Y + X_n) \geqslant E(Y + \underline{X})$$

and

$$\liminf_{n\to\infty} E(Y - X_n) \geqslant E(Y - \overline{X}),$$

which gives at once the relation (R). Further, if

$$o\text{-}\lim_{n\to\infty} X_n = X = \underline{X} = \overline{X},$$

then $X \in \mathscr{L}_1$, so that by (R)

$$\liminf_{n\to\infty} E(X_n) \geqslant E(X) \geqslant \limsup_{n\to\infty} E(X_n),$$

which gives the equality (S).

2.3. We shall prove the following theorem:

Theorem 2.9.

If $X_n \in \mathscr{L}_1$ *with* $0 \leqslant X_n \leqslant X_{n+1}$, $n = 1, 2, \ldots$, *and* $o\text{-}\lim_{n\to\infty} X_n = X \in \mathscr{V}$, *then* $X \in \mathscr{L}_1$ *if and only if* $\lim E(X_n) < +\infty$. *We then have*

$$E(X) = \lim_{n\to\infty} E(X_n).$$

In order to prove this theorem, we shall extend the function E from $\mathscr{L}_1^{\,+}$ to $\mathscr{V}^+$. We know that $\mathscr{S}$ is o-dense in $\mathscr{V}$, therefore $\mathscr{S}^+$ is o-dense in $\mathscr{V}^+$. Hence, every $X \in \mathscr{V}^+$ may be represented in the form

$$X = o\text{-}\lim_{n\to\infty} X_n,$$

where $0 \leqslant X_n \in \mathscr{S}$, i.e., $X_n \in \mathscr{S}^+$, $n = 1, 2, \ldots$, and $X_n \leqslant X_{n+1}$, $n = 1, 2, \ldots$. We then have $E(X_n) \leqslant E(X_{n+1})$, $n = 1, 2, \ldots$. Hence $\lim_{n\to\infty} E(X_n)$ exists and is either a non-negative real number or $+\infty$. If another sequence $0 \leqslant Y_n \in \mathscr{S}^+$, $Y_n \leqslant Y_{n+1}$, $n = 1, 2, \ldots$, o-converges to $X \in \mathscr{V}^+$, then according to Lemma 2.1,

$$\lim_{n\to\infty} E(Y_n) = \lim_{n\to\infty} E(X_n).$$

If $X \in \mathscr{L}_1^+$, then $\lim_{n\to\infty} E(X_n) = E(X)$.

Now, we define

$$E^*(X) = \lim_{n\to\infty} E(X_n);$$

then E^* is defined for every $X \in \mathscr{V}^+$ and the value $E^*(X)$ is uniquely determined and $0 \leqslant E^*(X) \leqslant +\infty$; if $X \in \mathscr{L}_1^+$ then $E^*(X) = E(X)$; therefore E^* may be considered as an extension of E from $\mathscr{L}_1^+$ to $\mathscr{V}^+$. We obviously have:

$$\text{if} \quad X \leqslant Y \quad \text{then} \quad E^*(X) \leqslant E^*(Y)$$

$$E^*(X+Y) = E^*(X)+E^*(Y), \; E^*(\lambda X) = \lambda E^*(X), \; \lambda \geqslant 0.$$

Further, the following lemma is true:

Lemma 2.3.

If $X_n \in \mathscr{V}^+$ *with* $X_n \leqslant X_{n+1}$, $n = 1, 2, \ldots$, *and*

$$X = o\text{-}\lim_{n\to\infty} X_n \in \mathscr{V}^+,$$

then

$$E^*(X) = \lim_{n\to\infty} E^*(X_n).$$

Proof. Let us write

$$X_n = o\text{-}\lim_{k\to\infty} X_{nk} \text{ with } X_{nk} \in \mathscr{S} \text{ and } X_{nk} \leqslant X_{n,k+1}, \quad k = 1, 2, \ldots.$$

Further, let

$$U_n = X_{1n} \vee X_{2n} \vee \ldots \vee X_{nn}, \quad n = 1, 2, \ldots;$$

then $U_n \in \mathscr{S}$, $U_n \leqslant X_n$ and $U_n \leqslant U_{n+1}$, $n = 1, 2, \ldots$; therefore, the $o\text{-}\lim_{n\to\infty} U_n$ exists in $\mathscr{V}^+$ and is $\leqslant o\text{-}\lim_{n\to\infty} X_n = X$, i.e.,

$$o\text{-}\lim_{n\to\infty} U_n = X.$$

We can prove

$$o\text{-}\lim_{n\to\infty} U_n \geqslant X,$$

i.e.,

$$o\text{-}\lim_{n\to\infty} U_n = X.$$

In fact, for any fixed index i, we have

$$X_{ik} \leqslant U_k \quad \text{if} \quad i \leqslant k; \tag{2}$$

therefore,

$$X_i = o\text{-}\lim_{k\to\infty} X_{ik} \leqslant o\text{-}\lim_{k\to\infty} U_k, \quad i = 1, 2, \ldots;$$

hence
$$X = o\text{-}\lim_{i\to\infty} X_i \leqslant o\text{-}\lim_{k\to\infty} U_k;$$
thus
$$X = o\text{-}\lim_{n\to\infty} U_n$$
and, according to the definition of E^*, we have
$$E^*(X) = \lim_{n\to\infty} E(U_n). \tag{3}$$
We shall prove:
$$\lim_{n\to\infty} E(U_n) = \lim_{n\to\infty} E^*(X_n).$$
In order to prove it, we note that since $U_n \leqslant X_n$, we have
$$\lim_{n\to\infty} E(U_n) \leqslant \lim_{n\to\infty} E(X_n). \tag{4}$$
Since (2) is true, we have $E(X_{in}) \leqslant E(U_n)$, if $i \leqslant n$, hence
$$E^*(X_i) = \lim_{n\to\infty} E(X_{in}) \leqslant \lim_{n\to\infty} E(U_n), \quad i = 1, 2, \ldots;$$
thus
$$\lim_{i\to\infty} E^*(X_i) \leqslant \lim_{n\to\infty} E(U_n). \tag{5}$$
Now (3), (4) and (5) imply
$$\lim_{n\to\infty} E^*(X_n) = E^*(X).$$

Proof of Theorem 2.9. If $X \in \mathscr{L}_1^+$, then
$$\lim_{n\to\infty} E(X_n) = E(X) < +\infty \text{ (see Theorem 2.7).}$$
Suppose $\lim_{n\to\infty} E(X_n) < +\infty$; since every $X_n \in \mathscr{L}_1^+ \subseteq \mathscr{V}^+$, we have
$$E^*(X_n) = E(X_n), \quad n = 1, 2, \ldots,$$
and by Lemma 2.3,
$$E^*(X) = \lim_{n\to\infty} E(X_n) < +\infty, \text{ i.e., } E^*(X) < +\infty.$$
Suppose $X \in \mathscr{V}^+ - \mathscr{L}_1^+$; since $\mathscr{E}^+$ is u-dense in $\mathscr{V}^+$ there exists a representation
$$X = u\text{-}\lim_{n\to\infty} Y_n \quad \text{with} \quad Y_n \leqslant Y_{n+1}, \ Y_n \in \mathscr{E}^+, \quad n = 1, 2, \ldots .$$
The elements Y_n, $n = 1, 2, \ldots$, do not belong to $\mathscr{K}^+$, for every $n \geqslant$ a certain index n_0, for, if a subsequence Y_n belongs to $\mathscr{K}^+$, then by the u-convergence of this sequence, we have $X \in \mathscr{L}_1^+$, which contradicts the assumption $X \in \mathscr{V}^+ - \mathscr{L}_1^+$. Now, let n be an index with

$Y_n \in \mathscr{E}^+ - \mathscr{K}^+$ and let

$$Y_n = \sum_{i \geqslant 1} \xi_i I_{b_i}$$

be the reduced representation of Y_n by indicators; then

$$\sum_{i \geqslant 1} \xi_i p(b_i) = +\infty$$

and

$$o\text{-}\lim_{k \to \infty} \sum_{i=1}^{k} \xi_i I_{b_i} = Y_n$$

with

$$\sum_{i=1}^{k} \xi_i I_{b_i} \in \mathscr{S};$$

hence

$$E^*(Y_n) = \lim_{k \to \infty} \sum_{i=1}^{k} \xi_i p(b_i) = +\infty.$$

On the other hand, $Y_n \leqslant X$, so that $E^*(Y_n) \leqslant E^*(X) = +\infty$, which contradicts the assumption $E^*(X) < +\infty$.

Exercise 1. Prove: if $X_n \in \mathscr{L}_1{}^+$, $n = 1, 2, \ldots$, and

$$\liminf_{n \to \infty} E(X_n) < +\infty,$$

then

$$o\text{-}\liminf_{n \to \infty} X_n \in \mathscr{L}_1{}^+$$

and

$$E(o\text{-}\liminf X_n) \leqslant \liminf_{n \to \infty} E(X_n).$$

Exercise 2. Prove: if

$$X_n \in \mathscr{L}_1{}^+ \text{ and } \limsup_{n \to \infty} E(X_n) < +\infty,$$

then

$$o\text{-}\liminf_{n \to \infty} X_n \in \mathscr{L}_1{}^+.$$

and

$$E(o\text{-}\liminf_{n \to \infty} X_n) \leqslant \liminf_{n \to \infty} E(X_n) \leqslant \limsup_{n \to \infty} E(X_n) \leqslant E(o\text{-}\limsup_{n \to \infty} X_n),$$

if

$$o\text{-}\limsup_{n \to \infty} X_n \in \mathscr{L}_1{}^+.$$

3. SIGNED MEASURES

3.1. Let $\mathfrak{B}$ be a Boolean σ-algebra; then a finite real-valued function ψ on $\mathfrak{B}$, which is σ-additive, i.e., for which

$$\sum_{n \geqslant 1} \psi(a_n) = \psi\Big(\bigvee_{n \geqslant 1} a_n\Big),$$

if $a_1, a_2, \ldots,$ are pairwise disjoint, is said to be a *signed measure* ψ on $\mathfrak{B}$. If $(\mathfrak{B}, p)$ is a pr σ-algebra and X a rv over $\mathfrak{B}$, with $X \in \mathscr{L}_1(\mathfrak{B})$, then the indefinite integral $\psi(b) = E(I_b X)$, $b \in \mathfrak{B}$, is a signed measure on $\mathfrak{B}$ (see Section 1.4). If ψ_1 and ψ_2 are signed measures on $\mathfrak{B}$, then

$$\psi = \lambda_1 \psi_1 + \lambda_2 \psi_2$$

where λ_1 and λ_2 are real numbers, is also a signed measure, i.e., the signed measures on $\mathfrak{B}$ form a vector space $\mathscr{M}(\mathfrak{B})$. Obviously every signed measure $\psi \in \mathscr{M}(\mathfrak{B})$ is continuous, i.e., if $a_n \in \mathfrak{B}$, $n = 1, 2, \ldots,$ with $a_n \downarrow \emptyset$, then $\psi(a_n) \to 0$.

3.2. An element $a \in \mathfrak{B}$ is said to be *positive* with respect to a signed measure $\psi \in \mathscr{M}(\mathfrak{B})$ if and only if, for every $x \in \mathfrak{B}$, $\psi(x \wedge a) \geqslant 0$. Similarly, an element $a \in \mathfrak{B}$ is said to be *negative* with respect to ψ if and only if, for every $x \in \mathfrak{B}$, $\psi(x \wedge a) \leqslant 0$. The zero element $\emptyset$ is both positive and negative in this sense. We do not assert that there necessarily exists any other, non-trivial, positive element or negative element with respect to a signed measure ψ. But the following theorem is true:

Theorem 3.1.

If ψ is a signed measure on a Boolean σ-algebra $\mathfrak{B}$, then there exist two disjoint elements $a \in \mathfrak{B}$, $b \in \mathfrak{B}$ such that $a \vee b = e$ and a is positive and b is negative with respect to ψ. We shall then say that the elements a and b form a Hahn-decomposition of the unit $e \in \mathfrak{B}$ with respect to $\psi \in \mathscr{M}(\mathfrak{B})$.

Proof. The set of all negative elements of $\mathfrak{B}$ with respect to ψ is non-empty, for $\emptyset$ belongs to this set. It is easy to see that a countable join of negative elements is also a negative element. We write

$$\xi = \inf \psi(x)$$

for all negative elements $x \in \mathfrak{B}$; then there exists a sequence b_n, $n = 1, 2, \ldots,$ of negative elements in $\mathfrak{B}$ such that

$$\xi = \lim_{n \to \infty} \psi(b_n).$$

We write

$$b = \bigvee_{n=1}^{\infty} b_n.$$

Then b is a negative element in $\mathfrak{B}$ such that $\psi(b) \leqslant \psi(x)$, for every negative element x in $\mathfrak{B}$. We shall prove that the element $a = b^c = e - b$ is a positive element. In order to prove it, we suppose that, on the contrary, a is not positive. Then there exists an element x_0 in $\mathfrak{B}$, such that $x_0 \leqslant a$ and $\psi(x_0) < 0$. The element x_0 cannot be a negative element in $\mathfrak{B}$, for then $b \vee x_0$ would be a negative element with

$$\psi(b \vee x_0) = \psi(b) + \psi(x_0) < \psi(b),$$

which is impossible. Let k_1 be the smallest positive integer with the property that x_0 contains an element x_1 in $\mathfrak{B}$ for which $\psi(x_1) \geqslant 1/k_1$.† Since

$$\psi(x_0 - x_1) = \psi(x_0) - \psi(x_1) \leqslant \psi(x_0) - (1/k_1) < 0,$$

and moreover $x_0 - x_1 \leqslant a$, the argument just applied to x_0 is applicable to the element $x_0 - x_1$. Let k_2 be the smallest positive integer with the property that $x_0 - x_1$ contains an element x_2 with $\psi(x_2) \geqslant 1/k_2$ and so on. We have, obviously, $\lim_{n \to \infty} 1/k_n = 0$. It follows that, for every $y \in \mathfrak{B}$ with

$$y \leqslant y_0 = x_0 - \bigvee_{n=1}^{\infty} x_n,$$

we have $\psi(y) \leqslant 0$, i.e., y_0 is a negative element. Since y_0 is disjoint from b and since

$$\psi(y_0) = \psi(x_0) - \sum_{n=1}^{\infty} \psi(x_n) \leqslant \psi(x_0) < 0,$$

i.e., $\psi(y_0) \leqslant 0$, this contradicts the minimality of b (for $\psi(y_0 \vee b) < \psi(b)$ and $y_0 \vee b$ is negative) and we conclude that the assumption $\psi(x_0) < 0$ is untenable, hence the elements $a \in \mathfrak{B}$ and $b \in \mathfrak{B}$ form a Hahn decomposition of e with respect to ψ.

3.3. One can easily construct examples to show that a Hahn-decomposition of e with respect to ψ is not unique. But, if $e = a_1 \vee b_1$ and $e = a_2 \vee b_2$ are two Hahn decompositions of e with respect to ψ, then one can prove that $\psi(a_1 + a_2) = 0$ and $\psi(b_1 + b_2) = 0$.

In fact, we observe that $a_1 - a_2 \leqslant a_1$, so that $\psi(a_1 - a_2) \geqslant 0$ and $a_1 - a_2 \leqslant b_2$, so that $\psi(a_1 - a_2) \leqslant 0$. Hence $\psi(a_1 - a_2) = 0$ and, by symmetry, $\psi(a_2 - a_1) = 0$. Hence, if we define, for every $x \in \mathfrak{B}$, $\psi^+(x) = \psi(x \wedge a)$ and $\psi^-(x) = -\psi(x \wedge b)$, where $e = a \vee b$, is any Hahn decomposition of e with respect to ψ, then ψ^+ and ψ^- are two uniquely determined functions on $\mathfrak{B}$, called the *upper variation* and the *lower variation* of ψ, respectively. The function $|\psi|$, defined by

$$|\psi|(x) = \psi^+(x) + \psi^-(x), \quad x \in \mathfrak{B},$$

† Since x_0 is not negative, there exist such elements.

is called the *total variation* of ψ (observe that $|\psi|(x)$ and $|\psi(x)|$ are distinct notations).

Obviously, the functions ψ^+, ψ^-, $|\psi|$ are non-negative measures on $\mathfrak{B}$ and we have $\psi = \psi^+ - \psi^-$, i.e., every signed measure ψ on $\mathfrak{B}$ can be uniquely represented as the diffeience of its upper and lower variation. We call $\psi = \psi^+ - \psi^-$ the *Jordan decomposition* of ψ. For the signed measure ψ_X, which corresponds to a rv $X \in \mathscr{L}_1$ (see Section 1.4), ψ_{X^+} and ψ_{X^-} are the upper and lower variations of ψ_X respectively, i.e., $\psi_X = \psi_{X^+} - \psi_{X^-}$ is the Jordan decomposition of ψ_X.

3.4. Let $(\mathfrak{B}, p)$ be a pr σ-algebra and $\mathscr{M}(\mathfrak{B})$ the set of all signed measures on $\mathfrak{B}$; then $p \in \mathscr{M}(\mathfrak{B})$. Let ψ, $\hat{\psi}$ be two signed measures on $\mathfrak{B}$; then we shall say that $\hat{\psi}$ is absolutely continuous with respect to ψ, in symbols $\hat{\psi} \ll \psi$ if and only if $\hat{\psi}(x) = 0$ for every x with $|\psi|(x) = 0$.

Theorem 3.2.

Let $\psi \in \mathscr{M}(\mathfrak{B})$, $\hat{\psi} \in \mathscr{M}(\mathfrak{B})$. Then the following statements are equivalent:

(a) $\hat{\psi} \ll \psi$

(b) $\hat{\psi}^+ \ll \psi$ *and* $\hat{\psi}^- \ll \psi$

(c) $|\hat{\psi}| \ll |\psi|$

(d) *for every* $\varepsilon > 0$, *there exists a positive number* $\delta(\varepsilon)$ *such that* $|\hat{\psi}|(x) < \varepsilon$, *for every* x *for which* $|\psi|(x) < \delta(\varepsilon)$.

Proof. It is easy to see that, if (a) is valid, then (b) is also valid. In fact, we have $\hat{\psi}(x) = 0$, whenever $|\psi|(x) = 0$. If $e = a \vee b$ is a Hahn decomposition with respect to $\hat{\psi}$, then we have, whenever $|\psi|(x) = 0$,

$$0 \leqslant |\psi|(x \wedge a) \leqslant |\psi|(x) = 0$$

and

$$0 \leqslant |\psi|(x \wedge b) \leqslant |\psi|(x) = 0,$$

and therefore

$$\hat{\psi}^+(x) = \hat{\psi}(x \wedge a) = 0$$

and

$$\hat{\psi}^-(x) = -\hat{\psi}(x \wedge b) = 0;$$

this proves the validity of (b). The fact that (b) implies (c), and (c) implies (a) follows from the relations

$$|\hat{\psi}|(x) = \hat{\psi}^+(x) + \hat{\psi}^-(x) \quad \text{and} \quad 0 \leqslant |\hat{\psi}(x)| \leqslant |\hat{\psi}|(x),$$

respectively.

That (d) implies (a) is obvious; we shall prove that (a) implies (d):

Suppose the relation (d) is not true. Then it is possible, for some $\varepsilon > 0$, to find a sequence x_n, $n = 1, 2, \ldots$, such that

$$|\psi|(x_n) < \frac{1}{2^n} \quad \text{and} \quad |\hat{\psi}|(x_n) \geqslant \varepsilon, \quad n = 1, 2, \ldots .$$

We write

$$x = o\text{-}\lim_n \sup x_n,$$

then

$$|\psi|(x) \leqslant \sum_{i=n}^{\infty} |\psi|(x_n) < \frac{1}{2^{n-1}}, \quad n = 1, 2, \ldots,$$

and therefore
$$|\psi|(x) = 0.$$

On the other hand,

$$|\hat{\psi}|(x) = \lim_{n\to\infty} |\hat{\psi}|(x_n \vee x_{n+1} \vee \ldots) \geqslant \lim_n \sup |\hat{\psi}|(x_n) \geqslant \varepsilon.$$

But this contradicts (a). Hence if (a), then (d).

The relation $\ll$ in $\mathcal{M}(\mathfrak{B})$ is reflexive and transitive. We shall call two signed measures $\hat{\psi}$ and ψ equivalent in symbols $\hat{\psi} \equiv \psi$ if and only if $\psi \ll \hat{\psi}$ and $\psi \ll \hat{\psi}$. The following theorem is true:

Theorem 3.3.

For every $X \in \mathcal{L}_1$, $\psi_X \ll p$ and, moreover, for every $\psi \in \mathcal{M}(\mathfrak{B})$, $\psi \ll p$.

Proof. We have $p(x) = 0$ if and only if $x = \emptyset$, and $\psi(\emptyset) = 0$, for every $\psi \in \mathcal{M}(\mathfrak{B})$.

4. THE RADON–NIKODYM THEOREM

4.1. Let $(\mathfrak{B}, p)$ be a pr σ-algebra and $\mathcal{M}(\mathfrak{B})$ the set of all signed measures on $\mathfrak{B}$; then the indefinite integral $\psi_X(b) = E(I_b X)$, $b \in \mathfrak{B}$, for any rv $X \in \mathcal{L}_1$, belongs to $\mathcal{M}(\mathfrak{B})$. We shall prove that, conversely, every signed measure $\psi \in \mathcal{M}(\mathfrak{B})$ is the indefinite integral of a rv $X \in \mathcal{L}_1$. In order to prove this we need the following lemmas:

Lemma 4.1.

If $\hat{\psi}$ and ψ are non-negative elements in $\mathcal{M}(\mathfrak{B})$ such that $\hat{\psi} \ll \psi$ and $\hat{\psi}$ is not identically zero, then there exists a positive number ε and an element $a \in \mathfrak{B}$ such that $\psi(a) > 0$ and such that a is a positive element in $\mathfrak{B}$ for the signed measure $\hat{\psi} - \varepsilon\psi$.

Proof. Let $e = a_n \vee b_n$ be a Hahn decomposition with respect to the signed measure $\hat{\psi} - (1/n)\psi$, $n = 1, 2, \ldots$. We write

$$a_0 = \bigvee_{n=1}^{\infty} a_n, \quad b_0 = \bigwedge_{n=1}^{\infty} b_n;$$

then, since $b_0 \leqslant b_n$, we have $0 \leqslant \hat{\psi}(b_0) \leqslant (1/n)\psi(b_0)$, $n = 1, 2, \ldots$, and consequently $\hat{\psi}(b_0) = 0$. It follows that $\hat{\psi}(a_0) > 0$ and therefore, by $\hat{\psi} \ll \psi$, that $\psi(a_0) > 0$. Hence we must have $\psi(a_n) > 0$, for at least one value of n; if, for such a value of n, we write $a = a_n$ and $\varepsilon = 1/n$, the requirements of the theorem are all satisfied.

Lemma 4.2.

Let ψ be any non-negative element of $\mathcal{M}(\mathfrak{B})$; then there exists a uniquely determined rv $X \in \mathcal{L}_1$ such that $\psi(b) = \psi_X(b) = E(I_b X)$ for every $b \in \mathfrak{B}$.

Proof. Let $\mathcal{F}$ be the class of all non-negative rv's $X \in \mathcal{L}_1$ such that $\psi_X(b) = E(I_b X) \leqslant \psi(b)$, for every $b \in \mathfrak{B}$. The class $\mathcal{F}$ is non-empty, for the element 0 belongs to $\mathcal{F}$. We write $\xi = \sup \psi_X(e) = \sup E(X)$, for all $X \in \mathcal{F}$; then there exists a sequence $X_n \in \mathcal{F}$, $n = 1, 2, \ldots$, such that

$$\lim_{n\to\infty} \psi_{X_n}(e) = \lim_{n\to\infty} E(X_n) = \xi \leqslant \psi(e) < +\infty.$$

Let

$$X_n^* = \bigvee_{k=1}^{n} X_k,$$

so that

$$0 \leqslant X_n^* \leqslant X_{n+1}^*, \quad n = 1, 2, \ldots,$$

and

$$X_n^* \xrightarrow{o} X = \bigvee_{n=1}^{\infty} X_n;$$

then $X_n^* \in \mathcal{L}_1$ and $X \in \mathcal{L}_1$ and since $E(X_n^*) \leqslant E(X) \leqslant \xi$, we have $X_n^* \in \mathcal{F}$, $X \in \mathcal{F}$, and

$$\xi = E(X) = \lim_{n\to\infty} E(X_n).$$

We assert that $\psi_X(b) = E(I_b X) = \psi(b)$ for every $b \in \mathfrak{B}$. Suppose that $\psi(b)$ is not equal to $\psi_X(b)$ for every $b \in \mathfrak{B}$, i.e., $\hat{\psi} = \psi - \psi_X$ is not identically zero; then, since $\hat{\psi} \ll p$, by Lemma 4.1, there exists a positive number ε and an element $a \in \mathfrak{B}$ such that $p(a) > 0$, i.e., $a \neq \emptyset$ and such that

$$0 \leqslant \varepsilon p(x \wedge a) \leqslant \hat{\psi}(x \wedge a) = \psi(x \wedge a) - \psi_X(x \wedge a)$$

for every $x \in \mathfrak{B}$. Now we write $Y = X + \varepsilon I_a$; then

$$\begin{aligned}\psi_Y(x) &= E(I_x Y)\\ &= E(I_x X) + \varepsilon p(x \wedge a)\\ &\leqslant E(I_{x-a} X) + \psi(x \wedge a)\\ &\leqslant \psi(x-a) + \psi(x \wedge a)\\ &= \psi(x)\end{aligned}$$

for every $x \in \mathfrak{B}$, i.e., $Y \in \mathscr{F}$. Moreover, we have

$$E(Y) = E(X) + \varepsilon p(a) > \xi,$$

i.e., $E(Y) > \xi$, and this contradicts $Y \in \mathscr{F}$. Hence, the assertion $\psi_X(b) = \psi(b)$ for every $b \in \mathfrak{B}$ is true.

Now, the following theorem (known as the Radom–Nikodym theorem) is true:

Theorem 4.1.

If $(\mathfrak{B}, p)$ is a pr σ-algebra and ψ is any signed measure on $\mathfrak{B}$, then there exists a rv $X \in \mathscr{L}_1$ such that

$$\psi(b) = \psi_X(b) = E(I_b X) \quad \textit{for every} \quad b \in \mathfrak{B}.$$

The rv X is uniquely determined.

Proof. Let $\psi = \psi^+ - \psi^-$ be the Jordan decomposition of ψ; since ψ^+ and ψ^- are non-negative, there exist $X_1 \in \mathscr{L}_1$ and $X_2 \in \mathscr{L}_1$ such that $\psi^+(b) = \psi_{X_1}(b)$ and $\psi^-(b) = \psi_{X_2}(b)$ for every $b \in \mathfrak{B}$; then we have for every $b \in \mathfrak{B}$,

$$\begin{aligned}\psi(b) &= \psi^+(b) - \psi^-(b)\\ &= \psi_{X_1}(b) - \psi_{X_2}(b)\\ &= E(I_b X_1) - E(I_b X_2)\\ &= E(I_b(X_1 - X_2))\\ &= E(I_b X),\end{aligned}$$

i.e.,

$$\psi(b) = \psi_X(b) \text{ where } X = X_1 - X_2.$$

Obviously, $X \in \mathscr{L}_1$ is uniquely determined, for, if $Y \in \mathscr{L}_1$ and $\psi(b) = \psi_Y(b)$, then $\psi_X(b) = \psi_Y(b)$ for every $b \in \mathfrak{B}$, i.e., $X = Y$.

5. REMARKS

5.1. Let $(\mathfrak{B}, p)$ be a pr σ-algebra and $\mathscr{V}(\mathfrak{B})$ the stochastic space of all rv's over $\mathfrak{B}$. In Section 2.3, we defined, for each $X \in \mathscr{V}^+ = \mathscr{V}^+(\mathfrak{B})$, $E^*(X)$ as an extension of the expectation E from $\mathscr{L}_1^{\,+}$ to $\mathscr{V}^+$. Now, let $X \in \mathscr{V}^-$, i.e., $X \in \mathscr{V}$ with $X \leqslant 0$; then $-X \in \mathscr{V}^+$ and we may define $E^*(X) = -E^*(-X)$. The most general class of rv's X for which it is convenient to define $E^*(X)$ is the class of all rv's X for which X^+ or X^- belongs to $\mathscr{L}_1$. We then define:

$$E^*(X) = E^*(X^+) - E^*(X^-).$$

E^* is unambiguously defined on the set $\mathscr{L}^*$ of all $X \in \mathscr{V}$ for which $E^*(X^+)$ or $E^*(X^-)$ is finite and if $X \in \mathscr{L}_1 \subseteq \mathscr{L}^*$ then $E^*(X) = E(X)$. It is easy to prove that:

(1) If $X \in \mathscr{L}^*$, $Y \in \mathscr{L}^*$ and $X+Y \in \mathscr{L}^*$ then

$$E^*(X+Y) = E^*(X) + E^*(Y).$$

(2) If $0 \leqslant X_n \underset{o}{\uparrow} X$, then

$$E^*(X_n) \uparrow E^*(X).$$

(3) If $X_n \in \mathscr{V}^+$ and $\sum_{n=1}^{\infty} X_n = X$ in the sense of o-convergence then

$$E^*(X) = \sum_{n=1}^{\infty} E^*(X_n).$$

(4) If $X \in \mathscr{L}^*$ and $E^*(X)$ is finite, then

$$\lim_{p(a) \to 0} E^*(I_a |X|) = 0.\dagger$$

4.2. Let $X \in \mathscr{L}^*$. Then the function

$$\psi_X(b) = E^*(I_b X), \quad b \in \mathfrak{B}$$

is p-continuous (i.e., vanishes for $b = \emptyset$) and σ-additive (i.e., if $a = \bigvee_{n=1}^{\infty} a_n$ where $a_1, a_2, \ldots$ are pairwise disjoint, then

$$\psi_X(a) = \sum_{n=1}^{\infty} \psi_X(a_n)\Big).$$

Now we may consider signed measures with values in the extended real line $R^* = [-\infty, +\infty]$. An *extended signed measure* ψ on $\mathfrak{B}$ is a σ-additive function defined on $\mathfrak{B}$ and valued in R^*, such that $\psi(\emptyset) = 0$ and such that ψ assumes at most one of the values $+\infty$ and $-\infty$. If $\psi(b) \geqslant 0$, for every $b \in \mathfrak{B}$, then ψ is called simply "*measure* on $\mathfrak{B}$"; if $0 \leqslant \psi(b) < +\infty$, then ψ is called "*finite measure* on $\mathfrak{B}$"; and if there exists a partition $e = a_1 \vee a_2 \vee \ldots$, with $a_1, a_2, \ldots$ pairwise disjoint, and $0 \leqslant \psi(a_n) < +\infty$, for every $n = 1, 2, \ldots$, then ψ is called "*σ-finite*

† See Loève [1], Section 7.2.

measure on $\mathfrak{B}$". It is easy to prove that ψ_X with $X \in \mathscr{L}^*$ is an extended signed measure, and if $X \geqslant 0$ then ψ_X is a σ-finite measure on $\mathfrak{B}$. We have, for every $X \in \mathscr{L}^*$, $\psi_X = \psi_{X^+} - \psi_{X^-}$, and both ψ_{X^+} and ψ_{X^-} are σ-finite measures and at least one is always finite; $|\psi_X| = \psi_{X^+} + \psi_{X^-}$ is also a σ-finite measure.

There exists a Hahn decomposition $e = a \vee b$ with respect to an extended signed measure and ψ can be represented as the difference of its upper and lower variations. ψ is called a *σ-finite signed measure* if both its lower and upper variations are σ-finite measures on $\mathfrak{B}$.

5.2. Let μ be a measure on $\mathfrak{B}$; then we can define the concept of the integral $\int X\,d\mu$ of a non-negative srv $X \in \mathscr{S}^+(\mathfrak{B})$ with respect to μ by

$$\int X\,d\mu = \sum_{j=1}^{n} \xi_j\,\mu(a_j)$$

where

$$X = \sum_{j=1}^{n} \xi_j I_{a_j}$$

is the reduced representation of X by indicators. The integral $\int X\,d\mu$ of a non-negative rv $X \in \mathscr{V}^+$ with respect to μ is defined by

$$\int X\,d\mu = \lim_{n \to \infty} \int X_n\,d\mu,$$

where X_n, $n = 1, 2, \ldots$, is a non-decreasing sequence of non-negative simple rv's which o-converges to X. Now we say that a rv $X \in \mathscr{V}$ possesses an integral $\int X\,d\mu$ with respect to the measure μ if one of the integrals $\int X^+\,d\mu$ and $\int X^-\,d\mu$ is $< +\infty$; then we define

$$\int X\,d\mu = \int X^+\,d\mu - \int X^-\,d\mu.$$

Let $X \in \mathscr{V}$ possess an integral; then $\psi_{X,\mu}(b) = \int I_b X\,d\mu$ is defined for every $b \in \mathfrak{B}$ and is a signed measure on $\mathfrak{B}$. The following generalizations of the Radom–Nikodym theorem can be proved (*cf.* P. Halmos [2]):

Theorem 5.1.

Let ψ be a σ-finite signed measure on $\mathfrak{B}$. Then there exists a rv $X \in \mathscr{L}^*$ such that $\psi(b) = \psi_X(b)$ for every $b \in \mathfrak{B}$.

Theorem 5.2.

Let μ be a σ-finite measure on $\mathfrak{B}$ and ψ a σ-finite signed measure on $\mathfrak{B}$ which is absolutely continuous with respect to μ, i.e., $\psi(x) = 0$ for every

$x \in \mathfrak{B}$ with $\mu(x) = 0$. Then there exists a rv $X \in \mathscr{V}$ which possesses an integral $\int X \, d\mu$ with respect to μ such that

$$\psi(b) = \int I_b X \, d\mu \text{ for every } b \in \mathfrak{B}.$$

Theorem 5.1 can be proved by an easy modification of the proof of the corresponding Theorem B, p. 128, of Halmos [2]. In order to prove Theorem 5.2, observe that ψ and μ are comparable in the sense of Carathéodory [6], page 208, and Carathéodory's Theorem 2 of this page can be applied.

VI

MOMENTS, SPACES $\mathscr{L}_q$

1. POWERS OF RV'S

1.1. Let $(\mathfrak{B}, p)$ be a pr σ-algebra and X be an erv with

$$X = \sum_{j \geqslant 1} \xi_j I_{a_j}$$

as its reduced representation by indicators. Then, for every $q = 1, 2, \ldots,$ we define

$$X^q = \sum_{j \geqslant 1} \xi_j^q I_{a_j} \tag{1}$$

as the q-th power of X. Obviously X^q is an erv. If $X \geqslant 0$, then we can define by (1) the q-th power for every real number $q > 0$. Hence $|X|^q$ is defined for every real number $q > 0$ and for every erv X. Now let X be any rv in $\mathscr{V}(\mathfrak{B})$; then there always exists a sequence $X_n \in \mathscr{E}(\mathfrak{B})$, $n = 1, 2, \ldots,$ such that $X_n \overset{o}{\to} X$. It is easy to see that, for every $q = 1, 2, \ldots,$ the sequence X_n^q, $n = 1, 2, \ldots,$ is then defined in $\mathscr{E}(\mathfrak{B})$ and o-converges to a rv in $\mathscr{V}(\mathfrak{B})$, which we denote by X^q and call the *q-th power* of X. This definition is independent of the choice of the o-convergent sequence X_n, $n = 1, 2, \ldots,$ to X. In the same way, if $X \geqslant 0$, $X \in \mathscr{V}(\mathfrak{B})$, since there exists a sequence $X_n \in \mathscr{E}(\mathfrak{B})$, $X_n \geqslant 0$, $n = 1, 2, \ldots,$ such that $X_n \overset{o}{\to} X$ and X_n^q, $n = 1, 2, \ldots,$ is o-convergent, we can define for every real number $q > 0$, the q-th power X^q of X as the o-limit of the sequence X_n^q, $n = 1, 2, \ldots .$

2. MOMENTS OF A RANDOM VARIABLE

2.1. Expectations of powers of a rv $X \in \mathscr{V}(\mathfrak{B})$, if they exist, are called *moments*. More precisely, if, for any positive integer $k = 1, 2, \ldots$, $X^k \in \mathscr{L}_1(\mathfrak{B})$, then we shall call the expectation $E(X^k)$ the k-th moment of X. If, for any positive number q, $|X|^q \in \mathscr{L}_1(\mathfrak{B})$, then we shall call $E(|X|^q)$ the q-th absolute moment of X.

Theorem 2.1.

If $|X|^q \in \mathscr{L}_1$, for some $X \in \mathscr{V}$ and $q > 0$, then $|X|^{q'} \in \mathscr{L}_1$ for every q' with $0 < q' \leqslant q$, and, if k is any positive integer with $k \leqslant q$, then $X^k \in \mathscr{L}_1$.

Proof. We have

$$|X|^{q'} \leqslant 1 + |X|^q;$$

hence $|X|^{q'} \in \mathscr{L}_1$ and if k is a positive integer then, since

$$|X^k| = |X|^k \in \mathscr{L}_1,$$

we have $X^k \in \mathscr{L}_1$

2.2. It is easy to see that the following inequality is true

$$|\alpha + \beta|^q \leqslant \gamma_q |\alpha|^q + \gamma_q |\beta|^q,$$

where α and β are real numbers, q a real number with $q > 0$, and $\gamma_q = 1$, if $q \leqslant 1$ and $\gamma_q = 2^{q-1}$, if $q \geqslant 1$.

This inequality implies

Theorem 2.2.

If $X \in \mathscr{V}$, $Y \in \mathscr{V}$ with $|X|^q \in \mathscr{L}_1$ and $|Y|^q \in \mathscr{L}_1$ for some $q > 0$, then $|X + Y|^q \in \mathscr{L}_1$ and moreover

$$E(|X+Y|^q) \leqslant \gamma_q E(|X|^q) + \gamma_q E(|Y|^q), \tag{1}$$

where $\gamma_q = 1$ or 2^{q-1} according as $q \leqslant 1$ or $q \geqslant 1$.

The following elementary inequality

$$|\alpha\beta| \leqslant \frac{|\alpha|^q}{q} + \frac{|\beta|^s}{s}, \quad \text{with} \quad q > 1 \quad \text{and} \quad \frac{1}{q} + \frac{1}{s} = 1,$$

is true for any two real numbers α and β so that we have

Theorem 2.3.

If $X \in \mathscr{V}$, $Y \in \mathscr{V}$ with $|X|^q \in \mathscr{L}_1$, $|Y|^s \in \mathscr{L}_1$ where $q > 1$ and

$$\frac{1}{q} + \frac{1}{s} = 1,$$

then $XY \in \mathscr{L}_1$ *and the so-called Hölder inequality holds*

$$E(|XY|) \leqslant (E(|X|^q))^{1/q} . (E(|Y|^s))^{1/s}. \tag{2}$$

Inequality (2) *yields the so-called Minkowski inequality*:

$$(E(|X+Z|^q))^{1/q} \leqslant (E(|X|^q))^{1/q} + (E(|Z|^q))^{1/q}, \tag{3}$$

for every $q \geqslant 1$ *with* $|X|^q \in \mathscr{L}_1$ *and* $|Z|^q \in \mathscr{L}_1$.

Proof (see Loève [1], p. 156. See also Theorems 6.4 and 6.6, Chapter VII, where proofs in more general cases are given.

Hölder's inequality with $q = 2$, $s = 2$ is called the Schwarz inequality

$$(E(|XY|))^2 \leqslant E(|X|^2)\, E(|Y|^2). \tag{4}$$

The following theorem is true:

Theorem 2.4.

Let $X \in \mathscr{V}$ *with* $|X|^q \in \mathscr{L}_1$ *for any* $q > 0$ *and let* ξ *be a positive real number. Then*

$$\frac{E(|X|^q) - \xi^q}{\sup |X|^q} \leqslant p([|X| \geqslant \xi]) \leqslant \frac{E(|X|^q)}{\xi^q}.† \tag{5}$$

If $\sup |X|^q = +\infty$, *then the left-hand side is zero. The inequality on the right in* (5) *is called the Markov inequality and for* $q = 2$ *it reduces to the Tchebichev inequality*

$$p([|X| \geqslant \xi]) \leqslant \frac{E(|X|^2)}{\xi^2}, \qquad \xi > 0. \tag{5a}$$

Proof. Let $a = [|X| \geqslant \xi]$; then from the obvious relations:

$$E(|X|^q) = E(I_a|X|^q) + E(I_{a^c}|X|^q)$$

and

$$|\xi|^q p(a) \leqslant E(I_a|X|^q) \leqslant \sup |X|^q . p(a),$$

$$0 \leqslant E(I_{a^c}|X|^q) \leqslant |\xi|^q,$$

it follows that

$$|\xi|^q p(a) \leqslant E(|X|^q) \leqslant \sup |X|^q p(a) + |\xi|^q.$$

This proves the inequality.

---

† The sup Y, where $Y \in \mathscr{V}$, is defined as the infimum of all real numbers η with $Y \leqslant \eta$. If the set of all real numbers η with $Y \leqslant \eta$ is empty, then sup $Y = +\infty$.

3. THE SPACES $\mathscr{L}_q$

3.1. We shall denote by $\mathscr{L}_q = \mathscr{L}_q(\mathfrak{B})$ the set of all $X \in \mathscr{V}$ with $|X|^q \in \mathscr{L}_1$, for every real number $q > 0$. Then $\mathscr{L}_q$ is a vector sublattice of $\mathscr{V}$. It will be convenient to introduce two extreme cases. The first is the trivial space $\mathscr{L}_0$ of all rv's X, i.e., $\mathscr{L}_0 = \mathscr{V}$, since

$$E(|X|^0) = E(1) = 1$$

for every $X \in \mathscr{V}$. The second is the space $\mathscr{L}_\infty$ of all bounded rv's, i.e., $\mathscr{L}_\infty = \mathscr{M}$. Since

$$\lim_{q \to \infty} E(|X|^q) < +\infty$$

if and only if $|X| \leqslant 1$, it seems that only the set $\mathscr{L}_\infty'$ of all rv's with $|X| \leqslant 1$ ought to be introduced. However, we have

$$\lim_{q \to \infty} \left(E(|X|^q)\right)^{1/q} = \sup |X|. \tag{6}$$

Therefore

$$\|X\|_\infty = \lim_{q \to \infty} \left(E(|X|^q)\right)^{1/q} < +\infty$$

if and only if $X \in \mathscr{M}$, i.e., there exists a constant η such that $|X| \leqslant \eta$. In order to prove (6), we have

$$\sup |X| \geqslant \left(E(|X|^q)\right)^{1/q} \geqslant \left(E(|X|^q . I_{[|X| \geqslant \xi]})\right)^{1/q} \geqslant \xi\left(p([|X| \geqslant \xi])\right)^{1/q}$$

for every ξ with $0 > \xi - \sup |X|$, and the right-hand side converges to ξ as $q \to +\infty$. Letting $\xi \uparrow \sup |X|$ we infer (6).

Theroem 2.1 implies

Theorem 3.1.

$\mathscr{V} = \mathscr{L}_0 \supseteq \mathscr{L}_q \supseteq \mathscr{L}_s \supseteq \mathscr{L}_\infty = \mathscr{M} \supseteq \mathscr{L}_\infty'$, if $0 \leqslant q \leqslant s \leqslant +\infty$; *moreover* $\mathscr{L}_\infty \supseteq \mathscr{S}$.

The following theorem is true:

Theorem 3.2.

For every q with $0 < q < 1$, the space $\mathscr{L}_q$ is a linear metric space with metric defined by $\rho(X, Y) = E(|X - Y|^q)$. For every q with $1 \leqslant q \leqslant +\infty$ the space $\mathscr{L}_q$ is a normed linear space with norm defined by

$$\|X\|_q = \left(E(|X|^q)\right)^{1/q}, \qquad \|X\|_\infty = \sup |X| = \lim_{q \to \infty} \|X\|_q.$$

Proof. By the inequality (1), we can prove, for every q, with $0 < q < 1$, the inequality:

$$\rho(X, Z) \leqslant \rho(X, Y) + \rho(Y, Z)$$

and, by the inequality (3), we can prove for every q with $1 \leqslant q < +\infty$ the inequality:

$$\|X+Y\|_q \leqslant \|X\|_q + \|Y\|_q.$$

If $q = +\infty$, then

$$\|X\|_\infty = \sup |X| = \lim_{q\to\infty} \|X\|_q.$$

Hence
$$\lim_{q\to\infty} \|X+Y\|_q \leqslant \lim_{q\to\infty} \|X\|_q + \lim_{q\to\infty} \|Y\|_q;$$

i.e.,
$$\|X+Y\|_\infty \leqslant \|X\|_\infty + \|Y\|_\infty.$$

It is obvious that the other properties of the metric and the norm are true.

3.2. We first introduce the *convergence in the q-th mean.* We shall say that the sequence $X_n \in \mathscr{L}_q$, $n = 1, 2, \ldots$, converges to $X \in \mathscr{L}_q$ in the q-th mean and write $X_n \overset{q}{\to} X$ if and only if $E(|X_n - X|^q) \to 0$, when $0 < q < +\infty$, and if and only if $\|X_n - X\|_\infty \to 0$, when $q = +\infty$.

The following theorems are true:

Theorem 3.3.

If $X_n \overset{q}{\to} X$, *then* $E(|X_n|^q) \to E(|X|^q)$, $0 < q < +\infty$.

Proof. Let $X_n \overset{q}{\to} X$. If $q \leqslant 1$, then it follows from inequality (1), that

$$E(|X_n|^q) - E(|X|^q) \leqslant E(|X_n - X|^q),$$

i.e.,
$$E(|X_n|^q) \to E(|X|^q).$$

If $q > 1$, then it follows from inequality (3), that

$$|\, \|X_n\|_q - \|X\|_q| \leqslant \|X_n - X\|_q,$$

i.e.,
$$\|X_n\|_q \to \|X\|_q,$$

hence
$$E(|X_n|^q) \to E(|X|^q).$$

Theorem 3.4.

If $X_n \overset{q}{\to} X$, *then* $X_n \overset{\mathrm{p}}{\to} X$, $0 < q \leqslant +\infty$, i.e. convergence in the q-th mean implies convergence in probability.

Proof. By applying the inequality of Markov (on the right of (5)).

Theorem 3.5.

If $X_n \xrightarrow{\mathrm{P}} X$ *and the sequence* X_n *is bounded by a constant, then* $X_n \xrightarrow{q} X$, $0 < q < +\infty$.

Proof. By applying the inequality on the left of (5).

Exercise. Prove that

$$d(X, Y) = E\left(\frac{|X-Y|}{1+|X-Y|}\right)$$

defines a metric in $\mathscr{V}$ and convergence in this metric is equivalent to convergence in probability.

It is easy to prove:

Theorem 3.6.

Convergence in the distance ρ, *if* $0 < q < 1$, *and norm convergence, if* $1 \leqslant q \leqslant +\infty$, *respectively, are equivalent to convergence in the q-th mean in* $\mathscr{L}_q$.

We are now in a position to prove that the space $\mathscr{L}_q$ is closed relative to convergence in the q-th mean. More precisely, the following theorem is true:

Theorem 3.7.

A sequence $X_n \in \mathscr{L}_q$, $n = 1, 2, \ldots$, *converges in the q-th mean to a rv* X *in* $\mathscr{L}_q$ *if and only if* X_n, $n = 1, 2, \ldots$, *is a fundamental sequence in the q-th mean, i.e.,* $(X_m - X_k) \xrightarrow{q} 0$, $0 < q \leqslant +\infty$.

Proof. Let $q < +\infty$, and let $X_n \xrightarrow{q} X$; then $X_m - X_k \xrightarrow{q} 0$, since, by the inequality (1),

$$E(|X_m - X_k|^q) \leqslant \gamma_q\, E(|X_m - X|^q) + \gamma_q\, E(|X - X_k|^q).$$

Conversely, if $X_m - X_k \xrightarrow{q} 0$ then, by the Markov inequality, for every $\varepsilon > 0$,

$$p([|X_m - X_k| \geqslant \varepsilon]) \leqslant \frac{1}{\varepsilon^q}\, E(|X_m - X_k|^q),$$

so that $X_m - X_k \xrightarrow{\mathrm{P}} 0$. Therefore, there exists a subsequence X_{n_λ},

$\lambda = 1, 2, \ldots$, which o-converges to some $X \in \mathscr{V}$, i.e., $X = o\text{-}\lim_{\lambda \to \infty} X_{n_\lambda}$, in $\mathscr{V}$.

For every fixed m, we have $X_m - X_{n_\lambda} \xrightarrow{o} X_m - X$ in $\mathscr{V}$. Since

$$E(|X_m - X_{n_\lambda}|^q) \to 0,$$

it follows, by the Fatou–Lebesgue theorem and the hypothesis, that

$$\lim_m \sup E(|X_m - X|^q) \leqslant \lim_m \sup (\lim_{n_\lambda} \inf E(|X_m - X_{n_\lambda}|^q)) = 0.$$

Hence $X_m \xrightarrow{q} X$.

In the case $q = +\infty$, the theorem is true, because ∞-convergence in $\mathscr{L}_\infty$ is equivalent to uniform convergence.

4. CONVERGENCE IN MEAN AND EQUI-INTEGRABILITY

4.1. A family $X_i \in \mathscr{L}_1$, $i \in I$, is said to be *equi-integrable* if and only if

$$\sup_I E(I_{[|X_i| > \xi]} |X_i|) \downarrow 0 \quad \text{as} \quad \xi \uparrow +\infty.$$

It is easy to prove:

A family $x_i \in \mathscr{L}_1$, $i \in I$, is equi-integrable if and only if, for every $\varepsilon > 0$, there exists a $\delta(\varepsilon) > 0$, independent of i, such that

$$E(I_{[|X_i| > \xi]} |X_i|) < \varepsilon,$$

for every $\xi > \delta(\varepsilon)$ and $i \in I$, i.e., if and only if

$$E(I_{[|X_i| > \xi]} |X_i|) \to 0$$

uniformly in i, as $\xi \to +\infty$.

Theorem 4.1.

If the family $X_i \in \mathscr{L}_1$, $i \in I$, is absolutely bounded by a rv $X \in \mathscr{L}_1$, i.e., $|X_i| \leqslant X$, $i \in I$, then it is equi-integrable. In particular, any finite family in $\mathscr{L}_1$ is always equi-integrable.

Proof. Since the function $\psi_X(a) = E(I_a X)$, $a \in \mathfrak{B}$, is a measure on $\mathfrak{B}$, hence σ-additive and non-negative, i.e., continuous, we have

$$\lim_{\xi \to \infty} E(I_{[X > \xi]} X) = 0.$$

Since $|X_i| \leqslant X$, $i \in I$, we have

$$E(I_{[|X_i| > \xi]} |X_i|) \leqslant E(I_{[X > \xi]} X), \qquad i \in I,$$

i.e.,

$$\sup_I E(I_{[|X_i| > \xi]} |X_i|) \downarrow 0.$$

If I is finite, then obviously the family is absolutely bounded and the hypothesis of the theorem is satisfied.

4.2. A family $X_i \in \mathscr{L}_1$, $i \in I$, is said to be *equi-continuous* if

$$\sup_I E(I_a|X_i|) \to 0, \quad \text{as} \quad p(a) \to 0,$$

i.e., for every $\varepsilon > 0$ there exists a $\delta(\varepsilon) > 0$ such that

$$\sup_I E(I_a|X_i|) < \varepsilon$$

for every $a \in \mathfrak{B}$ with $p(a) < \delta(\varepsilon)$.

Theorem 4.2.

A family $X_i \in \mathscr{L}_1$, $i \in I$, is equi-integrable if, and only if, it satisfies the following two conditions:

(a) *The family X_i, $i \in I$, is equi-continuous.*

(b) $\sup_I E(|X_i|) < +\infty$.

Proof. Let X_i, $i \in I$, be equi-integrable; then, since, for $\xi \geqslant 0$ and any $a \in \mathfrak{B}$

$$E(I_a|X_i|) = E(I_{a \wedge [|X_i| \leqslant \xi]}|X_i|) + E(I_{a \wedge [|X_i| > \xi]}|X_i|),$$

i.e.,

$$E(I_a|X_i|) \leqslant \xi p(a) + E(I_{[|X_i| > \xi]}|X_i|),$$

we have

$$\sup_I E(I_a|X_i|) \leqslant \xi p(a) + \sup_I E(I_{[|X_i| > \xi]}|X_i|). \tag{1}$$

But, for every $\varepsilon > 0$, there exists a $\xi_\varepsilon > 0$ such that

$$\sup_I E(I_{[|X_i| > \xi_\varepsilon]}|X_i|) < \frac{\varepsilon}{2}.$$

Now we write

$$\delta(\varepsilon) = \frac{\varepsilon}{2\xi_\varepsilon};$$

then, if $a \in \mathfrak{B}$ with $p(a) < \delta(\varepsilon)$, we have

$$\sup_I E(I_a|X_i|) < \varepsilon;$$

thus X_i, $i \in I$, is equi-continuous.

We set in (1) $a = e$; then

$$\sup_I E(|X_i|) \leqslant \xi + \sup_I E(I_{[|X_i| > \xi]}|X_i|),$$

i.e.,
$$\sup_I E(|X_i|) < +\infty.$$

Thus the family X_i, $i \in I$ satisfies the two conditions (a) and (b).

Conversely, let the family X_i, $i \in I$, satisfy conditions (a) and (b); then the obvious inequality

$$E(|X_i|) \geqslant E(I_{[|X_i| \geqslant \xi]}|X_i|) \geqslant \xi p([|X_i| \geqslant \xi]), \qquad \xi > 0,$$

implies
$$\sup_I p([|X_i| \geqslant \xi]) \leqslant \frac{1}{\xi} \sup_I E(|X_i|).$$

Since, moreover, by condition (b), $\sup E(|X_i|) < +\infty$, we have,

$$\sup_I p([|X_i| \geqslant \xi]) \to 0, \quad \text{as} \quad \xi \to +\infty. \tag{2}$$

By condition (a), for every $\varepsilon > 0$, there exists $\delta(\varepsilon) > 0$ such that, if $a \in \mathfrak{B}$ with $p(a) < \delta(\varepsilon)$, then

$$\sup_I E(I_a|X_i|) < \varepsilon.$$

From (2), it follows that there exists $\eta(\varepsilon) > 0$, such that

$$p([|X_i| \geqslant \xi]) < \delta(\varepsilon),$$

if $\xi > \eta(\varepsilon)$; then we have

$$\sup_I E(I_{[|X_i| > \xi]}|X_i|) < \varepsilon.$$

Thus the family X_i, $i \in I$, is equi-integrable.

Theorems 4.1 and 4.2 imply:

Theorem 4.3.

If a family $X_i \in \mathscr{L}_1$, $i \in I$, is absolutely bounded by a rv $X \in \mathscr{L}_1$, then it is equi-continuous; in particular every finite family is equi-continuous.

The following theorem is true:

Theorem 4.4.

A sequence $X_n \in \mathscr{L}_1$, $n = 1, 2, \ldots$, converges in the mean (the first mean) to a rv $X \in \mathscr{L}_1$ if and only if $E(I_a X_n) \to E(I_a X)$ uniformly in $a \in \mathfrak{B}$, as $n \to \infty$; in particular, if X_n converges in the mean to X and

$$p\text{-}\lim a_n = a \quad \textit{in} \quad \mathfrak{B},$$

then
$$\lim_{n \to \infty} E(I_{a_n} X_n) = E(I_a X),$$

i.e., $$\psi_{X_n}(a_n) \to \psi_X(a).$$

Proof. From the obvious inequality

$$|E(I_a X_n) - E(I_a X)| \leqslant E(|I_a X_n - I_a X|) \leqslant E(|X_n - X|)$$

for every $a \in \mathfrak{B}$, it follows that, if X_n converges in the mean to X, then $E(I_a X_n)$ converges to $E(I_a X)$ uniformly in a. Conversely, the equality

$$E(|X_n - X|) = (E(I_{b_n} X_n) - E(I_{b_n} X)) - (E(I_{b_n{}^c} X_n) - E(I_{b_n{}^c} X)),$$

where $b_n = [X_n > X]$, implies that if $E(I_a X_n)$ converges to $E(I_a X)$ uniformly in a, then X_n converges in the mean to X. Now if X_n converges in the mean to X and $p\text{-}\lim_{n\to\infty} a_n = a$ in $\mathfrak{B}$ then, since

$$|E(I_{a_n} X_n) - E(I_a X)| \leqslant |E(I_{a_n} X_n) - E(I_{a_n} X)| + |E(I_{a_n} X) - E(I_a X)|$$

and the right-hand side converges to zero, as $n \to \infty$, we have

$$E(I_{a_n} X_n) \to E(I_a X).$$

Now we shall prove the following theorem.

Theorem 4.5.

Let $X_n \in \mathscr{L}_1$, $n = 1, 2, \ldots$, and $X \in \mathscr{V}$; then the following two statements are equivalent:

(a) *The sequence $X_n \in \mathscr{L}_1$, $n = 1, 2, \ldots$, is equi-integrable and $X_n \xrightarrow{\text{p}} X$.*

(b) *$X \in \mathscr{L}_1$ and X_n converges in the mean to X.*

Proof. Let (b) be true; then the sequence X_n, $n = 1, 2, \ldots$, is fundamental in the mean; i.e., for every $\varepsilon > 0$, there exists an index n_ε such that

$$E(|X_m - X_k|) < \frac{\varepsilon}{3} \quad \text{if} \quad m \geqslant n_\varepsilon, \qquad k \geqslant n_\varepsilon. \tag{1}$$

The obvious inequality

$$E(I_a|X_k|) \leqslant E(I_a|X_m|) + E(|X_k - X_m|), \qquad a \in \mathfrak{B}, \tag{2}$$

and (1) imply that, for every $a \in \mathfrak{B}$,

$$\sup_k E(I_a|X_k|) \leqslant \sup_{m \leqslant n_\varepsilon} E(I_a|X_m|) + \frac{\varepsilon}{3}\,; \tag{3}$$

since the finite family X_m, $m = 1, 2, \ldots, n_\varepsilon$, is, by Theorem 4.1, equi-integrable, we have by Theorem 4.2 that

$$\sup_{m \leqslant n_\varepsilon} E(|X_m|) < +\infty,$$

and that the sequence X_m, $m = 1, 2, \ldots, n$, is equi-continuous; i.e., there exists $\delta(\varepsilon) > 0$ such that

$$\sup_{m \leqslant n_\varepsilon} E(I_a|X_m|) \leqslant \frac{\varepsilon}{3},$$

if $a \in \mathfrak{B}$ with $p(a) < \delta(\varepsilon)$.

These and the inequality (3) imply:

$$\sup_k E(|X_k|) < +\infty, \tag{4}$$

and

$$\sup_k E(I_a|X_k|) < \varepsilon, \tag{5}$$

for every a with $p(a) < \delta(\varepsilon)$; i.e., that the sequence X_k, $k = 1, 2, \ldots$, is equi-continuous. Conditions (4) and (5) imply by Theorem 4.2 that the sequence X_k, $k = 1, 2, \ldots$, is equi-integrable. Furthermore, we shall prove that the sequence X_n, $n = 1, 2, \ldots$, converges in probability to X. In fact, by Theorem 2.4 we have the inequality

$$p([|X_n - X| \geqslant \varepsilon]) \leqslant \frac{1}{\varepsilon} E(|X_n - X|) \quad \text{for every} \quad \varepsilon > 0.$$

Since $E(|X_n - X|) \to 0$, we have $p([|X_n - X| \geqslant \varepsilon]) \to 0$ as $n \to \infty$ and for every $\varepsilon > 0$, i.e., the sequence X_n, $n = 1, 2, \ldots$, converges in probability to X.

Now, let (a) be true; then we shall prove that (b) is true.

Since $X_n \xrightarrow{p} X$, there exists a subsequence X_{n_k}, $k = 1, 2, \ldots$, such that X_{n_k} o-converges to X, as $k \to \infty$, i.e., $|X_{n_k}|$ also o-converges to $|X|$. Since the sequence X_n, $n = 1, 2, \ldots$, is equi-integrable, we have, by Theorem 4.2, $\sup E(|X_n|) < +\infty$, i.e.,

$$\lim_k \inf E(|X_{n_k}|) \leqslant \sup_k E(|X_{n_k}|) < +\infty.$$

By Exercise 1, Section 2.3, Chapter V,

$$o\text{-}\liminf_{k \to \infty} |X_{n_k}| = o\text{-}\lim_{k \to \infty} |X_{n_k}| = |X| \in \mathscr{L}_1;$$

i.e.,

$$X \in \mathscr{L}_1.$$

We shall prove $E(|X_n - X|) \to 0$, i.e., X_n converges in the mean to X.

In fact, we have

$$E(|X_n - X|) = E(I_{[|X_n - X| \leqslant \varepsilon]}|X_n - X|) + E(I_{[|X_n - X| > \varepsilon]}|X_n - X|)$$
$$\leqslant \varepsilon + E(I_{[|X_n - X| > \varepsilon]}|X_n|) + E(I_{[|X_n - X| > \varepsilon]}|X|).$$

Since

$$\lim_{n \to \infty} p([|X_n - X| > \varepsilon]) = 0,$$

for every $\varepsilon > 0$, we have

$$E(I_{[|X_n - X| > \varepsilon]}|X_n|) \to 0$$

and

$$E(I_{[|X_n - X| > \varepsilon]}|X|) \to 0$$

for every $\varepsilon > 0$,

i.e.,

$$E(|X_n - X|) \to 0.$$

Theorem 4.5 yields the following theorem:

Theorem 4.6.

Let $X_n \in \mathscr{L}_q$, $n = 1, 2, \ldots$, with $0 < q < +\infty$ and $X \in \mathscr{V}$; then the following two conditions are equivalent:

(a) *The sequence $|X_n|^q$, $n = 1, 2, \ldots$, is equi-integrable and the sequence X_n, $n = 1, 2, \ldots$, converges in probability to X.*

(b) *$X \in \mathscr{L}_q$ and $X_n \overset{q}{\to} X$.*

Proof. The proof is analogous to that of Theorem 4.5, which deals with the special case $q = 1$. It suffices to replace everywhere in that proof $|Y|$ by $|Y|^q$ and to apply instead of the inequality

$$|X + Y| \leqslant |X| + |Y|$$

the inequality

$$|X + Y|^q \leqslant \gamma_q(|X|^q + |Y|^q).$$

Theorems 3.6 and 3.7 and the fact that the space $\mathscr{L}_\infty$ is identical with the space $\mathscr{M} = \{X \in \mathscr{V}\colon$ there exists a constant η_X such that $|X| \leqslant \eta_X\}$ imply the following:

Theorem 4.7.

For every q with $1 \leqslant q \leqslant +\infty$ the space $\mathscr{L}_q$ is, with respect to the norm $\|X\|_q$, a complete normed vector lattice; i.e., a Banach lattice. For *every q with* $0 < q < 1$, *the space* $\mathscr{L}_q$ *is a metrically complete vector lattice.*

Furthermore, we obviously have

Theorem 4.8.

For every q with $0 \leqslant q \leqslant +\infty$ *the vector lattice* $\mathscr{L}_q$ *is (conditionally) complete with respect to the lattice operations.*

Proof. Obviously, if $X_1 \in \mathscr{L}_q$ and $X_2 \in \mathscr{L}_q$, then $X_1 \vee X_2 \in \mathscr{L}_q$ and $X_1 \wedge X_2 \in \mathscr{L}_q$. Moreover, if $X \in \mathscr{V}$ and $Y \in \mathscr{L}_q$ with $0 \leqslant X \leqslant Y$, then $X \in \mathscr{L}_q$; i.e., if $X_i \in \mathscr{L}_q$ with $0 \leqslant X_i \leqslant Y$, $i \in I$, and $Y \in \mathscr{L}_q$, then

$$\bigvee_{i \in I} X_i \in \mathscr{L}_q.$$

Corollary 4.1.

The space $\mathscr{L}_2$ *is a Hilbert space with* $\langle X, Y \rangle = E(XY)$ *as scalar product.*

Exercises. Prove that:

1. $X_n \xrightarrow{q} X$ implies $X_n \xrightarrow{q'} X$, for $q' < q$.

2. If $\sup_n E(|X_n|^q) < +\infty$ and X_n converges in probability to X, then $X_n \xrightarrow{q'} X$ for $q' < q$.

3. If $|X_n| \leqslant Y$, $n = 1, 2, \ldots$, with $Y \in \mathscr{L}_q$ and if X_n converges in probability to X, then $X_n \xrightarrow{q} X \in \mathscr{L}_q$, $0 < q < +\infty$.

VII

GENERALIZED RANDOM VARIABLES

(Random variables having values in any space)

1. PRELIMINARIES

1.1. Let $\mathscr{V}$ be the stochastic space (space of all real-valued random variables) over a probability σ-algebra $(\mathfrak{B}, p)$. Further, let $\mathbf{B}$ be the σ-field (Boolean σ-algebra) of all Borel subsets of the real line R. Then the set $\mathfrak{D}$ of all open intervals $\iota_\xi \equiv (-\infty, \xi)$, $\xi \in R$, is a chain in $\mathbf{B}$, which σ-generates $\mathbf{B}$. According to Section 5.6 Chap. IV, if X is a rv in $\mathscr{V}$, then there exists a uniquely determined Boolean σ-homomorphism h_X of $\mathbf{B}$ into $\mathfrak{B}$ such that $h_X(\iota_\xi) = [X<\xi] \in \mathfrak{B}$, $\xi \in R$. With the help of this σ-homomorphism h_X, a quasi-probability:

$$P_X(A) = p(h_X(A)), \qquad A \in \mathbf{B},$$

can be defined on $\mathbf{B}$. Then $(R, \mathbf{B}, P_X)$ is the so-called "sample probability space of X". Conversely, if h is any Boolean σ-homomorphism of $\mathbf{B}$ into $\mathfrak{B}$, then there always exists exactly a real-valued random variable $X \in \mathscr{V}$ such that:

$$h(A) = h_X(A) \text{ for every } A \in \mathbf{B}.$$

Under consideration of this connection between the concepts of real-valued rv's over $(\mathfrak{B}, p)$ and Boolean σ-homomorphisms of $\mathbf{B}$ into $\mathfrak{B}$, one could define a real-valued rv over $(\mathfrak{B}, p)$ as a Boolean σ-homomophism of $\mathbf{B}$ into $\mathfrak{B}$. The algebraic and topological structure of the space of all these σ-homomorphisms, considered as real-valued rv's, could also be studied directly according to this definition (see Olmsted [1]). In an analogous way, the concept of random variables over $(\mathfrak{B}, p)$ with values in any topological or in general in any abstract space Σ, respectively, may also be introduced in the probability theory (see Segal [1], Dubins [1]). In fact, let **B** be the Boolean σ-algebra of all Borel subset of Σ, if it is a topological space, or let **B** be any Boolean σ-subalgebra of the Boolean algebra $\mathfrak{P}(\Sigma)$ of all subsets of Σ, if Σ is any abstract space. Then a suitable Boolean σ-homomorphism h of **B** into $\mathfrak{B}$ defines also a quasi-probability:

$$P(A) = p(h(A)), \qquad A \in \mathbf{B}.$$

Hence, the pr space $(\Sigma, \mathbf{B}, P)$ can replace the concept of a sample pr space, corresponding to a certain rv having values in Σ. However, an important problem in the probability theory is the definition and study of an algebraic and topological structure in the space of all rv's over $(\mathfrak{B}, p)$ with values in Σ with the help of the algebraic and topological structure existing in the space Σ. For this reason, the Boolean σ-homomorphisms of **B** into $\mathfrak{B}$ are not always suitable (see Kappos [13], Dubins [1], Nedoma [1]). However, this question can be studied in a way which is analogous to that of Chapter IV. Furthermore, by this method, all problems related to the study of the algebraic and topological structure of the space of all rv's with values in any space Σ may be established more naturally and simply.

2. GENERALIZED ELEMENTARY RANDOM VARIABLES

2.1. Let $(\mathfrak{B}, p)$ be a pr σ-algebra and Σ be any space. A function X defined on an experiment $\mathbf{a} \in \mathscr{T}(\mathfrak{B})$ and having values in Σ, i.e., a map

$$X : \mathbf{a} \ni a_j \Rightarrow X(a_j) = \xi_j \in \Sigma, \qquad j \geqslant 1,$$

is said to be a *generalized elementary random variable* (erv) *over* $\mathfrak{B}$ *with respect to* Σ. The set of all generalized erv's over $\mathfrak{B}$ with respect to Σ will be denoted by $\mathscr{E}(\mathfrak{B}, \Sigma)$ and called a *generalized elementary stochastic space over* $\mathfrak{B}$ *with respect to* Σ. If $\Sigma = R$, i.e., the field of the real numbers, then we have the elementary stochastic space $\mathscr{E}(\mathfrak{B})$ over $\mathfrak{B}$ which we studied in Section 2, Chapter IV. By $\mathscr{S}(\mathfrak{B}, \Sigma)$ we will denote the generalized *simple* stochastic space over $\mathfrak{B}$ with respect to Σ. Let

$$X : \mathbf{a} \ni a_j \Rightarrow X(a_j) = \xi_j \in \Sigma, \qquad j \geqslant 1$$

$$Y : \mathbf{b} \ni b_i \Rightarrow Y(b_i) = \eta_i \in \Sigma, \qquad i \geqslant 1$$

be two erv's of $\mathscr{E}(\mathfrak{B}, \Sigma)$; then an equality $X = Y$ will be defined if and only if, for every pair (j, i) with $a_j \wedge b_i \neq \emptyset$, we have $\xi_j = \eta_i$ in Σ. Furthermore, any relation r, resp any operation $\otimes$, defined in Σ induces a relation r, resp an operation $\otimes$, in $\mathscr{E}(\mathfrak{B}, \Sigma)$ defined as follows:

(1) $X\ r\ Y$ in $\mathscr{E}(\mathfrak{B}, \Sigma)$ if and only if, for every pair (j, i) with $a_j \wedge b_i \neq \emptyset$, we have $\xi_j\, r\, \eta_i$ in Σ, resp,

(2) $X \otimes Y = Z : \mathbf{a} \wedge \mathbf{b} \ni a_j \wedge b_i \Rightarrow Z(a_j \wedge b_i) = X(a_j) \otimes Y(b_i) = \xi_j \otimes \eta_i$ in Σ, for every pair (j, i) with $a_j \wedge b_i \neq \emptyset$.

If a multiplication of a scalar λ and an element $\xi \in \Sigma$, i.e., $\lambda\xi \in \Sigma$, is defined in Σ, then a multiplication λX is induced also in $\mathscr{E}(\mathfrak{B}, \Sigma)$ defined as follows:

$$\lambda X = Z : \mathbf{a} \ni a_j \Rightarrow Z(a_j) = \lambda X(a_j) = \lambda \xi_j \in \Sigma, \qquad j \geqslant 1.$$

2.2. Let Σ be equipped with an algebraic, resp a topological, structure; then a corresponding algebraic, resp topological, structure can be induced in $\mathscr{E}(\mathfrak{B}, \Sigma)$ with the help of the definitions of Sections 2.1. Particularly, if Σ is equipped with a vector lattice or norm structure, then order and other convergences can be induced in $\mathscr{E}(\mathfrak{B}, \Sigma)$ and a completion process with respect to these convergences can be applied on $\mathscr{E}(\mathfrak{B}, \Sigma)$ such that the corresponding extended generalized stochatic space $\mathscr{V}(\mathfrak{B}, \Sigma)$ will be complete with respect to all convergences which are considered as important in the probability theory. In the following sections the above mentioned algebraic and topological questions will be discussed in the case in which Σ is a lattice group, resp vector lattice or normed space.

3. COMPLETION WITH RESPECT TO O-CONVERGENCE

3.1. Let Σ be a σ-saturated (i.e. conditionally σ-complete with respect to the lattice operations) lattice group; then for every sequence $\xi_n \in \Sigma$, $n = 1, 2, \ldots$, which is bounded from above (resp from below) there exists the supremum $\bigvee_{n=1}^{\infty} \xi_n$, resp the infimum $\bigwedge_{n=1}^{\infty} \xi_n$ in Σ. Furthermore, a σ-saturated l-group is always archimedean, and every archimedean l-group is commutative. It is easy to prove that the elementary stochastic space $\mathscr{E}(\mathfrak{B}, \Sigma)$ is also a lattice group, but not always σ-saturated unless the Boolean σ-algebra $\mathfrak{B}$ is atomic. The lattice group Σ can be considered as the lattice subgroup of all constant erv's. Obviously, Σ is then a regular lattice subgroup of $\mathscr{E} \equiv \mathscr{E}(\mathfrak{B}, \Sigma)$. Moreover, $\mathscr{E}$ is an archimedean l-group. In fact, let $X \in \mathscr{E}$, $Y \in \mathscr{E}$ with $nY \leqslant X, n = 1, 2, \ldots$; we can assume that X and Y are defined on the same experiment, i.e.,

$$X : \mathbf{a} \ni a_j \Rightarrow \xi_j \in \Sigma, \qquad j \geqslant 1$$

$$Y : \mathbf{a} \ni a_j \Rightarrow \eta_j \in \Sigma, \qquad j \geqslant 1;$$

then $nY \leqslant X$ is equivalent to $n\eta_j \leqslant \xi_j$, $j \geqslant 1$, $n = 1, 2, \ldots$. But $n\eta_j \leqslant \xi_j$, $n = 1, 2, \ldots$, implies $\eta_j \leqslant \theta$, and this holds for every $j \geqslant 1$; hence $Y \leqslant \theta$. (Here θ is the zero of Σ and $\mathscr{E}$).

3.2. For a real-valued erv X, i.e.,

$$X : \mathbf{a} \ni a_j \Rightarrow X(a_j) = \xi \in R, \qquad j \geqslant 1$$

we have used a representation by a series:

$$X = \sum_{j \geqslant 1} \xi_j I_{a_j}.$$

That is possible here also in the following way:

We denote by ξI_a for every $\xi \in \Sigma$ and $a \in \mathfrak{B}$ the erv

$$\xi I_a \begin{cases} a \Rightarrow \xi \\ a^c \Rightarrow \theta \end{cases} \qquad \text{where } \theta \text{ the zero of } \Sigma.$$

Particularly we define: $\theta I_a = \theta$, $\xi I_\varnothing = \theta$, i.e., the constant erv θ; and $\xi I_e = \xi$, i.e., the constant erv ξ. Let now X be an erv having values in Σ, i.e.,

$$X : \mathbf{a} \ni a_j \Rightarrow X(a_j) = \xi_j \in \Sigma, \qquad j \geqslant 1;$$

then we put

$$X = \sum_{j \geqslant 1} \xi_j I_{a_j}.$$

It is easily verified here also that:

$$X = o\text{-}\lim_{n \to \infty} \sum_{j=1}^{n} \xi_j I_{a_j}.$$

If Σ is a vector lattice and a commutative multiplication $\xi\eta$ is defined in it, i.e., Σ is a lattice algebra over a scalar field F, and $\mathbf{1}$ denotes the neutral element of the multiplication, the so-called unit of Σ, then the set of all $\mathbf{1} \cdot I_a \equiv I_a$, $a \in \mathfrak{B}$, may be considered as the Boolean kernel $\mathscr{I}(\mathfrak{B})$ of all indicators in $\mathscr{E}(\mathfrak{B}, \Sigma)$ (*cf.* Section 2, Chapter IV).

3.3. We remark that the space $\mathscr{E}(\mathfrak{B}, \Sigma)$, considered as an archimedean l-group, may always be embedded regularly in a saturated l-group, which may be constructed in a way that is analogous to that of the construction of the real numbers by Dedekind cuts of the rational numbers. MacNeille [1] has proved that this Dedekind process can be applied to embed regularly every partially ordered set into a saturated partially ordered set. Everett [1], Banaschewski [1], and others (see also Kantorovic, Vulikh and Pinsker [1]) have proved that the fact that an l-group is archimedean is not only sufficient but also necessary for assuming that the minimal saturated MacNeille extension in which it may be embedded regularly is also an l-group. In the general case this extension is a lattice-ordered semigroup. Let $\hat{\mathscr{E}} = \hat{\mathscr{E}}(\mathfrak{B}, \Sigma)$ be the MacNeille minimal saturated extension of $\mathscr{E} = \mathscr{E}(\mathfrak{B}, \Sigma)$. We then can consider $\mathscr{E}$ as a regular lattice-subgroup of $\hat{\mathscr{E}}$.

3.4. In the l-group $\mathscr{E}$ an o-convergence and the concept of o-fundamental sequences may be defined as in Section 2, Chapter IV; then theorems 3.1–3.3, 3.5, part of Theorem 3.4, Sections 4.1–4.5, and part of Section 4.6, Chapter VII, are also true here. An o-fundamental sequence in $\mathscr{E}$ is not always o-convergent, i.e., $\mathscr{E}$ is not always o-complete (i.e., complete with respect to the o-convergence). Obviously the extension $\hat{\mathscr{E}}$ of $\mathscr{E}$, as a saturated l-group, is also o-complete, i.e., for every o-fundamental sequence in $\hat{\mathscr{E}}$ there exists in $\hat{\mathscr{E}}$ an element to which the fundamental sequence o-converges. $\hat{\mathscr{E}}$ is moreover complete with respect to the o-convergence of nets. Papangelou [1, 2] (see also Everett and Ulam [1]), has established another process to complete directly a commutative l-group with respect to the o-convergence of sequences. This process is hence applicable to complete $\mathscr{E}$ and leads, if some additional conditions are fulfilled, to the minimal saturated extension $\hat{\mathscr{E}}$ of $\mathscr{E}$. This happens, for example, in the case $\Sigma = R$ (see Chapter IV, Section 5). The process which we have stated in the mentioned Section 5, Chapter IV, to define the extension space of $\mathscr{E}(\mathfrak{B}, R)$ is analogous to that of Papangelou. But the proofs given there are based on special properties of R as a linearly ordered l-group. On account of that, we shall state now briefly the process of Papangelou with the more general establishments of him. The extension space of Papangelou has the advantage that every element of it can be expressed as an o-limit of a sequence of erv's.

3.5. Let $\mathscr{M}$ be the set of all o-fundamental sequences in $\mathscr{E}$. If $\{X_n\}$ and $\{Y_n\}$ are o-fundamental sequences of $\mathscr{M}$, we define the following operations in $\mathscr{M}$:

$$\{X_n\} + \{Y_n\} = \{X_n + Y_n\}$$

$$-\{X_n\} = \{-X_n\}$$

$$\{X_n\} \vee \{Y_n\} = \{X_n \vee Y_n\} \quad \text{and dually.}$$

If $\mathscr{E}$ is, moreover, a vector lattice, then we define:

$$\lambda\{X_n\} = \{\lambda X_n\}, \qquad \lambda \in R.$$

3.6. It is easy to verify that $\mathscr{M}$ is a lattice group with respect to these operations, resp a vector lattice, if $\mathscr{E}$ is a vector lattice. Let $\{X\}$ denote the constant o-fundamental sequence $X_n = X \in \mathscr{E}$, $n = 1, 2, \ldots$; then the map:

$$\mathscr{E} \ni X \Rightarrow \{X\} \in \mathscr{M}$$

embeds $\mathscr{E}$ isomorphically and regularly in $\mathscr{M}$. The set $\mathscr{N}$ of all o-null sequences, i.e., sequences which o-converge to the zero element θ in $\mathscr{E}$, forms a lattice ideal in $\mathscr{M}$. Let $\mathscr{V} = \mathscr{V}(\mathfrak{B}, \Sigma)$ denote the quotient lattice group $\mathscr{M}/\mathscr{N}$ (of the l-group $\mathscr{M}$ modulo the l-ideal $\mathscr{N}$) and $\tilde{X}$ the class $\{X_n\}/\mathscr{N} \in \mathscr{M}/\mathscr{N}$. The map

$$\mathscr{E} \ni X \Rightarrow \{X\}/\mathscr{N} \in \mathscr{M}/\mathscr{N}$$

embeds, obviously, $\mathscr{E}$ isomorphically and regularly in $\mathscr{V} = \mathscr{M}/\mathscr{N}$. Hence, $\mathscr{E}$ can be considered as a regular lattice subgroup of $\mathscr{V}$ and the class $\{X\}/\mathscr{N}$ can be identified with the element $X \in \mathscr{E}$.

We shall call $\mathscr{V}(\mathfrak{B}, \Sigma)$ *the stochastic space of all rv's with values in* Σ over the pr σ-algebra $(\mathfrak{B}, p)$ and we shall regard $\mathscr{E}(\mathfrak{B}, \Sigma)$ as the stochastic subspace of $\mathscr{V}(\mathfrak{B}, \Sigma)$ constisting of all erv's.

3.7. A sequence $\tilde{X}_n \in \mathscr{V}$, $n = 1, 2, \ldots$, is said to be *$\mathscr{E}$-o-convergent* to $\tilde{X} \in \mathscr{V}$ if there exists a decreasing sequence $Y_n \in \mathscr{E}$, $n = 1, 2, \ldots$, such that:

$$|\tilde{X}_n - \tilde{X}| \leqslant Y_n, \qquad n = 1, 2, \ldots, \quad \text{and} \quad (\mathscr{E}) \bigwedge_{n=1}^{\infty} Y_n = \theta.$$

A sequence $\tilde{X}_n \in \mathscr{V}$, $n = 1, 2, \ldots$, is said to be $\mathscr{E}$-o-fundamental in $\mathscr{V}$, if and only if there is a decreasing sequence $Y_n \in \mathscr{E}$, $n = 1, 2, \ldots$, such that:

$$|\tilde{X}_{n+k} - \tilde{X}_n| \leqslant Y_n, \quad \text{equivalently} \quad \tilde{X}_n - Y_n \leqslant \tilde{X}_{n+k} \leqslant \tilde{X}_n + Y_n,$$

$n = 1, 2, \ldots$, $k = 1, 2, \ldots$, and $(\mathscr{E}) \bigwedge_{n=1}^{\infty} Y_n = \theta$.†

An analogous theorem to Theorem 5.1, Chapter IV, is also true here, i.e.,

---

† For these definitions compare also Section 4.5., Chapter IV.

Theorem 3.1.

If $X_n \in \mathscr{E}$, $n = 1, 2, \ldots$, *is an o fundamental sequence in* $\mathscr{E}$ *(hence* $\tilde{X} \equiv \{X_n\}/\mathscr{N}$ *an element in* $\mathscr{V}$*), then the sequence* X_n, $n = 1, 2, \ldots$, *considered as a sequence in* $\mathscr{V}$, *is* $\mathscr{E}$*-o-convergent to* $\tilde{X}$ *in* $\mathscr{V}$, *i.e.,* $\mathscr{E}$ *is* $\mathscr{E}$*-o-dense in* $\mathscr{V}$.

Proof. Since $X_n \in \mathscr{E}$, $n = 1, 2, \ldots$, is o-fundamental in $\mathscr{E}$, there are† two sequences, an increasing $Y_n \in \mathscr{E}$, $n = 1, 2, \ldots$, and a decreasing $Z_n \in \mathscr{E}$, $n = 1, 2, \ldots$, such that:

$$Y_n \leqslant X_n \leqslant Z_n, \qquad n = 1, 2, \ldots,$$

and

$$\bigwedge_{n=1}^{\infty} (Z_n - Y_n) = \theta \quad \text{in} \quad \mathscr{E},$$

i.e.,

$$o\text{-}\lim (Z_n - Y_n) = \theta \quad \text{in} \quad \mathscr{E}.$$

Since $\mathscr{E}$ is regular in $\mathscr{V}$, we have also

$$\bigwedge_{n=1}^{\infty} (Z_n - Y_n) = \theta \quad \text{in} \quad \mathscr{V}.$$

We have to prove:

$$(\mathscr{V}) \bigvee_{n=1}^{\infty} Y_n = (\mathscr{V}) \bigwedge_{n=1}^{\infty} Z_n = \tilde{X}, \tag{I}$$

Obviously, we have $\tilde{X} \geqslant Y_n$, $n = 1, 2, \ldots$. Let $\{X_n{}^*\}/\mathscr{N} = \tilde{X}^*$ be another upper bound of Y_n, $n = 1, 2, \ldots$; the sequence $X_n{}^*$, $n = 1, 2, \ldots$, can be chosen to be decreasing. Then $X_n{}^* \geqslant Y_n$, $n = 1, 2, \ldots$; hence

$$\{X_n{}^*\}/\mathscr{N} \geqslant \{Y_n\}/\mathscr{N} = \{X_n\}/\mathscr{N} \equiv \tilde{X},$$

i.e.,

$$\tilde{X}^* \geqslant \tilde{X}.$$

Hence

$$(\mathscr{V}) \bigvee_{n=1}^{\infty} Y_n = \tilde{X};$$

the dual can be established similarly. This shows that in $\mathscr{V}$, we have:

$$\tilde{X} = (\mathscr{V})\text{-}o\text{-}\lim Y_n = (\mathscr{V})\text{-}o\text{-}\lim Z_n = (\mathscr{V})\text{-}o\text{-}\lim X_n \tag{II}$$

and since X_n, $n = 1, 2, \ldots$, is $\mathscr{E}$-o-fundamental, we conclude

$$\mathscr{E}\text{-}o\text{-}\lim X_n = \tilde{X}.$$

---

† see Lemma 4.1., Chapter IV.

The lattice group $\mathscr{V}$ is, therefore, the $\mathscr{E}$-o-closure of the lattice subgroup $\mathscr{E}$ in $\mathscr{V}$, Moreover, (I) implies:

Corollary 3.1.

Every $\tilde{X} \in \mathscr{V}$, can be expressed as an $\mathscr{E}$-o-limit (hence, also as an $\mathscr{V}$-o-limit) of an increasing resp descreasing sequence of erv's.

We can now prove:

Theorem 3.2.

The l-group $\mathscr{V}$ is $\mathscr{E}$-o-complete, i.e., complete with respect to the $\mathscr{E}$-o-convergence in $\mathscr{V}$.

We shall prove first:

Lemma 3.1.

If $\tilde{X}_n \in \mathscr{V}$, $n = 1, 2, \ldots$ is decreasing, then there exists a decreasing sequence $Y_n \in \mathscr{E}$, $n = 1, 2, \ldots$, with $\tilde{X}_n \leqslant Y_n$, $n = 1, 2, \ldots$, and such that $\tilde{X}_n$, $n = 1, 2, \ldots$, and Y_n, $n = 1, 2, \ldots$, have the same lower bounds in $\mathscr{E}$; if, moreover, $\tilde{X}_n$, $n = 1, 2, \ldots$, is $\mathscr{E}$-o-fundamental, then Y_n, $n = 1, 2, \ldots$, can be chosen to be $(\mathscr{E})$-o-equivalent to the sequence $\tilde{X}_n$, $n = 1, 2, \ldots$ (i.e., *such that $(\mathscr{E})$-o-*lim $(\tilde{X}_n - Y_n) = \theta$), *hence $(\mathscr{E})$-o-fundamental.*†

Proof of Lemma 3.1. By Corollary 3.1, for each $\tilde{X}_n \in \mathscr{V}$, there is a decreasing sequence

$$Y_{n,\,i} \in \mathscr{E}, \qquad i = 1, 2, \ldots,$$

with

$$Y_{n,\,i} \geqslant \tilde{X}_n$$

and

$$(\mathscr{E})\text{-}o\text{-lim } Y_{n,\,i} = \tilde{X}_n;$$

hence there is a decreasing sequence

$$Z_{n,\,i} \in \mathscr{E}, \qquad i = 1, 2, \ldots,$$

such that

$$(\mathscr{E}) \bigwedge_{i=1}^{\infty} Z_{n,\,i} = \theta$$

and $|Y_{n,\,i} - \tilde{X}_n| \leqslant Z_{n,\,i}$, $i = 1, 2, \ldots$, $n = 1, 2, \ldots$. We put

$$Y_n = Y_{1,\,n} \wedge Y_{2,\,n} \wedge \ldots \wedge Y_{n,\,n}, \qquad n = 1, 2, \ldots;$$

clearly

$$Y_n \in \mathscr{E}, \qquad n = 1, 2, \ldots,$$

† A dual lemma is also true.

is decreasing and $\tilde{X}_n \leqslant Y_n$, since

$$\tilde{X}_n \leqslant \tilde{X}_k \leqslant Y_{k,n},$$

for all $k = 1, 2, \ldots, n$. Let now $V \in \mathscr{E}$ with $V \leqslant Y_n$, $n = 1, 2,$; then $V \leqslant Y_{k,n}$ for all k and n with $k \leqslant n$; hence

$$V \leqslant (\mathscr{V}) \bigwedge_{n=1}^{\infty} Y_{k,n} = \tilde{X}_k,$$

for every $k = 1, 2, \ldots$. This shows that $\tilde{X}_n$, $n = 1, 2, \ldots$, and Y_n, $n = 1, 2, \ldots$, have the same lower bounds in $\mathscr{E}$. Assume now that $\tilde{X}_n$, $n = 1, 2, \ldots$, is $(\mathscr{E})$-o-fundamental, i.e.,

$$\tilde{X}_n - \tilde{X}_{n+k} \leqslant V_n, \qquad n = 1, 2, \ldots, \quad k = 1, 2, \ldots,$$

with

$$V_n \in \mathscr{E}, \qquad n = 1, 2, \ldots,$$

and

$$(\mathscr{E}) \bigwedge_{n=1}^{\infty} V_n = \theta;$$

then

$$\begin{aligned} Y_n - \tilde{X}_n &= Y_{1,n} \wedge Y_{2,n} \wedge \ldots \wedge Y_{n,n} - \tilde{X}_n \\ &= (Y_{1,n} - \tilde{X}_n) \wedge (Y_{2,n} - \tilde{X}_n) \wedge \ldots \wedge (Y_{n,n} - \tilde{X}_n) \\ &= \{(Y_{1,n} - \tilde{X}_1) + (\tilde{X}_1 - \tilde{X}_n)\} \wedge \{(Y_{2,n} - \tilde{X}_2) + (\tilde{X}_2 - \tilde{X}_n)\} \wedge \ldots \\ &\qquad \wedge \{(Y_{n,n} - \tilde{X}_n) + (\tilde{X}_n - \tilde{X}_n)\} \\ &\leqslant (Z_{1,n} + V_1) \wedge (Z_{2,n} + V_2) \wedge \ldots \wedge (Z_{n,n} + V_n) = W_n, \quad n = 1, 2, \ldots . \end{aligned}$$

It is easy to verify that

$$W_n \in \mathscr{E}, \qquad n = 1, 2, \ldots,$$

is decreasing with

$$(\mathscr{E}) \bigwedge_{n=1}^{\infty} W_n = \theta.$$

In fact, let $Z \in \mathscr{E}$ with

$$Z \leqslant W_n, \qquad n = 1, 2, \ldots;$$

then

$$Z \leqslant Z_{k,n} + V_k$$

for all k and n with $n \geqslant k$, hence, for all k and n (since $Z_{k,n}$, $n = 1, 2, \ldots$, is decreasing); i.e.,

$$Z \leqslant (\mathscr{E}) \bigwedge_{n=1}^{\infty} (Z_{k,n} + V_k) = (\mathscr{E}) \bigwedge_{n=1}^{\infty} Z_{k,n} + V_k = \theta + V_k, \text{ hence } Z \leqslant V_k, \quad k = 1, 2, \ldots;$$

this implies

$$Z \leqslant (\mathscr{E}) \bigwedge_{n=1}^{\infty} V_k = \theta.$$

Thus we have

$$|Y_n - \tilde{X}_n| \leqslant W_n, \qquad n = 1, 2, \ldots,$$

with

$$W_n \in \mathscr{E}, \qquad n = 1, 2, \ldots,$$

decreasing, and

$$(\mathscr{E}) \bigwedge_{n=1}^{\infty} W_n = \theta,$$

i.e., the sequence

$$Y_n \in \mathscr{E}, \qquad n = 1, 2, \ldots$$

is $(\mathscr{E})$-o-equivalent to the sequence

$$\tilde{X}_n \in \mathscr{V}, \qquad n = 1, 2, \ldots;$$

hence

$$Y_n \in \mathscr{E}, \qquad n = 1, 2, \ldots,$$

is $(\mathscr{E})$-o-fundamental.

Proof of Theorem 3.2. According to Theorem 3.1, we have:

$$(\mathscr{E})\text{-}o\text{-}\lim Y_n = \tilde{Y} \equiv \{Y_n\}/\mathscr{N} \quad \text{in} \quad \mathscr{V};$$

hence, we have also:

$$(\mathscr{E})\text{-}o\text{-}\lim \tilde{X}_n = \tilde{Y} \quad \text{in} \quad \mathscr{V}.$$

This implies that Theorem 3.2 is true.

3.8. It is easily verified that in $\mathscr{V}$, $(\mathscr{E})$-o-convergence and $(\mathscr{V})$-o-convergence are equivalent; moreover, $(\mathscr{E})$-o-fundamentality and $(\mathscr{V})$-o-fundamentality are equivalent in $\mathscr{V}$. In fact, $\mathscr{E}$-o-convergence, resp $\mathscr{E}$-o-fundamentality, implies $(\mathscr{V})$-o-convergence, resp $(\mathscr{V})$-o-fundamentality, for $\mathscr{E}$ is regular in $\mathscr{V}$. In order to prove that $(\mathscr{V})$-o-convergence, resp $(\mathscr{V})$-o-fundamentality, implies $(\mathscr{E})$-o-convergence, resp $(\mathscr{E})$-o-fundamentality, it suffices to prove: If $\tilde{V}_n \in \mathscr{V}$, $n = 1, 2, \ldots$, is decreasing with

$$(\mathscr{V}) \bigwedge_{n=1}^{\infty} \tilde{V}_n = \theta,$$

then there is a decreasing sequence

$$Z_n \in \mathscr{E}, \qquad n = 1, 2, \ldots$$

with

$$(\mathscr{E}) \bigwedge_{n=1}^{\infty} Z_n = \theta$$

and

$$\tilde{V}_n \leqslant Z_n, \qquad n = 1, 2, \ldots.$$

That the latter is true follows by Lemma 3.1.

3.9. The lattice group $\mathscr{V} \equiv \mathscr{M}/\mathscr{N}$ is the minimal, with respect to the o-convergence, complete regular extension of $\mathscr{E}$. Hence $\mathscr{V}$ can be embedded regularly in the MacNeille minimal saturated regular extension $\hat{\mathscr{E}}$ of $\mathscr{E}$. One might suspect that $\mathscr{V}$ is at least σ-saturated. This is true in the case $\Sigma = R =$ the field of the real numbers; moreover, in this case $\mathscr{V}$ is saturated, hence isomorphic to $\hat{\mathscr{E}}$. In the general case this question is open. Papangelou [1, 2] has proved that:

A necessary and sufficient condition that the minimal with respect to the o-convergence regular completion $\mathscr{G}_{\mathscr{G}}$ of an archimedean lattice group $\mathscr{G}$ is σ-saturated is the following:

(*P*) *If* $X_n \in \mathscr{G}$, $n = 1, 2, \ldots$, *is an increasing bounded sequence, then there is a decreasing sequence*

$$Y_n \in \mathscr{G}, \qquad n = 1, 2, \ldots,$$

with

$$X_n \leqslant Y_n, \qquad n = 1, 2, \ldots,$$

and

$$(\mathscr{G}) \bigwedge_{n=1}^{\infty} (Y_n - X_n) = \theta.$$

Susan Papadopoulou has, moreover, proved that:

Theorem 3.3.

A necessary and sufficient condition that the Papangelou completion $\mathscr{G}_{\mathscr{G}}$ *is saturated, i.e., isomorphic to the MacNeille extension* $\mathscr{G}$, *is the following*:

(SP) *For every subset* $\mathscr{B}$ *of* $\mathscr{G}$ *which is upper bounded in* $\mathscr{G}$, *there is a sequence* $X_n \in \mathscr{G}$, $n = 1, 2, \ldots$, *such that this sequence and* $\mathscr{B}$ *have the same upper bounds in* $\mathscr{G}$.

Proof. Since we do not use the particular structure of $\mathscr{E}$ in the application of the extension process of Papangelou, in order to prove Theorem 3.3, we can put $\mathscr{G} = \mathscr{E}$; then $\mathscr{G}_{\mathscr{G}} = \mathscr{V}$.

(1) The condition is necessary: Let, namely, $\mathscr{V}$ be saturated and $\mathscr{B}$ be any subset of $\mathscr{E}$ upper bounded in $\mathscr{E}$; then there is $\bigvee \{X : X \in \mathscr{B}\} = Z$ in $\mathscr{V}$; then, by Corollary 3.1, there is an increasing sequence $X_n \in \mathscr{E}$, $n = 1, 2, \ldots$, such that:

$$Z = (\mathscr{V}) \bigvee_{n=1}^{\infty} X_n;$$

this sequence possesses the same upper bounds in $\mathscr{E}$ with $\mathscr{B}$.

(2) The condition is sufficient: Let $\mathscr{B}$ be upper bounded in $\mathscr{E}$ and consider $\mathscr{E}$ as a lattice subgroup of the MacNeille extension $\hat{\mathscr{E}}$; then there is

$$(\hat{\mathscr{E}})\bigvee\{X : X \in \mathscr{B}\} = Z \in \hat{\mathscr{E}}.$$

Let now $X_n \in \mathscr{E}$, $n = 1, 2, \ldots$, be the sequence which possesses the same upper bounds in $\mathscr{E}$ with $\mathscr{B}$. Then we have:

$$Z = (\hat{\mathscr{E}}) \bigvee_{n=1}^{\infty} X_n.$$

The sequence X_n can be supposed to be increasing, for, otherwise, we can consider the sequence: $X_1, X_1 \vee X_2, X_1 \vee X_2 \vee X_3, \ldots$. Then

$$(\hat{\mathscr{E}})\text{-}o\text{-}\lim X_n = Z.$$

In the same way it can be proved that there is a decreasing sequence

$$Y_n \in \mathscr{E}, \qquad n = 1, 2, \ldots,$$

such that

$$Z = (\hat{\mathscr{E}}) \bigwedge_{n=1}^{\infty} Y_n.$$

(This is derived from the fact that an increasing sequence

$$Z_n \in \mathscr{E}, \qquad n = 1, 2, \ldots,$$

exists such that: $(\hat{\mathscr{E}}) \bigvee_{n=1}^{\infty} Z_n = -Z$). Hence we have:

$$|Z - X_n| = Z - X_n \leqslant Y_n - X_n.$$

But

$$(\hat{\mathscr{E}}) \bigwedge_{n=1}^{\infty} (Y_n - X_n) = (\hat{\mathscr{E}}) \bigwedge_{n=1}^{\infty} Y_n - (\hat{\mathscr{E}}) \bigvee_{n=1}^{\infty} X_n = Z - Z = \theta.$$

Hence also

$$(\mathscr{E}) \bigwedge_{n=1}^{\infty} (Y_n - X_n) = \theta,$$

and moreover, $Y_n - X_n \in \mathscr{E}$, $n = 1, 2, \ldots$

is decreasing. This means that

$$\mathscr{E}\text{-}o\text{-}\lim X_n = Z.$$

Consider now $\mathscr{V}$ as a regular lattice subgroup of $\hat{\mathscr{E}}$; then we have also

$$(\mathscr{V})\text{-}o\text{-}\lim X_n = Z$$

and, since $\mathscr{V}$ is complete with respect to the o-convergence, we have $Z \in \mathscr{V}$; hence $\mathscr{V}$ is saturated and, since it is a lattice subgroup of $\hat{\mathscr{E}}$, equal to $\hat{\mathscr{E}}$.

3.10. In the particular case in which the range Σ of $\mathscr{E} \equiv \mathscr{E}(\mathfrak{B}, \Sigma)$ is a real archimeadean vector lattice, Susan Papadopoulou has given a sufficient condition that $\mathscr{V}(\mathfrak{B}, \Sigma)$ be saturated, namely:

(SP^*) *If for a subset* Ξ *of positive elements of* Σ (*i.e.*, $\Xi \subseteq \Sigma^+$) *there is an element* $\alpha \in \Sigma$ *such that for every finite subset* $\{\xi_1, \xi_2, \ldots, \xi_k\} \subseteq \Xi$ *we have*

$$\alpha \geqslant \xi_1 + \xi_2 + \ldots + \xi_k,$$

then Ξ *is at most countable.*

This condition for Σ implies the same for the $\mathscr{E}(\mathfrak{B}, \Sigma)$. Moreover, condition ($SP^*$) formulated for any archimedean lattice group $\mathscr{G}$ implies condition (SP), so that the Papangelou extension of $\mathscr{G}$ is saturated. In the following, this statement will be established. We notice first of all, that the set R of all real numbers, as also every archimedean vector lattice, which is finite-dimensional, satisfies condition (SP^*). The following lemma is true:

Lemma 3.2.

Let $\mathscr{G}$ *be an archimedean lattice group satisfying the condition* (SP^*); *then* $\mathscr{G}$ *satisfies also the condition* (SP).

Proof. In fact, let $[\mathscr{A}, \mathscr{B}]$ be a non-void cut in $\mathscr{G}$, that is, let $\mathscr{A}$ and $\mathscr{B}$ be non-void and let $\mathscr{A}$ be the set of all lower bounds of $\mathscr{B}$ and $\mathscr{B}$ the set of all upper bounds of $\mathscr{A}$. We can suppose that $b > \theta$, for every $b \in \mathscr{B}$ (for, otherwise, we may consider the cut $[\mathscr{A} - a, \mathscr{B} - a]$, where $a \in \mathscr{A}$ and $a \notin \mathscr{B}$). There is, hence, an $a \in \mathscr{A}$ with $a > \theta$. For, otherwise, we would have $\theta = (\mathscr{G})\bigwedge\{b : b \in \mathscr{B}\}$, therefore $\theta \in \mathscr{B}$. By transfinite induction we can define a family $x_i \in \mathscr{A}$, $i \in I$, with $x_i > \theta$, $i \in I$, and such that if

$$\Xi^* \equiv \{x_{i_1} + x_{i_2} + \ldots + x_{i_n} \colon\ i_k \in I, \quad k = 1, 2, \ldots, n, \text{ and } i_{k_1} \neq i_{k_2} \text{ if } k_1 \neq k_2\},$$

then
$$\Xi^* \subseteq \mathscr{A},$$

and there is no $x \in \mathscr{G}$ with $x > \theta$ and $x + \Xi^* \subseteq \mathscr{A}$. By ($SP^*$), the set of the values of the above family is at most countable; and since $\mathscr{G}$ is archimedean, each value can appear in the family finitely many times. So I is also at most countable, hence so is Ξ^*. Since $\Xi^* \subseteq \mathscr{A}$, the elements of $\mathscr{B}$ are upper bounds of Ξ^*. We assert that every bound of Ξ^* belongs to $\mathscr{B}$ In fact, let x be such a bound, and let us suppose that x does not belong to $\mathscr{B}$; then there is $a \in \mathscr{A}$ such that $a \leqslant x$ is not true, i.e.,

$$(a-x)^{+} > \theta.$$

Let y be an arbitrary element of Ξ^*; then we have

$$\begin{aligned}(a-x)^{+}+y &= [(a-x) \vee \theta]+y \\ &= (a-x+y) \vee (\theta+y) \\ &= [a-(x-y)] \vee y.\end{aligned}$$

Since $x \geqslant y$, we have $a - (x - y) \leqslant a$. Therefore: $(a - (x - y)) \in \mathscr{A}$. And, since $y \in \mathscr{A}$, we have $[a-(x-y)] \vee y \in \mathscr{A}$. Hence $(a-x)^{+} + \Xi^* \subseteq \mathscr{A}$, which is a contraction to the properties of Ξ^*. Hence Ξ^* and $\mathscr{A}$ have the same upper bounds, and since Ξ^* is at most countable, (SP) is fulfilled.

It can now be proved:

Theorem 3.4.

If Σ is a real vector lattice satisfying condition (SP^*), then the stochastic spce $\mathscr{E}(\mathfrak{B}, \Sigma)$ of all erv's on $\mathfrak{B}$ with values in Σ satisfies also condition (SP^*).

Proof. Let $\mathscr{S}(\mathfrak{B}, \Sigma)$ be the space of all simple random variables in $\mathfrak{B}$ with values in Σ; then fcr every $X \in \mathscr{S}(\mathfrak{B}, \Sigma)$ we have a representation of the form:

$$X = \sum_{i=1}^{n} \xi_i I_{a_i}, \quad \text{where} \quad \mathbf{a} = \{a_1, a_2, \dots, a_n\}$$

is a finite experiment in $\mathfrak{B}$ and $\xi_i \in \Sigma$, $i = 1, 2, \dots, n$. To every $X \in (\mathfrak{B}, \Sigma)$ we can now correspond a value

$$E(X) = \sum_{i=1}^{n} p(a_i)\, \xi_i \in \Sigma.$$

By this correspondence, a linear and strictly increasing function $E(X)$ is defined on $\mathscr{S}(\mathfrak{B}, \Sigma)$. Let $\Xi \subseteq \mathscr{E}(\mathfrak{B}, \Sigma)$ and Ξ such that $X > \theta$ for every $X \in \Xi$. Then to every $X \in \Xi$ there corresponds a representation of the form:

$$X = \sum_{i=1}^{\infty} \xi_{X,\, i} I_{a_{X,\, i}} \quad \text{with} \quad \xi_{X,\, i} > \theta, \qquad i = 1, 2, \dots .$$

Suppose, there is an erv

$$X_0 = \sum_{i=1}^{\infty} \xi_{0,\,i} I_{a_{0,\,i}}$$

where
$$a_{0,\,i} \wedge a_{0,\,j} = \emptyset, \qquad i \neq j,$$

which is an upper bound of all finite sums of pairwise different elements of Ξ. Now let us suppose that Ξ is uncountable; then there is an index i_0 and an uncountable subset Ξ_1 of Ξ such that: for every $Y \in \Xi_1$, there is an index $i(Y)$ such that $a_{Y,\,i(Y)} \wedge a_{0,\,i_0} \neq \emptyset$. We consider the family

$$X_Y \equiv \xi_{Y,\,i(Y)} I_{a_{Y,\,i(Y)} \wedge a_{0,\,i_0}}, \qquad Y \in \Xi_1.$$

We have $X_Y > \theta$, for every $Y \in \Xi_1$, and all finite sums of pairwise different terms of this family are bounded by $\xi_{0,\,i_0} I_{a_{0,\,i_0}}$. Hence, moreover, all finite sums of pairwise different terms of the family $E(X_Y)$, $Y \in \Xi_1$, are bounded by $p(a_{0,\,i_0})\,\xi_{0,\,i_0}$. Furthermore, we have $E(X_Y) > \theta$ for every $Y \in \Xi_1$. But Σ satisfies (SP^*) and is archimedean. In fact, if it is not archimeadean, there exists an $x > \theta$ and a $y \in \Sigma$ such that: $nx \leqslant y$, for every $n = 1, 2, \ldots$. But then the set $\{\lambda x, \lambda \in R\}$ is uncountable and all finite sums of pairwise different elements of it are bounded by y, which is a contradiction to the property (SP^*). Thus, as it was proved in the proof of Lemma 3.2, Ξ_1 must be at most countable, which is a contradiction. Hence the theorem is true.

3.11. A lattice group Σ is said to be of the *countable type* if and only if every bounded set of pairwise orthogonal elements is at most countable. The following theorem is true:

Theorem 3.5.

If Σ is a lattice group of the countable type, then the stochastic space $\mathscr{E}(\mathfrak{B}, \Sigma)$ of all erv's over $\mathfrak{B}$ with values in Σ is also a lattice group of the countable type.

We shall prove first the following lemmas:

Lemma 3.3.

Let $\mathfrak{A}$ be a subset of the Boolean σ-algebra $\mathfrak{B}$ with $|\mathfrak{A}| > \aleph_0$; then there is a subset $\mathfrak{A}_0 \subseteq \mathfrak{A}$ such that $|\mathfrak{A}_0| > \aleph_0$ and $a_1 \wedge a_2 \neq \emptyset$ for every pair $a_1 \in \mathfrak{A}$, $a_2 \in \mathfrak{A}_0$.

Proof. Since $|\mathfrak{A}| > \aleph_0$ and the probability is strictly positive, there is a natural number n_0 and a subset $\mathfrak{A}_1 \subseteq \mathfrak{A}$ with $|\mathfrak{A}_1| > \aleph_0$ and $p(a) > 1/n_0$ for every $a \in \mathfrak{A}_1$. Obviously the set $\mathfrak{A}_1$ satisfies the condition:

(C_1) Among any n_0 elements $a_1, a_2, \ldots, a_{n_0}$ in $\mathfrak{A}_1$ there are at least two which are not disjoint.

We shall now prove:

(C_2) There is a subset $\mathfrak{A}_0 \subseteq \mathfrak{A}_1$ such that $|\mathfrak{A}_0| > \aleph_0$ and for every pair $a_1 \in \mathfrak{A}_0$, $a_2 \in \mathfrak{A}_0$ we have $a_1 \wedge a_2 \neq \emptyset$.

We notice that, if $n_0 = 2$, then condition (C_2) follows immediately from condition (C_1) with $\mathfrak{A}_0 = \mathfrak{A}_1$. In the general case $n_0 \geqslant 3$, condition (C_1) implies the existence at least of two elements a_1, a_2 in $\mathfrak{A}_1$ with $a_1 \wedge a_2 \neq \emptyset$. If there is not a third element $a_3 \in \mathfrak{A} - \{a_1, a_2\}$ such that a_1, a_2, a_3 are pairwise not disjoint, we put $\mathfrak{A}_2 = \{a_1, a_2\}$. In the opposite case, we consider three elements a_1, a_2, a_3 in $\mathfrak{A}_1$, that are pairwise not disjoint and carry on this process to define $\mathfrak{A}_2$, in the following way: we define $\mathfrak{A}_2 = \{a_1, a_2, a_3\}$ if there is not a fourth element

$$a_4 \in \mathfrak{A}_1 - \{a_1, a_2, a_3\}$$

such that a_1, a_2, a_3, a_4 are pairwise not disjoint, otherwise the process to define $\mathfrak{A}_2$ will be carried on. In this way after a finite or transfinite number of steps a subset $\mathfrak{A}_2 \subseteq \mathfrak{A}_1$ will be defined, which must satisfy the following two conditions:

(I) for every pair $a_1 \in \mathfrak{A}_2$, $a_2 \in \mathfrak{A}_2$, we have $a_1 \wedge a_2 \neq \emptyset$, i.e., the elements of $\mathfrak{A}$ are pairwise not disjoint.

(II) there is not an element $a \in \mathfrak{A}_1 - \mathfrak{A}_2$ such that the set $\mathfrak{A}_2 \cup \{a\} \subseteq \mathfrak{A}_1$ satisfies also condition (I), i.e., $\mathfrak{A}_2$ is a maximal subset of $\mathfrak{A}_1$ satisfying condition (I).

Assume now that $|\mathfrak{A}_2| > \aleph_0$; then ($C_2$) is true with $\mathfrak{A}_0 = \mathfrak{A}_2$. If $|\mathfrak{A}_2| \leqslant \aleph_0$, i.e., $\mathfrak{A} = \{a_1, a_2, \ldots\}$, then condition (II) implies that: for every $a \in \mathfrak{A}_1 - \mathfrak{A}_2$ there is at least an $a_n \in \mathfrak{A}_2$ such that $a_n \wedge a = \emptyset$; thus if we define

$$\mathfrak{B}_n = \{b \in \mathfrak{A}_1 : b \wedge a_n = \emptyset\} \quad \text{for} \quad \text{every} \quad n = 1, 2, \ldots,$$

then, for every $a \in \mathfrak{A}_1 - \mathfrak{A}_2$, there is at least one $\mathfrak{B}_n$ such that $a \in \mathfrak{B}_n$, hence

$$\mathfrak{A}_1 - \mathfrak{A}_2 \subseteq \bigcup_{n \geqslant 1} \mathfrak{B}_n.$$

Since

$$|\mathfrak{A}_1| > \aleph_0 \text{ and } |\mathfrak{A}_2| \leqslant \aleph_0,$$

we have
$$\left|\bigcup_{n\geq 1} \mathfrak{B}_n\right| > \aleph_0;$$

hence there exists at least an index n_1 such that $|\mathfrak{B}_{n_1}| > \aleph_0$. Now, we shall prove that (C_2) is true for a subset $\mathfrak{A}_0 \subseteq \mathfrak{B}_{n_1}$. Let first be $n_0 = 3$, and consider any two elements $b_1, b_2 \in \mathfrak{B}_{n_1}$ and the element a_{n_1}; then condition (C_1) implies that for the three elements b_1, b_2, a_{n_1} in $\mathfrak{A}_1$ two of them are not disjoint; . but $b_1 \wedge a_1 = \emptyset$ and $b_2 \wedge a_1 = \emptyset$; hence b_1, b_2 ought not to be disjoint. Thus (C_2) is true if $n_0 = 3$ with $\mathfrak{A}_0 = \mathfrak{B}_{n_1}$.

Assume now (C_2) is true if $n_0 = p \geq 3$; then we shall prove: (C_2) is true if $n_0 = p+1$ with a subset $\mathfrak{A}_0 \subseteq \mathfrak{B}_{n_1}$. In order to prove it, let $b_1, b_2, \ldots, b_p$ b_ρ any elements in $\mathfrak{B}_{n_1}$; then condition (C_1), with $n_0 = \rho+1$, implies that among the $\rho+1$ elements $b_1, b_2, \ldots, b_\rho, a_{n_1}$ there are two not disjoint elements; but, according to the definition of $\mathfrak{B}_{n_1}$, we have $a_{n_1} \wedge b_j = \emptyset$, $j = 1, 2, \ldots, p$; hence the two not disjoint elements ought to be among the elements $b_1, b_2 \ldots, b_\rho$; thus the set $\mathfrak{B}_{n_1}$ instead of $\mathfrak{A}_1$, satisfies also, with $n_0 = \rho$, condition (C_1) and we have assumed that, if $n_0 = \rho$, condition (C_1) implies condition (C_2); hence there exists a subset $\mathfrak{A}_0\, \mathfrak{A}\, \mathfrak{B}_{n_1}$ with $|\mathfrak{A}_0| > \aleph_0$ such that the elements of $\mathfrak{A}_0$ are pairwise not disjoint. Thus the Lemma 3.3 is true.

Obviously, the following lemma is true:

Lemma 3.4.

Let $a_i \in \mathfrak{B}$, $i \in I$, with $|I| > \aleph_0$ and $a_i \neq \emptyset$, $i \in I$. Then there is a subset $J \subseteq I$ with $|J| > \aleph_0$ such that the members a_j of the subfamily a_j, $j \in J$, are pairwise not disjoint.

Proof of Theorem 3.5. Let $X_i \in \mathscr{E}(\mathfrak{B}, \Sigma)$ with $X_i \neq \theta$, $i \in I$, be any bounded family of pairwise orthogonal erv's. It can be assumed that $X_i > \theta$, $i \in I$ (for otherwise we can consider the family $|X_i|$, $i \in I$); then for every $i \in I$, there is an element $\xi_i \in \Sigma$, $\xi_i > \theta$ and an element $a_i \in \mathfrak{B}$, $a_i \neq \emptyset$, such that $\xi_i I_{a_i} \leq X_i$. We put $Y_i = \xi_i I_{a_i}$. Then the family $Y_i \in \mathscr{E}(\mathfrak{B}, \Sigma)$, $i \in I$, is bounded with pairwise orthogonal members. Assume that $|I| > \aleph_0$ and consider the family $a_i \in \mathfrak{B}$, $i \in I$; then Lemma 3.4 implies the existence of a subset $J \subseteq I$ with $|J| > \aleph_0$ and such that the members a_j of the subfamily $a_j \in \mathfrak{B}$, $j \in J$, are pairwise not disjoint. But then, for every $i_1 \in J$, $i_2 \in J$, with $i_1 \neq i_2$, $Y_{i_1} \wedge Y_{i_2} = \theta$ and $a_{i_1} \wedge a_{i_2} \neq \emptyset$ imply $\xi_{i_1} \wedge \xi_{i_2} = \theta$.

Let now
$$Y: \mathbf{a} \ni a_j \Rightarrow \eta_j \in \Sigma, \quad j \geq 1,$$

be an upper bound of Y_i, $i \in I$, in $\mathscr{E}(\mathfrak{B}, \Sigma)$. Then there exists a $j_0 \geqslant 1$, and a $J_1 \subseteq J$ with $|J_1| > \aleph_0$, such that: $a_{j_0} \wedge a_i \neq \emptyset$ for every $i \in J_1$. Then, obviously, the family $\xi_j \in \Sigma$, $j \in J_1$, is bounded by η_{j_0} and the elements ξ_j are pairwise disjoint, i.e., $|J_1|$ cannot be $> \aleph_0$ (contradiction!). The theorem is hence true.

3.12. R. Cristescu ([1], Chapter III, Theorem 2.43, compare also Vulikh [1], Theorem VI 2.2) has proved:

(C) If a vector lattice Σ is saturated and of the countable type, then every upper bounded (resp lower bounded) subset of Σ has a countable subset A' such that $\sup A' = \sup A$ (resp $\inf A' = \inf A$).

From this theorem, the following theorem can be derived:

Theorem 3.6.

If Σ is an archimedean vector lattice of the countable type, Σ_Σ the Papangelou extension of Σ, then Σ is saturated, i.e., Σ_Σ is isomorphic to the MacNeille extension $\hat{\Sigma}$ of Σ.

Proof. We notice that the MacNeille extension $\hat{\Sigma}$ of Σ is also of the countable type. In fact, let T be a bounded subset of $\hat{\Sigma}$ with pairwise orthogonal elements. Then, for every $\xi \in T$ with $\xi \neq \theta$, there exists an $\eta_\xi \in \Sigma$ with $\theta < \eta_\xi \leqslant |\xi|$. The set $\Gamma = \{\eta \in \Sigma \colon \eta = \eta_\xi, \xi \in T\}$ is a bounded subset of Σ with pairwise orthogonal elements; hence Γ is at most countable and the same is true for T. Thus the previous theorem of Cristescu can be applied to $\hat{\Sigma}$ and it follows that Σ satisfies the condition (SP); therefore Σ_Σ is saturated, i.e., isomorphic to $\hat{\Sigma}$.

The following theorem is now true:

Theorem 3.7.

Let Σ be an archimedean vector lattice of the countable type; then the Papangelou extension of the vector lattice $\mathscr{E}(\mathfrak{B}, \Sigma)$ of all erv's over $\mathfrak{B}$ with values in Σ, is a saturated vector lattice.

Proof. Since Σ is of the countable type, the vector lattice $\mathscr{E}(\mathfrak{B}, \Sigma)$ is also of the countable type. Therefore its Papangelou extension is saturated.

4. COMPLETION WITH RESPECT TO A NORM

4.1. Let Σ be a *normed vector space* over a scalar field F†. The norm in Σ will be denoted by $\|\xi\|$, $\xi \in \Sigma$, and possesses the usual properties:

N_1. $\|\xi\| \geqslant 0$, and $\|\xi\| = 0$ if and only if $\xi = \theta$,

N_2. $\|\lambda\xi\| = |\lambda|\,\|\xi\|$, where $|\lambda|$ denotes the absolute value of the scalar λ,

N_3. $\|\xi+\eta\| \leqslant \|\xi\|+\|\eta\|$.

Then $d(x, y) = \|\xi-\eta\|$ defines a metric in Σ. If Σ is complete in the resulting by this metric topology, then Σ is said to be a *Banach space*, *B-space* for short. From these properties it follows that the norm $\|\xi\|$, as a real function defined on Σ, is continuous. If a commutative multiplication $\xi\eta$ is defined in Σ with the usual properties:

(1) $(\xi\eta)\zeta = \xi(\eta\zeta)$,

(2) $\xi(\eta+\zeta) = \xi\eta+\xi\zeta$,

(3) $\lambda_1\xi_1 . \lambda_2\xi_2 = (\lambda_1\lambda_2)\xi_1\xi_2$,

(4) there exists a unit $\mathbf{1} \in \Sigma$ such that $\mathbf{1}.\xi = \xi$ for all $\xi \in \Sigma$,

then Σ is said to be a *normed commutative algebra with unit*; then the norm ought to possess moreover the properties:

N_4. $\|\xi\eta\| \leqslant \|\xi\|\,\|\eta\|$,

N_5. $\|\mathbf{1}\| = 1$.

If, moreover, Σ as a normed vector space is a B-space, then Σ is said to be a *B-algebra*.

It is easy to see that $\xi+\eta$ and $\xi\eta$ as functions of both variables together are metrically continuous. In particular, $\xi+\eta$ is uniformly continuous. The most known examples of B-spaces are also vector lattices. In a *normed vector lattice* we shall assume that the ordering and the norm are related by the following condition:

(O_1) $|\xi| \leqslant |\eta|$ implies $\|\xi\| \leqslant \|\eta\|$, where $|\xi|$ denotes the absolute value of ξ, i.e., $|\xi| = \xi^+ + \xi^-$.

The B-space will be called in this case a *B-lattice*. It is easy to see that:

(O_2) $\|\xi\| = \|\,|\xi|\,\|$ for all $\xi \in \Sigma$,

and moreover that the operations $\xi \wedge \eta$ and $\xi \vee \eta$, as functions of both variables together, are metrically uniformly continuous. In any Banach

† F is considered as the field R of the real numbers or the field C of the complex numbers.

lattice, metric convergence is equivalent to the ru^*-convergence.† In fact, we need only consider convergence to θ. Now, if $|\xi_n| \leqslant \lambda_n \eta$ with $\lambda_n \downarrow 0$ in R, then clearly

$$\|\xi_n\| \leqslant \|\lambda_n \eta\| = |\lambda_n| \, \|\eta\| \to 0.$$

Hence ru^*-convergence implies metric *-convergence and so metric convergence. Conversely, if $\|\xi_n\| \to 0$, we can choose a subsequence $n(k)$ so that

$$\|\xi_{n(k)}\| < \frac{1}{k^3},$$

and then define

$$\eta = \sum_{k=1}^{\infty} k|\xi_{n(k)}|,$$

where the series convergence is metric convergence; then

$$|\xi_{n(k)}| \leqslant \frac{1}{k} \eta,$$

i.e., the subsequence $\xi_{n(k)}$, $k = 1, 2, \ldots$, converges relatively uniformly to zero. Hence metric convergence implies ru^*-convergence.

But ru^*-convergence implies o^*-convergence. Hence, in any B-lattice, if a sequence ξ_n, $n = 1, 2, \ldots$, metric-converges to ξ, then there exists a subsequence which o-converges to ξ.

A normed vector lattice is always archimedean. In fact, it is sufficient to prove $(1/\nu)\xi \overset{o}{\to} \theta$. But obviously, we have $1/\nu \, \|\xi\| \to 0$, and this implies $(1/\nu)\xi \overset{o}{\to} \theta$. For, more generally, we have:

(I) $\xi_n \in \Sigma$, $n = 1, 2, \ldots$, and monotone decreasing with $\|\xi_n\| \to 0$ implies always the existence of $\bigwedge_{n=1}^{\infty} \xi_n = \theta$, i.e., $\xi_n \overset{o}{\downarrow} \theta$ in Σ. Because the two other possibilities: there not exists $\bigwedge_{n=1}^{\infty} \xi_n$ in Σ or there exists in Σ and is $\neq \theta$, lead to the contradiction that $\inf \|\xi_n\| > 0$. Therefore we define:

A normed vector lattice Σ is called *continuous* if and only if:

(C) $\xi_n \in \Sigma$, $n = 1, 2, \ldots$, and monotone decreasing with $\bigwedge_{n=1}^{\infty} \xi_n = \theta$ implies $\|\xi_n\| \to 0$.

---

† About this convergence see Section 9.2., Chapter IV.

Order-boundedness in a normed vector lattice clearly implies metric boundedness. The converse is not however usually true. A σ-saturated normed and continuous vector lattice Σ is said to be a *K-B-lattice* if and only if Σ satisfies the following condition:

(K) if $\xi_n \in \Sigma$, $n = 1, 2, \ldots$, is increasing and there is an $\alpha \in R$ such that $\|\xi_n\| \leqslant \alpha$, then the sequence ξ_n is bounded in Σ.

It is easy to prove that a K-B-lattice is always a B-lattice. But there are Banach lattices which are not σ-saturated; for example the Banach lattice of all continuous real functions defined on the closed interval $[\alpha, \beta]$ with respect to the norm $\|f\| = \sup_{x \in [\alpha, \beta]} |f(x)|$. We shall prove now:

Theorem 4.1.

A σ-saturated normed and continuous vector lattice Σ is of the countable type.

Proof. Let T be a bounded subset of Σ with pairwise orthogonal non-zero elements. We put

$$T_n = \left\{\xi \in T : \|\xi\| > \frac{1}{n}\right\}.$$

Obviously

$$T = \bigcup_{n=1}^{\infty} T_n.$$

Let $|T| > \aleph_0$; then there is an index n_0, such that $|T_{n_0}| \geqslant \aleph_0$. Then there is a subset of T_{n_0} of the form: $\{\tau_1, \tau_2, \ldots\}$; but then the series $\sum_{n=1}^{\infty} \tau_n$ o-converges, and this implies $\tau_n \xrightarrow{o} \theta$; the continuity of Σ implies then $\|\tau_n\| \to 0$, i.e., a contradiction; for $\|\tau_n\| > 1/n_0$, $n = 1, 2, \ldots$. Hence every T_n is finite and then T is countable.

Theorem 4.2.

In a K-B-lattice Σ the following statements are equivalent:

(I) $o\text{-}\lim \xi_n = \theta$ *in* Σ,

(II) $\lim_{n, m \to \infty} \| |\xi_n| \vee |\xi_{n+1}| \vee \ldots \vee |\xi_m| \| = 0.$

Proof. Suppose $\xi_n \overset{o}{\to} \theta$; then

$$\eta_n = \sup_{k \geqslant n} |\xi_k|$$

is decreasing and

$$o\text{-}\lim \eta_n = \theta,$$

hence $$\|\eta_n\| \to 0.$$

But since the norm is monotone we have

$$\| \, |\xi_n| \vee |\xi_{n+1}| \vee \ldots \vee |\xi_m| \, \| \leqslant \|\eta_n\|,$$

hence (II) is true. Conversely, if (II) is satisfied, then for any $\varepsilon > 0$ there is an index N_ε such that for $n \geqslant N_\varepsilon$, we have

$$\| \, |\xi_n| \vee |\xi_{n+1}| \vee \ldots \vee |\xi_m| \, \| < \varepsilon$$

for every $m \geqslant n$; hence, by condition (K), it follows that η_n exists; since the norm is continuous with respect to o-convergence, it follows that $\|\eta_n\| \leqslant \varepsilon$ for $n \geqslant N_\varepsilon$. Thus $\|\eta_n\| \to 0$. But $\eta_n \overset{o}{\downarrow} o\text{-}\limsup |\xi_n|$ and therefore by the monotony of the norm we have

$$o\text{-}\limsup |\xi_n| = \theta,$$

i.e., $$o\text{-}\lim \xi_n = \theta.$$

4.2. Let Σ be a Banach space over the scalar field F; then the space $\mathscr{E} = \mathscr{E}(\mathfrak{B}, \Sigma)$ of all erv's over $\mathfrak{B}$ with values in Σ is a vector space. To every element $X \in \mathscr{E}$, i.e.,

$$X : \mathbf{a} \ni a_j \Rightarrow X(a_j) = \xi_j \in \Sigma, \qquad j \geqslant 1,$$

there corresponds a non-negative real-valued erv:

$$\phi(X) : \mathbf{a} \ni a_j \Rightarrow \|X(a_j)\| = \|\xi_j\| \in R, \qquad j \geqslant 1,$$

called *norm-valuation* of X in $\mathscr{E}(\mathfrak{B}, R)$.

Obviously, we have:

(I) $\phi(X) \geqslant 0$, and $\phi(X) = 0$ if and only if $X = \theta$,

(II) $\phi(\lambda X) = |\lambda| \, \phi(X)$, if $\lambda \in F$,

(III) $\phi(X + Y) \leqslant \phi(X) + \phi(Y)$.

If Σ is a B-algebra, then $\mathscr{E}(\mathfrak{B}, \Sigma)$ is an algebra over the scalar field F and we have:

(IV) $\phi(XY) \leqslant \phi(X) \, \phi(Y)$.

If Σ is a B-lattice, then $\mathscr{E}(\mathfrak{B}, \Sigma)$ is a vector lattice and an absolute value $X = X^+ + X^-$ can be defined for every $X \in \mathscr{E}(\mathfrak{B}, \Sigma)$; then the norm valuation and the ordering relation $\leqslant$ are related by the following condition:

(V) $|X| \leqslant |Y|$ implies $\phi(X) \leqslant \phi(Y)$.

Moreover we have

(VI) $\phi(X) = \phi(|X|)$,

(VII) $\phi(X \vee Y) \leqslant \phi(X) + \phi(Y)$.

4.3. The so-called *norm o-convergence* (briefly n-o-convergence) may be introduced in $\mathscr{E}(\mathfrak{B}, \Sigma)$ with the aid of the norm valuation:

A sequence $X_n \in \mathscr{E}(\mathfrak{B}, \Sigma)$, $n = 1, 2, \ldots$, is called *n-o-convergent* to X in $\mathscr{E}(\mathfrak{B}, \Sigma)$, denoted by

$$X_n \underset{\mathscr{E}(\mathfrak{B},\, \Sigma)}{\overset{n\text{-}o}{\rightarrow}} X \quad \text{if and only if} \quad \phi(X_n - X) \underset{\mathscr{E}(\mathfrak{B},\, R)}{\overset{o}{\rightarrow}} 0.$$

A sequence $X_n \in \mathscr{E}(\mathfrak{B}, \Sigma)$, $n = 1, 2, \ldots$, is called *n-o-fundamental* if and only if

$$\phi(X_n - X_k) \underset{\mathscr{E}(\mathfrak{B},\, R)}{\overset{o}{\rightarrow}} 0.$$

4.4. We may now establish with the aid of the n-o-convergence a completion process similar to that of Section 3, leading to a space of rv's $\mathscr{V}^*(\mathfrak{B}, \Sigma)$, which is complete with respect to all convergences used in the probability theory. In order to do it, let $\mathscr{M}$, $\mathscr{K}$, $\mathscr{N}$ be the set of all n-o-fundamental, all n-o-convergent and all sequences which n-o-converge to the zero element of $\mathscr{E}(\mathfrak{B}, \Sigma)$ respectively; then we have $\mathscr{M} \supseteq \mathscr{K} \supseteq \mathscr{N}$. Let now $\{X_n\}$ and $\{Y_n\}$ belong to $\mathscr{M}$ or $\mathscr{K}$ or $\mathscr{N}$ respectively; then $\{X_n + Y_n\}$, $\{\lambda X_n\}$ with $\lambda \in F$, and if Σ is a Banach algebra $\{X_n\, Y_n\}$ belong to $\mathscr{M}$, or $\mathscr{K}$, or $\mathscr{N}$ respectively. Moreover, if Σ is a Banach algebra and only one of $\{X_n\}$ and $\{Y_n\}$ belongs to $\mathscr{N}$, then $\{X_n\, Y_n\}$ belongs also to $\mathscr{N}$; hence we can define the following operations:

$$\{X_n\} + \{Y_n\} = \{X_n + Y_n\}, \quad \lambda\{X_n\} = \{\lambda X_n\},$$

if Σ a Banach space, moreover

$$\{X_n\}\{Y_n\} = \{X_n\, Y_n\},$$

if Σ is a Banach algebra, and prove that $\mathscr{M}$ is a vector space, moreover an algebra respectively. $\mathscr{K}$ is a vector subspace and $\mathscr{N}$ is an ideal in $\mathscr{M}$ and in $\mathscr{K}$. The quotient space $\mathscr{V}^*(\mathfrak{B}, \Sigma) = \mathscr{M}/\mathscr{N}$ of all classes $\{X_n\}/\mathscr{N}$ constitutes a vector space or an algebra, over the scalar field F. $\mathscr{K}/\mathscr{N}$ is then a vector subspace of $\mathscr{M}/\mathscr{N}$ and can be identified with $\mathscr{E}(\mathfrak{B}, \Sigma)$. We shall call $\mathscr{V}^*(\mathfrak{B}, \Sigma)$ the stochastic space of all rv's valued in Σ over the pr σ-algebra $(\mathfrak{B}, p)$ and we shall regard $\mathscr{E}(\mathfrak{B}, \Sigma)$ as the stochastic subspace of $\mathscr{V}^*(\mathfrak{B}, \Sigma)$ consisting of all erv's. We shall denote the elements of $\mathscr{V}^*(\mathfrak{B}, \Sigma)$ also by capital letters $X, Y, Z, \ldots$.

The norm valuation ϕ can be extended from $\mathscr{E}(\mathfrak{B}, \Sigma)$ on $\mathscr{V}^*(\mathfrak{B}, \Sigma)$. Namely, since the inequality

$$|\phi(X)-\phi(Y)| \leqslant \phi(X-Y)$$

is true in $\mathscr{E}(\mathfrak{B}, \Sigma)$, it can be proved that, if $\{X_n\} \in \mathscr{M}$, then $\phi(X_n)$, $n = 1, 2, \ldots$, is an o-fundamental sequence in $\mathscr{E}(\mathfrak{B}, R)$ and defines a real valued rv in $\mathscr{V}(\mathfrak{B}, R)$. If, moreover, $\{Y_n\} \in \{X_n\}/\mathscr{N}$, i.e.,

$$X_n - Y_n \underset{\mathscr{E}(\mathfrak{B}, \Sigma)}{\overset{n\text{-}o}{\rightarrow}} \theta,$$

then
$$o\text{-}\lim \phi(X_n) = o\text{-}\lim \phi(Y_n) \quad \text{in} \quad \mathscr{V}(\mathfrak{B}, R).$$

Hence we can correspond to every class $\{X_n\}/\mathscr{N}$ a uniquely defined by

$$\phi(\{X_n\}/\mathscr{N}) \equiv o\text{-}\lim \phi(X_n)$$

norm valuation, which is an extension of the norm valuation ϕ from $\mathscr{E}(\mathfrak{B}, \Sigma)$ on $\mathscr{V}^*(\mathfrak{B}, \Sigma)$ and preserves all properties (I)–(VII) of Section 4.2. Hence, we can define an n-o-convergence in $\mathscr{V}^*(\mathfrak{B}, \Sigma)$ and prove that $\mathscr{V}^*(\mathfrak{B}, \Sigma)$ is then closed (complete) with respect to it, i.e., every n-o-fundamental sequence in $\mathscr{V}^*(\mathfrak{B}, \Sigma)$ is n-o-convergent. The stochastic subspace $\mathscr{E}(\mathfrak{B}, \Sigma)$ is n-o-dense in $\mathscr{V}^*(\mathfrak{B}, \Sigma)$. The inequality

$$|\phi(X)-\phi(Y)| \leqslant \phi(X-Y),$$

which is also true in $\mathscr{V}^*(\mathfrak{B}, \Sigma)$ implies: if

$$X_n \overset{n\text{-}o}{\rightarrow} X \quad \text{in} \quad \mathscr{V}^*(\mathfrak{B}, \Sigma)$$

then

$$\phi(X_n) \underset{\mathscr{V}(\mathfrak{B}, R)}{\overset{o}{\rightarrow}} \phi(X).$$

4.5. We may now introduce in $\mathscr{V}^*(\mathfrak{B}, \Sigma)$ the following kinds of convergence:

(u) *Norm uniform convergence* (*n-u-convergence* for short): We say the sequence X_n, $n = 1, 2, \ldots$, *n-u*-converges to X in $\mathscr{V}^*(\mathfrak{B}, \Sigma)$ and denote it by

$$X_n \overset{n\text{-}u}{\rightarrow} X \quad \text{in} \quad \mathscr{V}^*$$

if and only if

$$\phi(X_n - X) \overset{u}{\rightarrow} 0 \quad \text{in} \quad \mathscr{V}(\mathfrak{B}, R).$$

(p) *Norm probability convergence* (*n-p-convergence* for short): We say the sequence X_n, $n = 1, 2, \ldots$, *n-p*-converges to X in $\mathscr{V}^*$ and denote by

$$X_n \overset{n\text{-}p}{\rightarrow} X \quad \text{in} \quad \mathscr{V}^*$$

if and only if

$$\phi(X_n - X) \overset{p}{\rightarrow} 0 \quad \text{in} \quad \mathscr{V}(\mathfrak{B}, R).$$

(a) *Norm almost uniform convergence* (*n-au-convergence* for short): We say the sequence X_n, $n = 1, 2, \ldots$, *n-au*-converges to X in $\mathscr{V}^*$ and denote it by

$$X_n \overset{n\text{-}au}{\rightarrow} X \quad \text{in} \quad \mathscr{V}^*$$

if and only if

$$\phi(X_n - X) \overset{au}{\rightarrow} 0 \quad \text{in} \quad \mathscr{V}(\mathfrak{B}, R).$$

(r) *Norm relative uniform convergence* (*n-ru-convergence* for short): We say the sequence X_n, $n = 1, 2, \ldots$, *n-ru*-converges to X in $\mathscr{V}^*$ and denote it by

$$X_n \overset{n\text{-}ru}{\rightarrow} X \quad \text{in} \quad \mathscr{V}^*$$

if and only if

$$\phi(X_n - X) \overset{n\text{-}ru}{\rightarrow} 0 \quad \text{in} \quad \mathscr{V}(\mathfrak{B}, R).$$

In the usual way, we may now define for all these types of convergence, the correspondent so-called star convergence, i.e., *n-o**-convergence, *n-u**-convergence, *n-p**-convergence, *n-au**-convergence and *n-ru**-convergence.

On can prove in an analogous way as in Section 7, Chapter IV, that the previous types of convergence are related as follows:

(1) The *n-o*-convergence is equivalent to the *n-au*-convergence and implies the *n-p*-convergence.

(2) The n-p-convergence does not imply n-o-convergence; however, if $X_n \xrightarrow{n\text{-}p} X$, then there exists a subsequence X_{k_n}, $n = 1, 2, \ldots$ such that

$$X_{k_n} \xrightarrow{n\text{-}o} X.$$

It is easy to prove that the set $\mathscr{S}(\mathfrak{B}, \Sigma)$ of all simple rv's with values in Σ is n-o-dense in $\mathscr{V}^*(\mathfrak{B}, \Sigma)$ but not always n-u-dense in $\mathscr{V}^*(\mathfrak{B}, \Sigma)$.

4.6. If the space Σ of the values is a Banach lattice, in which case $\mathscr{E}(\mathfrak{B}, \Sigma)$ is a vector lattice, then an o-convergence can be introduced in $\mathscr{E}(\mathfrak{B}, \Sigma)$. Assume that the scalar field F is the field R of real numbers, and let $\eta > \theta$ be an element of Σ with $\|\eta\| = 1$; then the set

$$R\eta \equiv \{\xi \in \Sigma : \xi = \lambda\eta, \quad \lambda \in R\}$$

is a Banach sublattice of Σ and moreover regular with respect to the ordering relation; obviously $R\eta$ is isomorphic to R. Hence $\mathscr{E}(\mathfrak{B}, R\eta)$ is isomorphic to the stochastic space $\mathscr{E}(\mathfrak{B}, R)$ of all real valued erv's over $(\mathfrak{B}, p)$ and even a regular, with respect to the ordering relation, vector sublattice of $\mathscr{E}(\mathfrak{B}, \Sigma)$. Hence the vector lattice $\mathscr{E}(\mathfrak{B}, R)$ may be embedded regularly in $\mathscr{E}(\mathfrak{B}, \Sigma)$ and be considered as a vector sublattice of $\mathscr{E}(\mathfrak{B}, \Sigma)$. In this case, we can complete $\mathscr{E}(\mathfrak{B}, \Sigma)$ with respect to the o-convergence and define its completion $\mathscr{V}(\mathfrak{B}, \Sigma)$ (see Section 3.5). If the norm in Σ is continuous and Σ is σ-saturated, then Σ is of the countable type (see Theorem 4.1); moreover Σ is archimedean, hence (see Theorem 3.7) $\mathscr{V}(\mathfrak{B}, \Sigma)$ is a saturated vector lattice.

A comparison of the n-o-convergence with the o-convergence in $\mathscr{E}(\mathfrak{B}, \Sigma)$, if Σ is a Banach lattice, and the discovery of a relation between the two completion processes of $\mathscr{E}(\mathfrak{B}, \Sigma)$ with respect to these two convergences must be an interesting problem but seem to be still open. A completion process of $\mathscr{E}(\mathfrak{B}, \Sigma)$ in the general case in which Σ is a topological vector space could also be stated here. But to this purpose, we need advanced knowledge of general topology and the theory of topological vector spaces. Moreover, a comparison of such a completion to the completion with respect to the o-convergence, when Σ is a topological vector lattice, seems also to be an open problem. In the following sections we shall be restricted to the introduction of the concepts of expectation and moments of rv's with values in a Banach space with the help of the Bochner integration theory.

5. EXPECTATION OF RV'S WITH VALUES IN A BANACH SPACE

5.1. The theory of expectation and moments of real-valued rv's, that is stated in Chapters V and VI, may be generalized on rv's with values in a Banach space. Let $(\mathfrak{B}, p)$ be a pr σ-algebra and Σ a Banach space. Let $\mathscr{E}(\mathfrak{B}, \Sigma)$ be the stochastic space of all erv's over $\mathfrak{B}$ with values in Σ and $\mathscr{V}^*(\mathfrak{B}, \Sigma)$ the completion of $\mathscr{E}(\mathfrak{B}, \Sigma)$ with respect to the n-o-convergence. $\mathscr{S}(\mathfrak{B}, \Sigma)$ will denote the stochastic space of all simple rv's over $\mathfrak{B}$ with values in Σ. We shall consider $\mathscr{S}(\mathfrak{B}, \Sigma)$ and $\mathscr{E}(\mathfrak{B}, \Sigma)$ as vector subspaces of $\mathscr{V}^*(\mathfrak{B}, \Sigma)$. $\mathscr{S}(\mathfrak{B}, R)$ resp $\mathscr{E}(\mathfrak{B}, R)$ resp $\mathscr{V}(\mathfrak{B}, R)$ will denote the stochastic spaces of the real-valued simple resp elementary resp all rv's over $\mathfrak{B}$. In order to define the expectation and the moments, we shall follow the Bochner integration theory; the theory stated in Chapters V and VI is then a particular case of this general theory.

5.2. Let X be a simple rv, i.e.,

$$X \in \mathscr{S}(\mathfrak{B}, \Sigma) \quad \text{and} \quad X = \sum_{i=1}^{k} \xi_i I_{a_i}$$

be a representation of X by indicators (see Section 3.2); then we shall call *expectation* $E(X)$ of X in Σ the element of Σ defined by the equality

$$E(X) = \sum_{i=1}^{k} p(a_i)\, \xi_i.$$

It is easy to prove that the so defined expectation of X in Σ is independent of the particular representation of X by indicators.

According to this definition, we have

(I) $E(\xi I_a) = p(a)\, \xi$, for every $a \in \mathfrak{B}$ and $\xi \in \Sigma$.

The following properties may easily be proved:

1. The expectation E is a linear mapping of $\mathscr{S}(\mathfrak{B}, \Sigma)$ into Σ, i.e., we have

$$E(\alpha X + \beta Y) = \alpha E(X) + \beta E(Y)$$

with α and β scalars and $X \in \mathscr{S}(\mathfrak{B}, \Sigma)$, $Y \in \mathscr{S}(\mathfrak{B}, \Sigma)$.

2. If Σ is a Banach lattice, hence $\mathscr{S}(\mathfrak{B}, \Sigma)$ is a vector lattice, then E is strictly monotone, i.e., if $X > Y$, then $E(X) > E(Y)$.

3. For every

$$X \in \mathscr{S}(\mathfrak{B}, \Sigma) \quad \text{with} \quad X = \sum_{i=1}^{k} \xi_i I_{a_i},$$

we have

$$\phi(X) = \sum_{i=1}^{k} \|\xi_i\| \, I_{a_i} \in \mathscr{S}(\mathfrak{B}, R^+),$$

hence

$$E(\phi(X)) = \sum_{i=1}^{k} p(a_i) \, \|\xi_i\| \in R^+,$$

and obviously we have:

(II) $\|E(X)\| \leqslant E(\phi(X))$.

5.3. We define:

(III) $\|X\|_1 = E(\phi(X))$, for every $X \in \mathscr{S}(\mathfrak{B}, \Sigma)$.

Then $\| \; \|_1$ is a norm on $\mathscr{S}(\mathfrak{B}, \Sigma)$. In fact we have

(α) $\|X\|_1 \geqslant 0$, and $\|X\|_1 = 0$ if and only if $X = \theta$.

(β) $\|X+Y\|_1 \leqslant \|X\|_1 + \|Y\|_1$.

(γ) $\|\lambda X\|_1 = |\lambda| \, \|X\|_1$,

and if Σ is a Banach lattice then

(δ) $\|X\|_1 \leqslant \|Y\|_1$ if $|X| \leqslant |Y|$.

The inequality (II) can now be written as follows:

($\widetilde{\text{II}}$) $\|E(X)\| \leqslant \|X\|_1$.

The norm $\| \; \|_1$ defines a distance

$$\delta(X, Y) = \|X - Y\|_1, \qquad X \in \mathscr{S}(\mathfrak{B}, \Sigma), \qquad Y \in \mathscr{S}(\mathfrak{B}, \Sigma),$$

which induces in $\mathscr{S}(\mathfrak{B}, \Sigma)$ a metric convergence, i.e., a topology, called the topology of the *convergence in mean.* Instead of convergence in mean, we shall say also *norm*$_1$*-convergence* (briefly N_1*-convergence*) and denote by

$$X_n \overset{N_1}{\rightarrow} X \quad \text{in} \quad \mathscr{S}(\mathfrak{B}, \Sigma);$$

in other words we define

$$X_n \overset{N_1}{\rightarrow} X \quad \text{in} \quad \mathscr{S}(\mathfrak{B}, \Sigma)$$

if and only if $\|X_n - X\|_1 \rightarrow 0$.

The inequality ($\widetilde{\text{II}}$) implies:

(IV) *The expectation* $E(X)$, $X \in \mathscr{S}(\mathfrak{B}, \Sigma)$, *is continuous* for the convergence in mean.

5.4. A sequence $X_n \in \mathscr{S}(\mathfrak{B}, \Sigma)$, $n = 1, 2, \ldots$, is N_1-fundamental if and only if the double sequence $\|X_n - X_m\|_1$ converges to zero, i.e., if

$$\lim_{(n, m) \to (\infty, \infty)} E(\phi(X_n - X_m)) = 0.$$

Theorem 5.1.

If the sequence $X_n \in \mathscr{S}(\mathfrak{B}, \Sigma)$, $n = 1, 2, \ldots$, *is* N_1*-fundamental, then*

(1) *the sequence* $\phi(X_n) \in \mathscr{S}(\mathfrak{B}, \Sigma)$, $n = 1, 2, \ldots$, *is* N_1*-fundamental,*

(2) *the sequence* $E(X_n) \in \Sigma$, $n = 1, 2, \ldots$, *is fundamental with respect to the norm convergence in* Σ.

Proof. We have

$$|\phi(X_n) - \phi(X_m)| \leqslant \phi(X_n - X_m),$$

hence

$$\| \, |\phi(X_n) - \phi(X_m)| \, \|_1 \leqslant E(\phi(X_n - X_m)) = \|X_n - X_m\|_1.$$

This inequality implies (1). From

$$\|E(X_n) - E(X_m)\| = \|E(X_n - X_m)\| \leqslant E(\phi(X_n - X_m)) = \|X_n - X_m\|_1$$

follows then (2).

5.5. Let

$$X \in \mathscr{S}(\mathfrak{B}, \Sigma) \quad \text{and} \quad X = \sum_{i=1}^{k} \xi_i I_{b_i}$$

be a representation of X by indicators; then XI_a is defined by the equality

$$XI_a = \sum_{i=1}^{k} \xi_i I_{b_i \wedge a} \quad \text{for every} \quad a \in \mathfrak{B}.$$

For every $X \in \mathscr{S}(\mathfrak{B}, \Sigma)$ we may define:

$$\psi_X(a) = E(XI_a), \qquad a \in \mathfrak{B}.$$

It is easy to prove that:

(V) The function $\psi_X(a)$ is additive, i.e.,

$$\psi_X(a \vee b) = \psi_X(a) + \psi_X(b) \quad \text{if} \quad a \wedge b = \emptyset.$$

Theorem 5.2.

If $X_n \in \mathscr{S}(\mathfrak{B}, \Sigma)$, $n = 1, 2, \ldots$, *is an* N_1*-fundamental sequence then*

(1) *The sequence* $X_n I_a$, $n = 1, 2, \ldots$, *is* N_1*-fundamental for every* $a \in \mathfrak{B}$,

(2) *The sequence* $\psi_{X_n}(a)$, $n = 1, 2, \ldots$, *is a fundamental sequence with respect to the norm convergence in* Σ *and defines by*:

$\psi(a) = \lim \psi_{X_n}(a)$, *for every* $a \in \mathfrak{B}$, *an additive function* ψ *on* $\mathfrak{B}$ *with values in* Σ.

Proof. We have

$$\phi(X_n I_a - X_m I_a) = \phi((X_n - X_m) I_a) \leqslant \phi(X_n - X_m);$$

hence

$$\|(X_n I_a - X_m I_a)\|_1 \leqslant E(\phi(X_n - X_m)) = \|X_n - X_m\|_1.$$

This inequality implies (1). From (1) of this theorem and (2) of Theorem 5.1, it follows that $\psi_{X_n}(a)$, $n = 1, 2, \ldots$, is norm-fundamental; hence there exists

$$\lim_{n \to \infty} \psi_{X_n}(a) = \psi(a) \in \Sigma$$

for every $a \in \mathfrak{B}$, and we have for $a, b \in \mathfrak{B}$, with $a \wedge b = \emptyset$:

$$\begin{aligned} \psi(a \vee b) &= \lim_{n \to \infty} \psi_{X_n}(a \vee b) \\ &= \lim_{n \to \infty} \{\psi_{X_n}(a) + \psi_{X_n}(b)\} \\ &= \psi(a) + \psi(b); \end{aligned}$$

i.e., (2) is also true.

Theorem 5.3.

For every $X \in \mathscr{S}(\mathfrak{B}, \Sigma)$ *the function* $\psi_X(a) = E(XI_a)$, $a \in \mathfrak{B}$, *is* σ-*additive, i.e., for every sequence of pairwise disjoint elements* $a_i \in \mathfrak{B}$, $i = 1, 2, \ldots$, *we have*

$$\psi_X\left(\bigvee_{i=1}^{\infty} a_i\right) = \sum_{i=1}^{\infty} \psi_X(a_i) \quad in \quad \Sigma.$$

Proof. Let

$$X = \sum_{j=1}^{k} \xi_j I_{b_j}$$

be a representation of X by indicators and put

$$a = \bigvee_{i=1}^{\infty} a_i;$$

then

$$XI_a = \sum_{j=1}^{k} \xi_j I_{b_j \wedge a}$$

and, for every $i = 1, 2, \ldots$, we have

$$XI_{a_i} = \sum_{j=1}^{k} \xi_j I_{b_j \wedge a_i};$$

therefore

$$\begin{aligned}
\psi_X(a) = E(XI_a) &= \sum_{j=1}^{k} p(a \wedge b_j)\, \xi_j \\
&= \sum_{j=1}^{k} p\left(\left(\bigvee_{i=1}^{\infty} a_i\right) \wedge b_j\right) \xi_j \\
&= \sum_{j=1}^{k} p\left(\bigvee_{i=1}^{\infty} (a_i \wedge b_j)\right) \xi_j \\
&= \sum_{j=1}^{k} \sum_{i=1}^{\infty} p(a_i \wedge b_j)\, \xi_j \\
&= \sum_{i=1}^{\infty} \sum_{j=1}^{k} p(a_i \wedge b_j)\, \xi_j \\
&= \sum_{i=1}^{\infty} E(XI_{a_i}) \\
&= \sum_{i=1}^{\infty} \psi_X(a_i),
\end{aligned}$$

i.e.,

$$\psi_X(a) = \sum_{i=1}^{\infty} \psi_X(a_i).$$

5.6. A σ-additive function ψ defined on the Boolean σ-algebra $\mathfrak{B}$ and with values in a Banach space Σ is said to be a *measure* (or signed measure) ψ on $\mathfrak{B}$ with values in Σ (compare, for the case of real-valued measures on $\mathfrak{B}$, Section 3, Chapter V).

Theorem 5.3 implies:

Corollary 5.1.

For every $X \in \mathcal{S}(\mathfrak{B}, \Sigma)$ the function ψ_X is a measure on $\mathfrak{B}$ with values in Σ.

The following theorem is true:

Theorem 5.4.

If $$X_n \in \mathscr{S}(\mathfrak{B}, \Sigma), \qquad n = 1, 2, \ldots,$$
is an N_1*-fundamental sequence, then the function*

$$\psi(a) = \lim_{n\to\infty} \psi_{X_n}(a)$$

for every $a \in \mathfrak{B}$ *is a measure in* $\mathfrak{B}$ *with values in* Σ.

Proof. Let a_i, $i = 1, 2, \ldots$, be a sequence of pairwise disjoint elements. We put

$$a = \bigvee_{i=1}^{\infty} a_i;$$

then for $\varepsilon > 0$ there exists a natural number $n_0(\varepsilon)$ such that, for a fixed $n \geqslant n_0(\varepsilon)$, we have

$$\|\psi(a) - \psi_{X_n}(a)\| < \frac{\varepsilon}{3} \quad \text{and} \quad \left\|\psi\left(\bigvee_{i=1}^{k} a_i\right) - \psi_{X_n}\left(\bigvee_{i=1}^{k} a_i\right)\right\| < \frac{\varepsilon}{3},$$

for every $k = 1, 2, \ldots$. The additivity of every ψ_{X_n}, $n = 1, 2, \ldots$, and ψ implies

$$\left\|\sum_{i=1}^{k} \psi(a_i) - \sum_{i=1}^{k} \psi_{X_n}(a_i)\right\| < \frac{\varepsilon}{3}.$$

For the fixed number n, we have (Theorem 5.3)

$$\psi_{X_n}(a) = \sum_{i=1}^{\infty} \psi_{X_n}(a_i).$$

This implies the existence of a natural number $k_0(\varepsilon)$ such that for $k \geqslant k_0(\varepsilon)$ we have:

$$\left\|\psi_{X_n}(a) - \sum_{i=1}^{k} \psi_{X_n}(a_i)\right\| < \frac{\varepsilon}{3}.$$

Then for $k \geqslant k_0(\varepsilon)$, we have

$$\left\|\psi(a) - \sum_{i=1}^{k} \psi(a_i)\right\| \leqslant \left\|\psi(a) - \psi_{X_n}(a)\right\| + \left\|\psi_{X_n}(a) - \sum_{i=1}^{k} \psi_{X_n}(a_i)\right\| +$$
$$+ \left\|\sum_{i=1}^{k} \psi_{X_n}(a_i) - \sum_{i=1}^{k} \psi(a_i)\right\| < \frac{\varepsilon}{3} + \frac{\varepsilon}{3} + \frac{\varepsilon}{3} = \varepsilon;$$

therefore we have

$$\psi(a) = \lim_{k\to\infty} \sum_{i=1}^{k} \psi(a_i) = \sum_{i=1}^{\infty} \psi(a_i),$$

i.e., the σ-additivity of ψ.

Theorem 5.5.

Let $X_n \in \mathcal{S}(\mathfrak{B}, \Sigma)$, $n = 1, 2, \ldots$, be an N_1-fundamental sequence; then for every $\varepsilon > 0$ there exists a $\delta(\varepsilon) > 0$, such that

$$\psi_{\phi(X_n)}(a) = E(\phi(X_n I_a)) = \|X_n I_a\|_1 < \varepsilon,$$

for every $n = 1, 2, \ldots$, and every $a \in \mathfrak{B}$ with $p(a) < \delta(\varepsilon)$, i.e., the real-valued measures $\psi_{\phi(X_n)}$, $n = 1, 2, \ldots$, defined on $\mathfrak{B}$ are uniformly, relative to $n = 1, 2, \ldots$, absolutely continuous with respect to the probability p.

Proof. Since X_n, $n = 1, 2, \ldots$, is N_1-fundamental, for every $\varepsilon > 0$, there is a natural number $n_0 = n_0(\varepsilon)$ such that

$$E(\phi(X_n - X_m)) < \frac{\varepsilon}{2} \quad \text{for} \quad n \geqslant n_0, \qquad m \geqslant n_0.$$

Let

$$X_n = \sum_{i=1}^{k_n} \xi_{ni} I_{a_{ni}}$$

be a representation of X_n by indicators, $n = 1, 2, \ldots$, and put

$$s_n = \sup_{1 \leqslant i \leqslant k_n} \|\xi_{ni}\|;$$

choose then $\delta(\varepsilon) > 0$ such that

$$\delta(\varepsilon) . s_n < \frac{\varepsilon}{2} \quad \text{for} \quad n \leqslant n_0.$$

Namely, if $s_n = 0$, for every $n \leqslant n_0$, we choose $\delta(\varepsilon) = 1$, and if $s_n \neq 0$ for some $n \leqslant n_0$, we put

$$\delta(\varepsilon) = \inf\left(\frac{\varepsilon}{2s_n} : \quad s_n \neq 0, \quad n \leqslant n_0\right).$$

Let now $a \in \mathfrak{B}$ with $p(a) < \delta(\varepsilon)$; then for $n \leqslant n_0$ we have

$$E(\phi(X_n I_a)) \leqslant s_n p(a) \leqslant s_n . \delta(\varepsilon) < \frac{\varepsilon}{2} < \varepsilon,$$

and for $n > n_0$ we have

$$E(\phi(X_n I_a)) \leqslant E(\phi(X_n - X_{n_0}) I_a) + E(\phi(X_{n_0} I_a)) < \frac{\varepsilon}{2} + \frac{\varepsilon}{2} = \varepsilon.$$

Hence we have

$$\psi_{\phi(X_n)}(a) = E(\phi(X_n I_a)) < \varepsilon$$

for every a with $p(a) < \delta(\varepsilon)$ and every $n = 1, 2, \ldots$.

Theorem 5.6.

Let $X_n \in \mathscr{S}(\mathfrak{B}, \Sigma)$, $n = 1, 2, \ldots$, *be an* N_1*-fundamental sequence. Then there is a subsequence* $X_{n_k} \in \mathscr{S}(\mathfrak{B}, \Sigma)$, $k = 1, 2, \ldots$, *which is n-o-fundamental.*

Proof. Since *n-o*-convergence is equivalent to *n-au*-convergence, it is enough to prove the existence of a *n-au*-fundamental subsequence of X_n, $n = 1, 2, \ldots$.

Let ε be any number with $0 < \varepsilon < 1$. Then we put

$$b_{m,n}(\varepsilon) = [\phi(X_n - X_m) \geqslant \varepsilon] = s_{\phi}{}^{c}{}_{(X_n - X_m)}(\varepsilon) \in \mathfrak{B}.$$

(*cf.* Section 3.2, Chapter IV). Obviously, we have

(1) $E(\phi(X_n - X_m)) \geqslant \varepsilon . p(b_{m,n}(\varepsilon)).$

Since X_n, $n = 1, 2, \ldots$, is N_1-fundamental, there is a natural number $N(\varepsilon)$ such that

(2) $E(\phi(X_n - X_m)) < \varepsilon^2 < \varepsilon$, for $n \geqslant N(\varepsilon)$, $m \geqslant N(\varepsilon)$.

Relations (1) and (2) imply:

$$p(b_{m,n}(\varepsilon)) < \varepsilon, \quad \text{for} \quad n \geqslant N(\varepsilon), \quad m \geqslant N(\varepsilon).$$

For

$$\varepsilon = \frac{1}{2^k}, \qquad k = 1, 2, \ldots,$$

we obtain in the previous way a sequence of natural numbers

$$N_1 = N(1/2), \qquad N_2 = N(1/2^2), \ldots, \qquad N_k = N(1/2^k), \ldots$$

with the property

$$E(\phi(X_n - X_m)) < \frac{1}{2^k} \quad \text{and} \quad p(b_{m,n}(1/2^k)) < \frac{1}{2^k}$$

for every $n \geqslant N_k$, $m \geqslant N_k$. Put now

$$n_1 = N_1, \qquad n_2 = \max\{n_1 + 1, N_2\}, \ldots, \qquad n_k = \max\{n_{k-1} + 1, N_k\}, \ldots,$$

and consider the subsequence $X_{n_1}, X_{n_2}, \ldots, X_{n_k}, \ldots$. Put moreover for every k:

$$b_k = b_{n_k, n_{k+1}}(1/2^k) \in \mathfrak{B}, \qquad k = 1, 2, \ldots$$

and

$$d_k{}^c = \bigvee_{i=k}^{\infty} b_i \in \mathfrak{B}, \qquad k = 1, 2, \ldots .$$

For $i \geqslant k$, we have, obviously,

$$\phi(X_{n_i} - X_{n_{i+1}}) I_{d_k} < \frac{1}{2^i}, \qquad i \geqslant k;$$

then, for every $j > i \geqslant k$, we have

$$\phi(X_{n_i} - X_{n_j}) I_{d_k} \leqslant \{\phi(X_{n_i} - X_{n_{i+1}}) + \ldots + \phi(X_{n_{j-1}} - X_{n_j})\} I_{d_k}$$

$$< \frac{1}{2^i} + \frac{1}{2^{i+1}} + \ldots + \frac{1}{2^{j-1}} < \frac{1}{2^{i-1}} \leqslant \frac{1}{2^{k-1}}.$$

Let now any $\delta > 0$; choose k such that

$$\frac{1}{2^{k-1}} < \delta$$

and put $a_\delta{}^c = d_k{}^c$; then for $\delta > 0$ there is $a_\delta = e - a_\delta{}^c \in \mathfrak{B}$ such that $p(a_\delta{}^c) < \delta$ and $\phi(X_{n_i} - X_{n_j}) I_{a_\delta}$ *u*-converges to zero, i.e., the subsequence

$$X_{n_k} \in \mathscr{S}(\mathfrak{B}, \Sigma), \qquad k = 1, 2, \ldots,$$

is *n-au*-fundamental.

Theorem 5.7.

Let $X_n \in \mathscr{S}(\mathfrak{B}, \Sigma)$, $n = 1, 2, \ldots$, be an N_1-fundamental sequence which n-o-converges to zero; then X_n, $n = 1, 2, \ldots$, converges to zero in mean, i.e., $\lim_{n \to \infty} \|X_n\|_1 = 0$.

Proof. We shall prove first that

$$\lim_{n \to \infty} \psi_{\phi(X_n)}(a) = \lim_{n \to \infty} E(\phi(X_n) I_a) = 0 \quad \text{for every} \quad a \in \mathfrak{B}.$$

According to Theorem 5.5, for every $\varepsilon > 0$, hence for

$$\varepsilon^* = \frac{\varepsilon}{p(a) + 1} > 0,$$

there exists $\delta = \delta(\varepsilon^*) >$ such that

$$\text{(L)} \quad 0 \leqslant \psi_{\phi(X_n)}(b) = E(\phi(X_n I_b)) < \varepsilon^*$$

for every $b \in \mathfrak{B}$ with $p(b) < \delta$. Since X_n, $n = 1, 2, \ldots$, is *n-o*-convergent to zero, it is also *n-au*-convergent to zero, i.e., for $\delta > 0$ there is $a_\delta \in \mathfrak{B}$ such that $p(a_\delta{}^c) < \delta$ and

$$\phi(X_n I_{a_\delta}) \overset{u}{\to} 0.$$

Hence for $\varepsilon^* > 0$ there is an $N(\varepsilon) = N(\varepsilon^*)$ such that

$$0 \leqslant \phi(X_k) I_{a_\delta} = \phi(X_k I_{a_\delta}) < \varepsilon^* \quad \text{for every} \quad k \geqslant N(\varepsilon).$$

For each k with $k \geqslant N(\varepsilon)$ we put

$$b_k = [\phi(X_k) \geqslant \varepsilon^*].$$

Then $b_k \leqslant a_\delta{}^c$, hence $p(b_k) < \delta$; according to (L), we have:

$$\psi_{\phi(X_k)}(b_k) = E(\phi(X_k I_{b_k})) < \varepsilon^*.$$

Now we have for every $k \geqslant N(\varepsilon)$

$$\psi_{\phi(X_k)}(a) \leqslant \psi_{\phi(X_k)}(a - b_k) + \psi_{\phi(X_k)}(b_k) \leqslant \varepsilon^* . p(a - b_k) + \varepsilon^* \leqslant \varepsilon^* . p(a) + \varepsilon^* = \varepsilon.$$

Thus

$$\lim \psi_{\phi(X_k)}(a) = 0 \quad \text{for every} \quad a \in \mathfrak{B}.$$

But

$$\psi_{\phi(X_n)}(e) = E(\phi(X_n)) = \|X_n\|_1.$$

Hence

$$\lim_{n \to \infty} \psi_{(X_n)}(e) = \lim_{n \to \infty} \|X_n\|_1 = 0,$$

i.e., the sequence X_n, $n = 1, 2, \ldots$, converges in mean to zero.

Theorem 5.8.

If X_n and $Y_n \in \mathscr{S}(\mathfrak{B}, \Sigma)$, $n = 1, 2, \ldots$, are two N_1-fundamental sequences which n-o-converge to the same X in $\mathscr{V}^*(\mathfrak{B}, \Sigma)$, then we have

$$\lim_{n \to \infty} E(X_n) = \lim_{n \to \infty} E(Y_n).$$

Proof. Obviously $X_n - Y_n$, $n = 1, 2, \ldots$, is an N_1-fundamental sequence and *n-o*-converges to zero, hence it converges to zero in mean. Then $\lim_{n \to \infty} E(X_n - Y_n) = 0$, hence

$$\lim_{n \to \infty} E(X_n) = \lim_{n \to \infty} E(Y_n).$$

5.7. We can now define an expectation for certain rv's $X \in \mathscr{V}^*(\mathfrak{B}, \Sigma)$. We shall say a rv $X \in \mathscr{V}^*(\mathfrak{B}, \Sigma)$ *possesses an expectation in* Σ if and only if there exists an N_1-fundamental sequence

$$X_n \in \mathscr{S}(\mathfrak{B}, \Sigma), \qquad n = 1, 2, \ldots,$$

n-o-converging to X. The expectation of X is then the element $E(X) \in \Sigma$ defined by

$$E(X) = \lim_{n \to \infty} E(X_n).$$

According to Theorem 5.8, the expectation is independent of the particular N_1-fundamental sequence $X_n \in \mathscr{S}(\mathfrak{B}, \Sigma)$, $n = 1, 2, \ldots$, which *n-o*-converges to X. The set of all rv's $X \in \mathscr{V}^*(\mathfrak{B}, \Sigma)$, which possess an expectation $E(X) \in \Sigma$ will be denoted by $\mathscr{L}_1(\mathfrak{B}, \Sigma)$. Obviously $\mathscr{S}(\mathfrak{B}, \Sigma)$ is a subset of $\mathscr{L}_1(\mathfrak{B}, \Sigma)$. The set of all real-valued rv's possessing an expectation defined as above in the particular case $\Sigma = R$ will be denoted by $\mathscr{L}_1(\mathfrak{B}, R)$. We shall prove below that $\mathscr{L}_1(\mathfrak{B}, R)$ is isometric to $\mathscr{L}_1$ defined in Section 2, Chapter V.

It is now easy to prove that $\mathscr{L}_1(\mathfrak{B}, \Sigma)$ is a vector subspace of $\mathscr{V}^*(\mathfrak{B}, \Sigma)$ and the properties (1) and (2) of Section 5.2 are valid for the expectation E defined on $\mathscr{L}_1(\mathfrak{B}, \Sigma)$, i.e., E is linear and if Σ is a B-lattice then E is strictly monotone on the vector lattice $\mathscr{L}_1(\mathfrak{B}, \Sigma)$. Moreover, we have here:

If $X \in \mathscr{L}_1(\mathfrak{B}, \Sigma)$, then $\phi(X) \in \mathscr{L}_1(\mathfrak{B}, R)$ and

(II*) $\quad \|E(X)\| \leqslant E(\phi(X))$;

and, if we define

(III*) $\quad \|X\|_1 = E(\phi(X))$, for every $X \in \mathscr{L}_1(\mathfrak{B}, \Sigma)$,

then $\| \;\|_1$ is a norm on the vector space $\mathscr{L}_1(\mathfrak{B}, \Sigma)$; i.e., properties ($\alpha$), ($\beta$), ($\gamma$) and ($\delta$) of Section 5.3 are also valid for the expectation E on $\mathscr{L}_1(\mathfrak{B}, \Sigma)$. The norm $\| \;\|_1$ defines a distance, which induces a metric convergence in mean and denoted by

$$X_n \overset{N_1}{\to} X \quad \text{in} \quad \mathscr{L}_1(\mathfrak{B}, \Sigma).$$

The inequality (II*) can be written as follows:

($\widetilde{\text{II}}$*) $\quad \|E(X)\| \leqslant \|X\|_1$,

which implies that the expectation E is continuous for the convergence in mean. The concept of an N_1-fundamental sequence can be also defined in $\mathscr{L}_1(\mathfrak{B}, \Sigma)$ and the following theorem is true.

Theorem 5.9.

If $X_n \in \mathscr{S}(\mathfrak{B}, \Sigma)$, $n = 1, 2, \ldots$, is N_1-fundamental and n-o-converging to a rv $X \in \mathscr{L}_1(\mathfrak{B}, \Sigma)$, then X_n, $n = 1, 2, \ldots$, converges in mean to X, i.e.,

$$X_n \overset{N_1}{\to} X.$$

Proof. For every $\varepsilon > 0$ there is a natural number $N(\varepsilon)$ such that, for $n \geqslant N(\varepsilon)$, $m \geqslant N(\varepsilon)$, we have

$$\|X_n - X_m\|_1 = E(\phi(X_n - X_m)) < \varepsilon; \tag{1}$$

for fixed $n \geqslant N(\varepsilon)$, the sequence $X_n - X_m$, $m = 1, 2, \ldots$, is N_1-fundamental and n-o-converging to $X_n - X$; hence $\phi(X_n - X_m)$, $m = 1, 2, \ldots$, is an N_1-fundamental sequence of positive real-valued rv's in $\mathscr{S}(\mathfrak{B}, R)$ o-converging to $\phi(X_n - X)$; thus, according to (1), we have

$$E(\phi(X_n - X)) = \lim_{m \to \infty} E(\phi(X_n - X_m)) \leqslant \varepsilon$$

i.e., $\|X_n - X\|_1 \leqslant \varepsilon$ for $n \geqslant N(\varepsilon)$, hence X_n, $n = 1, 2, \ldots$, converges in mean to X.

Corollary 5.2.

The space $\mathscr{S}(\mathfrak{B}, \Sigma)$ is, with respect to the convergence in mean, dense in $\mathscr{L}_1(\mathfrak{B}, \Sigma)$.

Exercise. Let $\mathfrak{A}$ be any Boolean subalgebra of $\mathfrak{B}$, which σ-generates $\mathfrak{B}$, and consider the space $\mathscr{S}(\mathfrak{A}, \Sigma)$; then to, every $X \in \mathscr{L}_1(\mathfrak{B}, \Sigma)$ there exists an N_1-fundamental sequence $X_n \in \mathscr{S}(\mathfrak{A}, \Sigma)$ n-o-converging to X.

Theorem 5.10.

The normed vector space $\mathscr{L}_1(\mathfrak{B}, \Sigma)$ is complete with respect to the N_1-convergence, i.e., $\mathscr{L}_1(\mathfrak{B}, \Sigma)$ is a Banach space.

Proof. Let

$$X_n \in \mathscr{L}_1(\mathfrak{B}, \Sigma), \qquad n = 1, 2, \ldots,$$

be a Cauchy sequence, i.e., let the double sequence

$$\|X_n - X_m\|_1 = E(\phi(X_n - X_m))$$

converge to zero. For every n there exists a rv $Y_n \in \mathscr{S}(\mathfrak{B}, \Sigma)$ such that

$$\|X_n - Y_n\| = E(\phi(X_n - Y_n)) < \frac{1}{n}.$$

But

$$\|Y_n - Y_m\|_1 \leqslant \|Y_n - X_n\|_1 + \|X_n - X_m\|_1 + \|X_m - Y_m\|_1;$$

hence the double sequence $\|Y_n - Y_m\|_1$ converges to zero; i.e., the sequence

$$Y_n \in \mathscr{S}(\mathfrak{B}, \Sigma), \qquad n = 1, 2, \ldots,$$

is N_1-fundamental. Hence, by Theorem 5.6, it contains a subsequence Y_{n_k}, $k = 1, 2, \ldots$, n-o-converging to a rv $X \in \mathscr{V}^*(\mathfrak{B}, \Sigma)$, which by definition belongs to $\mathscr{L}_1(\mathfrak{B}, \Sigma)$. According to Theorem 5.9,

$$Y_{n_k} \overset{N_1}{\rightarrow} X,$$

hence also

$$Y_n \overset{N_1}{\rightarrow} X,$$

i.e.,

$$\| Y_n - X \|_1 \rightarrow 0.$$

Then we have

$$\| X_n - X \|_1 \leqslant \| X_n - Y_n \|_1 + \| Y_n - X \|_1,$$

hence

$$\| X_n - X \|_1 \rightarrow 0,$$

i.e.,

$$X_n \overset{N_1}{\rightarrow} X \quad \text{in} \quad \mathscr{L}_1(\mathfrak{B}, \Sigma).$$

5.8. Let $X \in \mathscr{L}_1(\mathfrak{B}, \Sigma)$; then there is an N_1-fundamental sequence

$$Y_n \in \mathscr{S}(\mathfrak{B}, \Sigma), \qquad n = 1, 2, \ldots,$$

n-o-converging to X; then for every $a \in \mathfrak{B}$, the sequence $Y_n I_a \in \mathscr{S}(\mathfrak{B}, \Sigma)$ is also N_1-fundamental and n-o-converges to a rv in $\mathscr{L}_1(\mathfrak{B}, \Sigma)$, which we denote by XI_a; obviously this definition of XI_a does not depend on the choice of the sequence Y_n, $n = 1, 2, \ldots$. Hence we can define for any $X \in \mathscr{L}_1(\mathfrak{B}, \Sigma)$ a function

$$\psi_X(a) = E(XI_a), \qquad a \in \mathfrak{B}$$

on $\mathfrak{B}$ with values in Σ.

Theorem 5.11.

For every $X \in \mathscr{L}_1(\mathfrak{B}, \Sigma)$ the function

$$\psi_X(a) = E(XI_a), \qquad a \in \mathfrak{B},$$

is a measure on $\mathfrak{B}$ with values in Σ.

Proof. Let

$$Y_n \in \mathscr{S}(\mathfrak{B}, \Sigma), \qquad n = 1, 2, \ldots,$$

be N_1-fundamental and n-o-converging to X; then for every $a \in \mathfrak{B}$, we have

$$E(XI_a) = \lim_{n \to \infty} E(Y_n I_a) = \lim_{n \to \infty} \psi_{Y_n}(a).$$

According to Theorem 5.4, the function $\psi_X(a)$ defined by

$$\lim_{n\to\infty} \psi_{Y_n}(a) = \psi_X(a) = E(XI_a), \qquad a \in \mathfrak{B},$$

is a measure in $\mathfrak{B}$ with values in Σ.

5.9. Let $X \in \mathscr{L}_1(\mathfrak{B}, R)$ and $\xi \in \Sigma$; then there exists an N_1-fundamental sequence $Y_n \in \mathscr{S}(\mathfrak{B}, R)$, $n = 1, 2, \ldots$, o-converging to X. Let

$$X_n = \sum_{i=1}^{k_n} \lambda_{ni} I_{b_{ni}}$$

be a representation of $X_n \in \mathscr{S}(\mathfrak{B}, R)$ by indicators; then λ_{ni} is a real number, hence $\lambda_{ni}.\xi \in \Sigma$, and we can define

$$X_n \xi = \sum_{i=1}^{k_n} \lambda_{ni}.\xi.I_{b_{ni}}.$$

Obviously $X_n \xi \in \mathscr{S}(\mathfrak{B}, \Sigma)$, $n = 1, 2, \ldots$, and

$$E(X_n \xi) = E(X_n)\,\xi, \qquad n = 1, 2, \ldots .$$

The sequence $X_n \xi$, $n = 1, 2, \ldots$, is N_1-fundamental and n-o-converges to a rv in $\mathscr{L}_1(\mathfrak{B}, \Sigma)$, which we denote by $X\xi$. This definition does not depend on the choice of the sequence Y_n, $n = 1, 2, \ldots$, in $\mathscr{S}(\mathfrak{B}, R)$. Obviously we have

$$E(X\xi) = \xi E(X).$$

Exercise. Let Σ and T be Banach spaces and $v : \Sigma \to T$ be any continuous linear operator of Σ into T. Let $X \in \mathscr{L}_1(\mathfrak{B}, \Sigma)$; define

$$v \circ X \in \mathscr{L}_1(\mathfrak{B}, T)$$

and prove

$$E(v \circ X) = v \circ E(X).$$

5.10. Theorem 5.6 possesses the following analogous in $\mathscr{L}_1(\mathfrak{B}, \Sigma)$:

Theorem 5.12.

If the sequence $X_n \in \mathscr{L}_1(\mathfrak{B}, \Sigma)$, $n = 1, 2, \ldots$, converges in mean to $X \in \mathscr{L}_1(\mathfrak{B}, \Sigma)$, then there is a subsequence X_{n_k}, $k = 1, 2, \ldots$, n-o-converging to X.

Proof. For every $k = 1, 2, \ldots$, there exists a natural number n_k such that

$$\|X_{n_k} - X\|_1 = E(\phi(X_{n_k} - X)) < \frac{1}{2^{2k}}.$$

We put

$$a_k = \left[\phi(X_{n_k} - X) \geqslant \frac{1}{2^k}\right] \in \mathfrak{B};$$

then

$$\frac{1}{2^{2k}} > E(\phi(X_{n_k} - X)) \geqslant E(\phi(X_{n_k} - X)\, I_{a_k}) \geqslant \frac{1}{2^k}\, p(a_k),$$

hence

$$p(a_k) < \frac{1}{2^k}.$$

We put

$$b_k = \bigvee_{i=k}^{\infty} a_i;$$

then

$$\phi(X_{n_i} - X)\, I_{b_k^c} < \frac{1}{2^i} \quad \text{for} \quad i \geqslant k;$$

hence the sequence $X_{n_i} I_{b_k^c}$, $i = 1, 2, \ldots$, *n-u*-converges to $X I_{b_k^c}$. We put

$$b = \bigwedge_{k=1}^{\infty} b_k;$$

then

$$p(b) \leqslant p(b_k) \leqslant \sum_{i=k}^{\infty} p(a_i) < \frac{1}{2^{k-1}} \quad \text{for every} \quad k = 1, 2, \ldots,$$

i.e., $p(b) = 0$, hence $b = \emptyset$;
Now for every $\varepsilon > 0$ there exists a natural number $k(\varepsilon)$ such that, if we put $b_\varepsilon = b_{k(\varepsilon)}$, we have

$$p(b_\varepsilon) < \varepsilon \quad \text{and} \quad X_{n_i} I_{b_\varepsilon^c} \overset{n\text{-}u}{\rightarrow} X I_{b_\varepsilon^c},$$

i.e., the sequence X_{n_i}, $i = 1, 2, \ldots$, *n-au*-converges to X and that is equivalent to the *n-o*-convergence of X_{n_i}, $i = 1, 2, \ldots$, to X.

Theorem 5.5 possesses the following analogous in $\mathscr{L}_1(\mathfrak{B}, \Sigma)$.

Theorem 5.13.

If $X_n \in \mathscr{L}_1(\mathfrak{B}, \Sigma)$, $n = 1, 2, \ldots$, *is* N_1*-fundamental, then, for every* $\varepsilon > 0$, *there exists a number* $\delta(\varepsilon)$ *such that*

$$E(\phi(X_n)\, I_a) = \psi_{\phi(X_n)}(a) < \varepsilon$$

for every $n = 1, 2, \ldots$, *and every* $a \in \mathfrak{B}$ *with* $p(a) < \delta(\varepsilon)$.

Proof. Let $\varepsilon > 0$; since $\mathscr{S}(\mathfrak{B}, \Sigma)$ is dense with respect to the convergence in mean in $\mathscr{L}_1(\mathfrak{B}, \Sigma)$, for every n there exists a $Y_n \in \mathscr{S}(\mathfrak{B}, \Sigma)$ such that

$$\|X_n - Y_n\|_1 = E(\phi(X_n - Y_n)) < \frac{\varepsilon}{2n};$$

then we have for every $a \in \mathfrak{B}$

$$E(\phi(X_n - Y_n) I_a) < \frac{\varepsilon}{2n}.$$

Thus for every $a \in \mathfrak{B}$ we have

$$E(\phi(X_n) I_a) < E(\phi(Y_n) I_a) + \frac{\varepsilon}{2n}$$

i.e.,

$$\psi_{\phi(X_n)}(a) < \psi_{\phi(Y_n)}(a) + \frac{\varepsilon}{2n}. \tag{2}$$

But we have the inequality

$$E(\phi(Y_n - Y_m)) \leqslant E(\phi(Y_n - X_n)) + E(\phi(X_n - X_m)) + E(\phi(X_m - Y_m));$$

i.e., the sequence Y_n, $n = 1, 2, \ldots$, is N_1-fundamental. Theorem 5.5 implies now the existence of a number $\delta(\varepsilon)$ such that

$$\psi_{\phi(Y_n)}(a) < \frac{\varepsilon}{2}, \quad \text{for every} \quad n = 1, 2, \ldots,$$

and every $a \in \mathfrak{B}$ with $p(a) < \delta(\varepsilon)$.

The inequality (2) implies

$$\psi_{\phi(X_n)}(a) < \frac{\varepsilon}{2} + \frac{\varepsilon}{2n} < \varepsilon, \quad \text{for every} \quad n = 1, 2, \ldots$$

and every $a \in \mathfrak{B}$ with $p(a) < \delta(\varepsilon)$.

5.11. The following theorem shows that, if we consider $\mathscr{L}_1(\mathfrak{B}, \Sigma)$ as start space and try to apply the process of Section 5.7 to extend the expectation to other rv's in $\mathscr{V}^*(\mathfrak{B}, \Sigma)$, we do not obtain new rv's.

Theorem 5.14.

If

$$X_n \in \mathscr{L}_1(\mathfrak{B}, \Sigma), \qquad n = 1, 2, \ldots,$$

is N_1-fundamental and n-o-converging to a rv $X \in \mathscr{V}^*(\mathfrak{B}, \Sigma)$, then $X \in \mathscr{L}_1(\mathfrak{B}, \Sigma)$. Moreover in this case we have

$$E(X) = \lim_{n\to\infty} E(X_n).$$

Proof. Since $\mathscr{L}_1(\mathfrak{B}, \Sigma)$ is complete with respect to the convergence in mean, there exists a $Y \in \mathscr{L}_1(\mathfrak{B}, \Sigma)$ such that $X_n \overset{N_1}{\to} Y$. Theorem 5.12 implies the existence of a subsequence X_{n_k}, $k = 1, 2, \ldots$, such that $X_{n_k} \overset{n\text{-}o}{\to} Y$; this and the assumption $X_n \overset{n\text{-}o}{\to} X$ imply $X = Y \in \mathscr{L}_1(\mathfrak{B}, \Sigma)$. Obviously, we have moreover

$$E(X) = \lim_{n \to \infty} E(X_n).$$

5.12. Let Σ be the set R of all real numbers, considered as a Banach space; then the norm $\|\xi\|$ is equal to the absolute value $|\xi|$ of ξ and the n-o-convergence is equivalent to the o-convergence. Moreover, the n-u-convergence is equivalent to the u-convergence. In Chapter V, we followed another process in order to define the concept of an expectation for certain real-valued rv's. One can prove that this process is equivalent to the one given here in the particular case $\Sigma = R$. The process of Chapter V may be applied also for rv's with values in any Banach space Σ, as follows: An erv $X \in \mathscr{E}(\mathfrak{B}, \Sigma)$ with a representation

$$X = \sum_{j \geqslant 1} \xi_j I_{a_j}$$

by indicators possesses an expectation if and only if the series

$$\sum_{j \geqslant 1} p(a_j)\, \xi_j$$

is absolutely convergent, i.e.,

$$\sum_{j \geqslant 1} p(a_j)\, \|\xi_j\| < +\infty;$$

then the series

$$\sum_{j \geqslant 1} p(a_j)\, \xi_j$$

converges in Σ and defines the expectation

$$E(X) = \sum_{j \geqslant 1} p(a_j)\, \xi_j \in \Sigma.$$

It is easy to prove that the sequence

$$X_n = \sum_{j=1}^{n} \xi_j I_{a_j} \in \mathscr{S}(\mathfrak{B}, \Sigma), \qquad n = 1, 2, \ldots,$$

n-o-converges to X and, moreover, X_n, $n = 1, 2, \ldots$, is N_1-fundamental, i.e., $X \in \mathscr{L}_1(\mathfrak{B}, \Sigma)$. Since

$$\lim_{n\to\infty} \sum_{j=1}^{n} p(a_j)\,\xi_j = \sum_{j\geqslant 1} p(a_j)\,\xi_j \quad \text{in} \quad \Sigma,$$

we conclude that both definitions lead to the same value of expectation $E(X)$ in Σ. Let now denote by $\mathscr{K}(\mathfrak{B}, \Sigma)$ the set of all erv's in $\mathscr{E}(\mathfrak{B}, \Sigma)$, which possess an expectation; then

$$\mathscr{K}(\mathfrak{B}, \Sigma) \subseteq \mathscr{L}_1(\mathfrak{B}, \Sigma).$$

Let

$$X_n \in \mathscr{K}(\mathfrak{B}, \Sigma), \qquad n = 1, 2, \ldots,$$

be an n-u-fundamental sequence; then the double sequence

$$\phi(X_n - X_m) \in \mathscr{K}(\mathfrak{B}, R)$$

u-converges to zero and obviously $E(\phi(X_n - X_m))$ converges also to zero in R (compare Theorem 1.6, Chapter V). We have

$$\|E(X_n) - E(X_m)\| = \|E(X_n - X_m)\| \leqslant E(\phi(X_n - X_m));$$

hence $E(X_n) \in \Sigma$, $n = 1, 2, \ldots$, is a fundamental sequence with respect to the norm in Σ, i.e., there exists

$$\lim_{n\to\infty} E(X_n) \in \Sigma.$$

Thus we can define, analogously to the Chapter V, Section 2: an element $X \in \mathscr{V}^*(\mathfrak{B}, \Sigma)$ possesses an expectation in Σ if and only if there exists a sequence

$$X_n \in \mathscr{K}(\mathfrak{B}, \Sigma), \qquad n = 1, 2, \ldots,$$

which n-u-converges to X; then there exists

$$\lim_{n\to\infty} E(X_n) \in \Sigma$$

and can be defined as the expectation $E(X)$ of X. It is easy to prove that this sequence

$$X_n \in \mathscr{K}(\mathfrak{B}, \Sigma) \subseteq \mathscr{L}_1(\mathfrak{B}, \Sigma), \qquad n = 1, 2, \ldots,$$

is N_1-fundamental and obviously n-o-converges to X, hence $X \in \mathscr{L}_1(\mathfrak{B}, \Sigma)$. Let $\tilde{\mathscr{K}}(\mathfrak{B}, \Sigma)$ denote the set of all rv's in $\mathscr{V}^*(\mathfrak{B}, \Sigma)$ which possess an expectation according to the last definition; then $\tilde{\mathscr{K}}(\mathfrak{B}, \Sigma) \subseteq \mathscr{L}_1(\mathfrak{B}, \Sigma)$. Since, moreover, the two definitions of the expectation coincide on $\tilde{\mathscr{K}}(\mathfrak{B}, \Sigma)$, in order to prove the equivalence of the two definitions, it is sufficient to show that every $X \in \mathscr{L}_1(\mathfrak{B}, \Sigma)$ belongs to $\tilde{\mathscr{K}}(\mathfrak{B}, \Sigma)$.

We shall use the following:

Lemma 5.1.

If $X \in \mathscr{L}_1(\mathfrak{B}, \Sigma)$ *and* $Y \in \mathscr{E}(\mathfrak{B}, \Sigma)$ *with* $\phi(X-Y) < \varepsilon$, *then* $Y \in \mathscr{K}(\mathfrak{B}, \Sigma)$.

Proof. Let

$$Y = \sum_{i \geqslant 1} \xi_i I_{a_i}$$

be a representation of Y by indicators, where $a_i \wedge a_j = \emptyset$, $i \neq j$; we put

$$d_k = \bigvee_{i=1}^{k} a_i;$$

then

$$(\phi(Y) - \phi(X)) I_{d_k} \leqslant \phi(Y-X) I_{d_k} \leqslant \phi(Y-X) < \varepsilon.$$

Hence

$$\phi(Y) I_{d_k} < \varepsilon + \phi(X) I_{d_k}, \qquad k = 1, 2, \ldots .$$

From this it follows that:

$$\sum_{i=1}^{k} p(a_i)\,|\xi_i| = E(\phi(Y) I_{d_k}) < \varepsilon + E(\phi(X) I_{d_k}) \leqslant \varepsilon + E(\phi(X))$$

for every $k = 1, 2, \ldots$; hence

$$\sum_{i \geqslant 1} p(a_i)\,|\xi_i| \leqslant E(\phi(X)) + \varepsilon < +\infty,$$

which implies $Y \in \mathscr{K}(\mathfrak{B}, \Sigma)$.

In order to prove now that

$$\mathscr{L}_1(\mathfrak{B}, \Sigma) \subseteq \tilde{\mathscr{K}}(\mathfrak{B}, \Sigma),$$

which implies

$$\mathscr{L}_1(\mathfrak{B}, \Sigma) = \tilde{\mathscr{K}}(\mathfrak{B}, \Sigma),$$

we first remark that for every

$$X \in \mathscr{L}_1(\mathfrak{B}, \Sigma) \cap \mathscr{E}(\mathfrak{B}, \Sigma),$$

the assumptions of the Lemma 5.1, with $X = Y$, are satisfied. Hence $X \in \mathscr{K}(\mathfrak{B}, \Sigma)$. Let now X be any element of $\mathscr{L}_1(\mathfrak{B}, \Sigma)$. We shall prove: there exists

$$X_n \in \mathscr{E}(\mathfrak{B}, \Sigma), \qquad n = 1, 2, \ldots, \quad \text{such that} \quad X_n \overset{n-u}{\rightarrow} X.$$

In fact, there exists a sequence

$$Y_n \in \mathscr{S}(\mathfrak{B}, \Sigma), \qquad n = 1, 2, \ldots,$$

n-o-converging to X, equivalently n-au-converging to X. Let $\varepsilon > 0$; then, for every $k = 1, 2, \ldots$, there exist a natural number n_k and an element $b_k \in \mathfrak{B}$ such that

$$p(b_k{}^c) < \frac{1}{k} \quad \text{and} \quad \phi((Y_{n_k} - X) I_{b_k}) < \varepsilon.$$

Obviously, we have

$$p\left(\bigwedge_{k=1}^{\infty} b_k{}^c\right) < \frac{1}{k}, \qquad k = 1, 2, \ldots,$$

hence

$$p\left(\bigwedge_{k=1}^{\infty} b_k{}^c\right) = 0,$$

i.e.,

$$\bigwedge_{k=1}^{\infty} b_k{}^c = \emptyset \quad \text{and} \quad \bigvee_{k=1}^{\infty} b_k = e.$$

We put

$$c_1 = b_1, \qquad c_k = b_k - \bigvee_{i=1}^{k-1} b_i, \qquad k = 2, 3, \ldots,$$

and

$$X_\varepsilon = \sum_{k=1}^{\infty} Y_{n_k} I_{c_k};$$

then

$$X_\varepsilon \in \mathscr{E}(\mathfrak{B}, \Sigma) \quad \text{and} \quad \phi(X_\varepsilon - X) < \varepsilon.$$

Thus, for every

$$\varepsilon = \frac{1}{n}, \qquad n = 1, 2, \ldots,$$

there exists an $X_n \in \mathscr{E}(\mathfrak{B}, \Sigma)$ such that

$$\phi(X_n - X) < \frac{1}{n}, \qquad n = 1, 2, \ldots. \tag{1}$$

This implies

$$X_n \overset{n-u}{\rightarrow} X.$$

Moreover, according to the Lemma 5.1, the inequality (1) implies

$$X_n \in \mathscr{K}(\mathfrak{B}, \Sigma), \qquad n = 1, 2, \ldots;$$

hence

$$X \in \tilde{\mathscr{K}}(\mathfrak{B}, \Sigma),$$

i.e.,

$$\mathscr{L}_1(\mathfrak{B}, \Sigma) \subseteq \tilde{\mathscr{K}}(\mathfrak{B}, \Sigma).$$

5.13. Lebesgue's theorem on term by term integration, which we proved in Chapter V, Section 2, possesses an analogous theorem in the general case of rv's with values in any Banach space Σ.

Theorem 5.15. (*Lebesgue's theorem on term by term integration.*)
Let

$$X_n \in \mathscr{L}_1(\mathfrak{B}, \Sigma), \qquad n = 1, 2, \ldots,$$

be an n-o-convergent sequence to a rv $X \in \mathscr{V}^*(\mathfrak{B}, \Sigma)$. *If there exists a positive real-valued rv* $Y \in \mathscr{L}_1(\mathfrak{B}, R)$ *such that*

$$\phi(X_n) \leqslant Y, \qquad n = 1, 2, \ldots,$$

then the limit rv

$$X = n\text{-}o\text{-}\lim_{n\to\infty} X_n \quad \textit{belongs to} \quad \mathscr{L}_1(\mathfrak{B}, \Sigma)$$

and we have

$$E(X) = \lim_{n\to\infty} E(X_n) \quad \textit{in} \quad \Sigma.$$

Proof. Since X_n, $n = 1, 2, \ldots$, is n-o-fundamental, the double sequence

$$\phi(X_n - X_m) \in \mathscr{L}_1(\mathfrak{B}, R), \qquad n = 1, 2, \ldots, \qquad m = 1, 2, \ldots,$$

o-converges to zero, hence

$$o\text{-}\limsup_{(n,m)\to(+\infty,+\infty)} \phi(X_n - X_m) = 0.$$

Theorem 2.8, Chapter V (which is also true for double sequences), implies:

$$\limsup_{(n,m)\to(+\infty,+\infty)} E(\phi(X_n - X_m)) \leqslant E(\limsup_{(n,m)\to(+\infty,+\infty)} \phi(X_n - X_m)) = 0.$$

It follows that X_n, $n = 1, 2, \ldots$, is an N_1-fundamental sequence in $\mathscr{L}_1(\mathfrak{B}, \Sigma)$, hence the limit

$$X = n\text{-}o\text{-}\lim_{n\to\infty} X_n \in \mathscr{L}_1(\mathfrak{B}, \Sigma)$$

and

$$E(X) = \lim_{n\to\infty} E(X_n)$$

(see Theorem 5.14).

Theorem 5.16.

If for a rv

$$X \in \mathscr{V}^*(\mathfrak{B}, \Sigma)$$

there exists a positive real-valued rv

$$Y \in \mathscr{L}_1(\mathfrak{B}, R)$$

such that
$$\phi(X) \leqslant Y,$$
then
$$X \in \mathscr{L}_1(\mathfrak{B}, \Sigma).$$

Proof. Since
$$X \in \mathscr{V}^*(\mathfrak{B}, \Sigma)$$
and $\mathscr{S}(\mathfrak{B}, \Sigma)$ is n-o-dense in $\mathscr{V}^*(\mathfrak{B}, \Sigma)$, there exists a sequence
$$X_n \in \mathscr{S}(\mathfrak{B}, \Sigma), \qquad n = 1, 2, \ldots,$$
such that $X_n \xrightarrow{n\text{-}o} X$ in $\mathscr{V}^*(\mathfrak{B}, \Sigma)$; moreover the sequence X_n, $n = 1, 2, \ldots,$ may be chosen so that
$$\phi(X_n) \leqslant Y;$$
then by Lebesgue's theorem
$$X \in \mathscr{L}_1(\mathfrak{B}, \Sigma).$$

Theorem 5.17.

A rv
$$X \in \mathscr{V}^*(\mathfrak{B}, \Sigma)$$
belongs to $\mathscr{L}_1(\mathfrak{B}, \Sigma)$ *if and only if*
$$\phi(X) \in \mathscr{L}_1(\mathfrak{B}, R).$$

Proof. If
$$X \in \mathscr{L}_1(\mathfrak{B}, \Sigma),$$
then
$$\phi(X) \in \mathscr{L}_1(\mathfrak{B}, R)$$
(see Section 5.7). Conversely, if
$$X \in \mathscr{V}^*(\mathfrak{B}, \Sigma)$$
and
$$\phi(X) \in \mathscr{L}_1(\mathfrak{B}, \Sigma),$$
then from the previous theorem with $Y = \phi(X)$, we deduce that
$$X \in \mathscr{L}_1(\mathfrak{B}, \Sigma).$$

5.14. *Remark.* An expectation of certain rv's with values in a Banach space Σ may be defined with the help of the Pettis integration theory. By this theory, the so-called conjugate space Σ^* of Σ plays an important role. We shall state here briefly this theory. Let Σ be a Banach space with R as scalar field. A mapping ξ^* of Σ into R is said to be a *linear functional* if and only if
$$\xi^*(\lambda\xi + \mu\eta) = \lambda\xi^*(\xi) + \mu\xi^*(\eta),$$

for all λ, $\mu \in R$ and ξ, $\eta \in \Sigma$. A linear functional ξ^* is said to be *bounded* if and only if there exists a real number $\delta > 0$ such that

$$|\xi^*(\xi)| \leqslant \delta \|\xi\| \quad \text{for all} \quad \xi \in \Sigma. \tag{1}$$

It is easy to prove that a linear functional is bounded if and only if it is continuous. In fact, the inequality (1) implies

$$|\xi^*(\xi) - \xi^*(\xi_0)| \leqslant \delta \|\xi - \xi_0\|, \qquad \xi \in \Sigma, \qquad \xi_0 \in \Sigma,$$

and this proves the continuity of ξ^* at the point $\xi_0 \in \Sigma$. Conversely, the continuity at the point zero of Σ implies that for $1 > 0$, there exists a real number $1/\delta > 0$ such that $|\xi^*(\xi)| \leqslant 1$ for every $\xi \in \Sigma$ with

$$\|\xi\| \leqslant \frac{1}{\delta}.$$

Let us now consider any $\eta \in \Sigma$, $\eta \neq \theta$, and put

$$\xi = \frac{\eta}{\delta \|\eta\|};$$

then

$$\|\xi\| \leqslant \frac{1}{\delta},$$

hence

$$|\xi^*(\xi)| = \frac{1}{\delta \|\eta\|} |\xi^*(\eta)| \leqslant 1;$$

this implies

$$|\xi^*(\eta)| \leqslant \delta \|\eta\|$$

for any $\eta \in \Sigma$, i.e., the boundedness of ξ^*.

Let now Σ^* be the set of all bounded linear functionals on Σ; then, if $\lambda_i \in R$ and $\xi_i^* \in \Sigma^*$, $i = 1, 2$, we have $\lambda_1 \xi_1^* + \lambda_2 \xi_2^* \in \Sigma^*$, i.e., Σ^* is also a vector space. The zero element of Σ^* is the functional which vanishes for all $\xi \in \Sigma$.

A norm may be introduced in Σ^* as follows:

$$\|\xi^*\| = \sup \{|\xi^*(\xi)| \colon \xi \in \Sigma, \ \|\xi\| \leqslant 1\}.$$

It is easy to prove that Σ^* is complete in this normed topology. Hence Σ^* is again a Banach space, called the *conjugate* or *adjoint* space of Σ. Let us now define:

$$\xi^*(\xi) = \langle \xi, \xi^* \rangle, \qquad \xi \in \Sigma, \qquad \xi^* \in \Sigma^*;$$

then the so defined function on $\Sigma \times \Sigma^*$ with values in R, can be considered as a continuous bilinear functional. In fact, we have

$$\langle \lambda\xi + \mu\eta, \xi^* \rangle = \lambda \langle \xi, \xi^* \rangle + \mu \langle \eta, \xi^* \rangle \tag{2}$$

$$\langle \xi, \lambda\xi^* + \mu\eta^* \rangle = \lambda \langle \xi, \xi^* \rangle + \mu \langle \eta, \xi^* \rangle, \tag{3}$$

and moreover we have

$$|\langle \xi, \xi^* \rangle| \leqslant \|\xi\| \, \|\xi^*\|. \tag{4}$$

The adjoint space Σ^*, considered as a Banach space, also possesses an adjoint space Σ^{**}. This consists of all bounded linear functionals defined on Σ^*. Formulas (3) and (4) show that, for fixed $\xi_0 \in \Sigma$,

$$\langle \xi_0, \xi^* \rangle = \xi_0^{**}(\xi^*),$$

for all $\xi^* \in \Sigma^*$, is a bounded linear functional on Σ^*, i.e., this functional $\xi_0^{**} = \langle \xi_0, . \rangle \in \Sigma^{**}$. By formula (4) we have $\|\xi_0^{**}\| \leqslant \|\xi_0\|$. Moreover it may be proved that the equality $\|\xi_0^{**}\| = \|\xi_0\|$ must hold.† In this way a mapping $\xi_0^{**} : \Sigma \to \Sigma^{**}$ of Σ into Σ^{**} can be defined, the so-called natural mapping of Σ into Σ^{**}, which is an isometric isomorphism of Σ on a subset Σ_0^{**} of Σ^{**}. Hence Σ can be considered as a Banach subspace of Σ^{**}. In the case $\Sigma = \Sigma^{**}$, the Banach space Σ is called *regular* or *reflexive*.

5.15. Let now

$$X \in \mathscr{E}(\mathfrak{B}, \Sigma) \quad \text{and} \quad X = \sum_{i \geqslant 1} \xi_i I_{a_i}$$

be a representation of X by indicators; then, for every $\xi^* \in \Sigma^*$, we have

$$\xi^*(X) = \sum_{i \geqslant 1} \xi^*(\xi_i) I_{a_i} = \sum_{i \geqslant 1} \langle \xi_i, \xi^* \rangle I_{a_i} \in \mathscr{E}(\mathfrak{B}, R).$$

It is easy to prove: If the sequence

$$X_n \in \mathscr{E}(\mathfrak{B}, \Sigma), \qquad n = 1, 2, \ldots,$$

is n-o-fundamental (resp n-u-fundamental), then

$$\xi^*(X_n) \in \mathscr{E}(\mathfrak{B}, R)$$

is o-fundamental (resp u-fundamental) for every $\xi^* \in \Sigma^*$. Since now every $X \in \mathscr{V}^*(\mathfrak{B}, \Sigma)$ is the o-limit of an o-convergent sequence

$$X_n \in \mathscr{E}(\mathfrak{B}, \Sigma)$$

and then for every $\xi^* \in \Sigma^*$, the sequence

$$\xi^*(X_n) \in \mathscr{E}(\mathfrak{B}, R), \qquad n = 1, 2, \ldots,$$

† see Hille–Phillips [1], p. 33.

is o-fundamental, i.e., o-convergent in $\mathscr{V}(\mathfrak{B}, R)$, we may define

$$\xi^*(X) = o\text{-}\lim_{n\to\infty} \xi^*(X_n).$$

It is easy to prove that this correspondence of an element $\xi^*(X) \in \mathscr{V}(\mathfrak{B}, R)$ to any $X \in \mathscr{V}^*(\mathfrak{B}, \Sigma)$, for every $\xi^* \in \Sigma^*$, is well-defined, i.e., independent of the choice of the sequence

$$X_n \in \mathscr{E}(\mathfrak{B}, \Sigma), \qquad n = 1, 2, \ldots,$$

with

$$o\text{-}\lim_{n\to\infty} X_n = X \quad \text{in} \quad \mathscr{V}(\mathfrak{B}, \Sigma).$$

The following theorem is true (see Hille–Phillips [1], p. 77, Theorem 3.7.1.).

Theorem 5.18.

Let $X \in \mathscr{V}^*(\mathfrak{B}, \Sigma)$ *with* $\xi^*(X) \in \mathscr{L}_1(\mathfrak{B}, R)$ *for every* $\xi^* \in \Sigma^*$; *then there exists a* $\xi^{**} \in \Sigma^{**}$ *such that*

$$\xi^{**}(\xi^*) = E(\xi^*(X)), \textit{ for every } \xi^* \in \Sigma^*.$$

Proof. We put

$$f(\xi^*) = E(\xi^*(X)) \quad \text{for every} \quad \xi^* \in \Sigma^*.$$

Obviously, f is a linear functional on Σ^*. It remains to prove that f is bounded. In order to prove that, one may consider the mapping

$$\Phi(\xi^*) = \xi^*(X) \in \mathscr{L}_1(\mathfrak{B}, R) \quad \text{for all} \quad \xi^* \in \Sigma^*,$$

which is linear as a mapping of the Banach space Σ^* into the Banach space $\mathscr{L}_1(\mathfrak{B}, R)$; it is easy to verify that Φ is bounded; then we have

$$|f(\xi^*)| = |E(\xi^*(X))| = |E(\Phi(\xi^*))| \leqslant E(|\Phi(\xi^*)|) \leqslant \|\Phi\| \, \|\xi^*\|;$$

hence

$$|f(\xi^*)| \leqslant \delta \|\xi^*\|$$

for every $\xi^* \in \Sigma$, i.e., f is bounded on Σ^*; thus

$$f = \xi^{**} \in \Sigma^{**}.$$

According to the above theorem one may define for every rv

$$X \in \mathscr{V}^*(\mathfrak{B}, \Sigma),$$

with

$$\xi^*(X) \in \mathscr{L}_1(\mathfrak{B}, R)$$

for all $\xi^* \in \Sigma^*$, the element $\xi^{**} \in \Sigma^{**}$ which exists and is uniquely determined such that

$$\xi^{**}(\xi^*) = E(\xi^*(X)),$$

as an expectation $E_P(X) = \xi^{**}$ in Σ^{**}; when, moreover ξ^{**} belongs to the range Σ_0^{**} of Σ in Σ^{**} by the natural mapping of Σ into Σ^{**}, then $E_P(X)$ can be replaced by an element of Σ and can be considered as an expectation of X in Σ. If the Banach space Σ is reflexive, i.e., $\Sigma = \Sigma^{**}$, then for any rv $X \in \mathscr{V}^*(\mathfrak{B}, \Sigma)$ the previously defined $E_P(X)$ exists in Σ if and only if $\xi^*(X) \in \mathscr{L}_1(\mathfrak{B}, R)$ for all $\xi^* \in \Sigma^*$. For more details about this definition of expectation, the reader can consult Chapter III of the book by Hille–Phillips. We only notice that it may be proved that for every $X \in \mathscr{L}_1(\mathfrak{B}, \Sigma)$ there exists also the expectation $E_P(X)$ and is equal to the Bochner expectation $E(X)$.

The topology defined in a Banach space with respect to the norm is often called the *strong topology*. In addition to the strong topology, another topology, called the *weak topology* is used in Banach spaces. A sequence $\xi_n \in \Sigma$, $n = 1, 2, \ldots$, is said to be *weakly* convergent to an element $\xi \in \Sigma$ if and only if, for every $\xi^* \in \Sigma^*$, the sequence $\xi^*(\xi_n)$ converges to $\xi^*(\xi)$ in R. Both these kinds of convergence defined in Σ induce a kind of convergence in $\mathscr{V}^*(\mathfrak{B}, \Sigma)$, the strong convergence, which is equivalent to the n-o-convergence and the weak convergence defined as follows:

A sequence
$$X_n \in \mathscr{V}^*(\mathfrak{B}, \Sigma), \qquad n = 1, 2, \ldots,$$
is said to be weakly convergent to a rv
$$X \in \mathscr{V}^*(\mathfrak{B}, \Sigma)$$
if and only if for every $\xi^* \in \Sigma^*$ the sequence
$$\xi^*(X_n) \in \mathscr{V}(\mathfrak{B}, R), \qquad n = 1, 2, \ldots,$$
o-converges to the rv $\quad \xi^*(X) \in \mathscr{V}(\mathfrak{B}, R)$.

It is easy to prove that strong convergence implies weak convergence and a strongly fundamental sequence is always a weakly fundamental sequence.

6. THE SPACES $\mathscr{L}_q$ OF RV'S HAVING VALUES IN A BANACH SPACE. MOMENTS.

6.1. Let q be any real positive number > 0; then we denote by
$$\mathscr{L}_q(\mathfrak{B}, \Sigma)$$
the set of all rv's $X \in \mathscr{V}^*(\mathfrak{B}, \Sigma)$ such that $\phi(X) \in \mathscr{L}_q(\mathfrak{B}, R)$, i.e. $(\phi(X))^q \in \mathscr{L}_1(\mathfrak{B}, R)$.† The positive real number $E\big((\phi(X))^q\big)$ is then said

† See Chapter VI.

to be the *q-th absolute moment* of X; we put

$$N_q(X) = \|X\|_q = \Big(E\big((\phi(X))^q\big)\Big)^{1/q} \quad \text{for every } X \in \mathscr{L}_q(\mathfrak{B}, \Sigma).$$

The case $\Sigma = R$ is stated in Chapter VI. In case $q = 1$, from Section 5, Theorem 5.17, it follows that the space $\mathscr{L}_1(\mathfrak{B}, \Sigma)$ defined here is equal to the space of all rv's $X \in \mathscr{V}^*(\mathfrak{B}, \Sigma)$, which possess an expectation in Σ defined in Section 5.

The following theorems are true:

Theorem 6.1.

The set $\mathscr{L}_q(\mathfrak{B}, \Sigma)$ is a vector space.

Proof. Let

$$X \in \mathscr{L}_q(\mathfrak{B}, \Sigma) \quad \text{and} \quad \lambda \in F;$$

then

$$(\phi(\lambda X))^q = |\lambda|^q\,(\phi(X))^q \in \mathscr{L}_1(\mathfrak{B}, R),$$

hence

$$\lambda X \in \mathscr{L}_q(\mathfrak{B}, \Sigma).$$

If

$$X,\ Y \in \mathscr{L}_q(\mathfrak{B}, \Sigma),$$

then

$$(\phi(X))^q \vee (\phi(Y))^q \in \mathscr{L}_1(\mathfrak{B}, R).$$

From the inequality

$$(\alpha+\beta)^q \leqslant 2^q \sup(\alpha^q, \beta^q),$$

which is true for real numbers $\alpha \geqslant 0$, $\beta \geqslant 0$, $q > 0$, it follows that

$$(\phi(X+Y))^q \leqslant (\phi(X)+\phi(Y))^q \leqslant 2^q\{(\phi(X))^q \vee (\phi(Y))^q\}.$$

But

$$2^q\{(\phi(X))^q \vee (\phi(Y))^q\} \in \mathscr{L}_1(\mathfrak{B}, R),$$

hence (Theorem 5.16)

$$(\phi(X+Y))^q \in \mathscr{L}_1(\mathfrak{B}, R),$$

i.e.,

$$X + Y \in \mathscr{L}_q(\mathfrak{B}, \Sigma).$$

Theorem 6.2.

$X \in \mathscr{L}_q(\mathfrak{B}, \Sigma)$ if and only if $\phi(X) \in \mathscr{L}_q(\mathfrak{B}, R)$, and then we have

$$\|X\|_q = \|\phi(X)\|_q.$$

The proof is obvious.

Theorem 6.3.

Let $X \in \mathscr{V}^*(\mathfrak{B}, \Sigma)$; if there exists a positive real-valued rv $Y \in \mathscr{L}_q(\mathfrak{B}, R)$ such that $\phi(X) \leqslant Y$ then $X \in \mathscr{L}_q(\mathfrak{B}, \Sigma)$.

Proof. We have

$$(\phi(X))^q \leqslant Y^q.$$

Since moreover

$$Y^q \in \mathscr{L}_1(\mathfrak{B}, R),$$

we deduce that

$$(\phi(X))^q \in \mathscr{L}_1(\mathfrak{B}, R),$$

i.e.,

$$X \in \mathscr{L}_q(\mathfrak{B}, \Sigma).$$

Obviously, we have

$$\mathscr{S}(\mathfrak{B}, \Sigma) \subseteq \mathscr{L}_q(\mathfrak{B}, \Sigma) \quad \text{for every} \quad q \quad \text{with} \quad 0 < q < +\infty.$$

6.2. Let q and s be two real numbers with $q > 1$, $s > 1$ and

$$\frac{1}{q} + \frac{1}{s} = 1;$$

then we say that q and s are conjugate to each other. In Chapter VI, Section 2, we have proved for real-valued rv's the so-called Holder's and Minkowski's inequalities. Both inequalities may be generalized here:

6.2.1. Let Σ_1, Σ_2, Σ_3 be three Banach spaces and let

$$b(\xi_1, \xi_2) \equiv \xi_1 \xi_2$$

be a bilinear mapping of $\Sigma_1 \times \Sigma_2$ into Σ_3 such that

$$\|\xi_1 \xi_2\| \leqslant \|\xi_1\| \, \|\xi_2\|.$$

This mapping induces a bilinear mapping

$$b(X_1, X_2) = X_1 X_2$$

of

$$\mathscr{V}^*(\mathfrak{B}, \Sigma_1) \times \mathscr{V}^*(\mathfrak{B}, \Sigma_2)$$

into

$$\mathscr{V}^*(\mathfrak{B}, \Sigma_3)$$

in the following way:

Let first

$$X_1 = \sum_{i \geqslant 1} \xi_{1i} I_{a_{1i}} \quad \text{and} \quad X_2 = \sum_{j \geqslant 1} \xi_{2j} I_{a_{2j}}$$

be representations by indicators of two rv's

$$X_1 \in \mathscr{E}(\mathfrak{B}, \Sigma_1) \quad \text{and} \quad X_2 \in \mathscr{E}(\mathfrak{B}, \Sigma_2);$$

then we define

$$b(X_1, X_2) = X_1 X_2 = \sum_{i \geqslant 1,\, j \geqslant 1} \xi_{1i} \xi_{2j} I_{a_{1i} \wedge a_{2j}}.$$

It is easy to see that

$$(1) \qquad \phi(X_1, X_2) \leqslant \phi(X_1) \phi(X_2).$$

Since the mapping $b(\xi_1, \xi_2)$ is continuous, we have: if

$$X_i \in \mathscr{V}^*(\mathfrak{B}, \Sigma_i) \quad \text{and} \quad X_{in} \in \mathscr{E}(\mathfrak{B}, \Sigma_i), \qquad n = 1, 2, \ldots,$$

with

$$X_{in} \overset{n\text{-}o}{\rightarrow} X_i, \qquad i = 1, 2,$$

then the sequence $X_{1n} X_{2n}$, $n = 1, 2, \ldots$, n-o-converges in $\mathscr{V}^*(\mathfrak{B}, \Sigma_3)$ and defines independently of the choice of the sequences

$$X_{in} \in \mathscr{E}(\mathfrak{B}, \Sigma_i), \qquad n = 1, 2, \ldots, \qquad i = 1, 2,$$

a rv

$$X_3 = n\text{-}o\text{-}\lim_{n \to \infty} X_{1n} X_{2n}$$

in $\mathscr{V}^*(\mathfrak{B}, \Sigma_3)$, which we put equal to

$$b(X_1, X_2) = X_1 X_2.$$

Obviously, we have

(2) $\phi(X_1 X_2) \leqslant \phi(X_1)\phi(X_2)$, for all $X_i \in \mathscr{V}^*(\mathfrak{B}, \Sigma_i)$, $i = 1, 2$.

Now the following theorem may be proved:

Theorem 6.4.

Let Σ_1, Σ_2, Σ_3 be three Banach spaces,

$$b(\xi_1, \xi_2) = \xi_1 \xi_2$$

be a bilinear mapping of $\Sigma_1 \times \Sigma_2$ into Σ_3 such that

$$\|\xi_1 \xi_2\| \leqslant \|\xi_1\| \, \|\xi_2\|$$

and q, s be real numbers conjugate to each other; then

$$X_1 \in \mathscr{L}_q(\mathfrak{B}, \Sigma_1) \quad \textit{and} \quad X_2 \in \mathscr{L}_s(\mathfrak{B}, \Sigma_2)$$

imply

$$b(X_1, X_2) = X_1 X_2 \in \mathscr{L}_1(\mathfrak{B}, \Sigma_3)$$

and the inequality

$$\|E(X_1 X_2)\| \leqslant \|X_1\|_q \, \|X_2\|_s,$$

i.e., the so-called generalized Hölder's inequality is true.

Proof. We have

$$\phi(X_1) \in \mathscr{L}_q(\mathfrak{B}, R) \quad \text{and} \quad \phi(X_2) \in \mathscr{L}_s(\mathfrak{B}, R);$$

hence (Theorem 2.3, Chapter VI)

$$\phi(X_1)\phi(X_2) \in \mathscr{L}_1(\mathfrak{B}, R).$$

This and the inequality (2) imply

$$\phi(X_1 X_2) \in \mathscr{L}_1(\mathfrak{B}, R),$$

i.e.,

$$X_1 X_2 \in \mathscr{L}_1(\mathfrak{B}, \Sigma).$$

Furthermore, we have

(3) $\|E(X_1 X_2)\| \leqslant E(\phi(X_1 X_2)) \leqslant E(\phi(X_1)\phi(X_2))$.

Hölder's inequality (Theorem 2.3, Chapter VI) gives now

(4) $$E(\phi(X_1)\phi(X_2)) = E(|\phi(X_1)\phi(X_2)|)$$
$$\leqslant \left(E\left((\phi(X_1))^q\right)\right)^{1/q} \left(E(\phi(X_2))^s\right)^{1/s}$$
$$= \|X_1\|_q \|X_2\|_q.$$

The inequalities (3) and (4) imply

$$\|E(X_1 X_2)\| \leqslant \|X_1\|_q \|X_2\|_s.$$

6.2.2. *Remark.* Let $\Sigma_1 = R$, $\Sigma_2 = \Sigma =$ any Banach space and $\Sigma_3 = \Sigma$; then the multiplication $\lambda\xi$, $\lambda \in R$, $\xi \in \Sigma$, is a bilinear mapping

$$R \times \Sigma \ni (\lambda, \xi) \Rightarrow \lambda\xi \in \Sigma \quad \text{with} \quad \|\lambda\xi\| = \|\lambda\| \, \|\xi\|;$$

hence a mapping

$$\mathscr{V}(\mathfrak{B}, R) \times \mathscr{V}^*(\mathfrak{B}, \Sigma) \ni (X, Y) \Rightarrow XY \in \mathscr{V}^*(\mathfrak{B}, \Sigma)$$

may be defined, the so-called multiplication of a real-valued rv X and a rv $Y \in \mathscr{V}^*(\mathfrak{B}, \Sigma)$. From Theorem 6.4 it follows then

Corollary 6.1.

If $X \in \mathscr{L}_q(\mathfrak{B}, R)$ *and* $Y \in \mathscr{L}_s(\mathfrak{B}, \Sigma)$, *then* $XY \in \mathscr{L}_1(\mathfrak{B}, \Sigma)$ *and the inequality*

$$\|E(XY)\| \leqslant \|X\|_q \, \|Y\|_s,$$

is true.

Theorem 6.5.

Let q and s be real numbers conjugate to each other; then

(1) $X \in \mathscr{L}_q(\mathfrak{B}, \Sigma)$ *implies* $(\phi(X))^{q-1} \in \mathscr{L}_s(\mathfrak{B}, R)$,

(2) $X \in \mathscr{L}_q(\mathfrak{B}, \Sigma)$ *if and only if* $(\phi(X))^{q-1} X \in \mathscr{L}_1(\mathfrak{B}, \Sigma)$.

Proof. (1) We have

$$((\phi(X))^{q-1})^s = (\phi(X))^q \in \mathscr{L}_1(\mathfrak{B}, R),$$

hence $$(\phi(X))^{q-1} \in \mathscr{L}_s(\mathfrak{B}, R).$$

(2) From $$X \in \mathscr{L}_q(\mathfrak{B}, \Sigma)$$

it follows $$(\phi(X))^{q-1} \in \mathscr{L}_s(\mathfrak{B}, R).$$

By Corollary 6.1 we have

$$(\phi(X))^{q-1} X \in \mathscr{L}_1(\mathfrak{B}, \Sigma).$$

Conversely, if

$$(\phi(X))^{q-1} X \in \mathscr{L}_1(\mathfrak{B}, \Sigma),$$

then $$(\phi(X))^q = \phi\big((\phi(X))^{q-1} X\big) \in \mathscr{L}_1(\mathfrak{B}, R),$$

hence $$X \in \mathscr{L}_q(\mathfrak{B}, \Sigma).$$

6.2.3. Minkowski's inequality may also be generalized.

Theorem 6.6.

Let q be a real number with $1 \leqslant q < +\infty$. If X and $Y \in \mathscr{L}_q(\mathfrak{B}, \Sigma)$, then the inequality

$$\| X + Y \|_q \leqslant \| X \|_q + \| Y \|_q$$

is true.

Proof. For $q = 1$, we have

$$\| X + Y \|_1 \leqslant \| X \|_1 + \| Y \|_1$$

(*cf.* Section 5.3, inequality (β)). Let us consider the case $1 < q < +\infty$ and the real number $s > 0$ such that

$$\frac{1}{q} + \frac{1}{s} \equiv 1.$$

We have

$$(1) \quad (\phi(X+Y))^q = (\phi(X+Y))^{q-1} \phi(X+Y)$$
$$\leqslant (\phi(X+Y))^{q-1} \phi(X) + (\phi(X+Y))^{q-1} \phi(Y).$$

Since $$X + Y \in \mathscr{L}_q(\mathfrak{B}, \Sigma),$$

we have $$(\phi(X+Y))^{q-1} \in \mathscr{L}_s(\mathfrak{B}, R)$$

and $$\phi(X) \in \mathscr{L}_q(\mathfrak{B}, R), \qquad \phi(Y) \in \mathscr{L}_q(\mathfrak{B}, R),$$

hence (by Theorem 2.3, Chapter VI)

$$(\phi(X+Y))^{q-1} \phi(X) \in \mathscr{L}_1(\mathfrak{B}, R)$$

and $$(\phi(X+Y))^{q-1} \phi(Y) \in \mathscr{L}_1(\mathfrak{B}, R).$$

From (1) it follows now

$$E\Big((\phi(X+Y))^q\Big) \leqslant E\Big(\phi(X)(\phi(X+Y))^{q-1}\Big)+E\Big(\phi(Y)(\phi(X+Y))^{q-1}\Big).$$

Using Hölder's inequality, we have

$$E\Big(\phi(X)(\phi(X+Y))^{q-1}\Big) \leqslant \|X\|_q \left(E\Big((\phi(X+Y))^q\Big)\right)^{1/s}$$

$$E\Big(\phi(Y)(\phi(X+Y))^{q-1}\Big) \leqslant \|Y\|_q \left(E\Big((\phi(X+Y))^q\Big)\right)^{1/s},$$

hence $$E\Big((\phi(X+Y))^q\Big) \leqslant \{\|X\|_q+\|Y\|_q\} \left(E\Big((\phi(X+Y))^q\Big)\right)^{1/s}.$$

Since $$\left(E\Big((\phi(X+Y))^q\Big)\right)^{1-(1/s)} = \left(E\Big((\phi(X+Y))^q\Big)\right)^{1/q} = \|X+Y\|_q,$$

we deduce $$\|X+Y\|_q \leqslant \|X\|_q+\|Y\|_q.$$

6.3. It is easy to see now that, for every q with $1 \leqslant q < +\infty$, the function $\|X\|_q$, $X \in \mathscr{L}_q(\mathfrak{B}, \Sigma)$ is a norm on $\mathscr{L}_q(\mathfrak{B}, \Sigma)$; for we have obviously:

(1) $\|X\|_q \geqslant 0$, and $\|X\|_q = 0$ if and only if $X = \theta$,

(2) $\|X+Y\|_q \leqslant \|X\|_q+\|Y\|_q$,

(3) $\|\lambda X\|_q = |\lambda|\ \|X\|_q$.

The topology defined on $\mathscr{L}_q(\mathfrak{B}, \Sigma)$ by the norm $\|\ \|_q$ will be called *the topology of the convergence in mean of order q* or *briefly N_q-convergence.*

The following theorems are true:

Theorem 6.7.

If the sequence

$$X_n \in \mathscr{L}_q(\mathfrak{B}, \Sigma), \qquad n = 1, 2, \ldots,$$

N_q-converges to $X \in \mathscr{L}_q(\mathfrak{B}, \Sigma)$, then

(I) $\lim_{n\to\infty} \|X_n\|_q = \|X\|_q$.

In fact the inequality

$$|\ \|X_n\|_q - \|X\|_q| \leqslant \|X_n - X\|_q$$

implies (I).

Theorem 6.8.

(*Lebesgue's theorem on term by term integration.*) *Let*

$$X_n \in \mathscr{L}_q(\mathfrak{B}, \Sigma), \qquad n = 1, 2, \ldots,$$

with
$$n\text{-}o\text{-}\lim_{n\to\infty} X_n = X.$$

Let
$$Y \in \mathscr{L}_q(\mathfrak{B}, R)$$

such that
$$Y \geqslant 0 \quad \text{and} \quad \phi(X_n) \leqslant Y, \qquad n = 1, 2, \ldots;$$

then
$$X \in \mathscr{L}_q(\mathfrak{B}, \Sigma)$$

and
$$N_q\text{-}\lim_{n\to\infty} X_n = X.$$

Proof. Obviously the sequence
$$(\phi(X_n))^q, \qquad n = 1, 2, \ldots,$$
n-o-converges to $|\phi(X)|^q$ and
$$(\phi(X_n))^q \leqslant Y^q, \qquad n = 1, 2, \ldots;$$
since
$$Y^q \in \mathscr{L}_1(\mathfrak{B}, R),$$
from Lebesgue's theorem for $\mathscr{L}_1(\mathfrak{B}, R)$ we deduce that
$$(\phi(X))^q \in \mathscr{L}_1(\mathfrak{B}, R),$$
and then
$$X \in \mathscr{L}_q(\mathfrak{B}, \Sigma).$$
The inequality
$$\phi(X_n) \leqslant Y, \qquad n = 1, 2, \ldots,$$
implies also that
$$\phi(X) \leqslant Y.$$

Therefore we have
$$\phi(X_n - X) \leqslant \phi(X_n) + \phi(X) \leqslant 2Y,$$
hence
$$(\phi(X_n - X))^q \leqslant 2^q\, Y^q.$$

Since moreover $X_n - X \overset{n\text{-}o}{\to} \theta$, we deduce again, from Lebesgue's theorem for $\mathscr{L}_1(\mathfrak{B}, R)$, that
$$(\phi(X_n - X))^q, \qquad n = 1, 2, \ldots,$$
N_1-converges to zero, i.e.,
$$\lim_{n\to\infty} \left(E(\phi(X_n - X))^q\right) = 0;$$

hence
$$\lim_{n\to\infty} \|X_n - X\|_q = 0,$$

i.e.,
$$N_q\text{-}\lim X_n = X.$$

6.4. *Exercises.* Let $X_n \in \mathscr{L}_q(\mathfrak{B}, \Sigma)$, $n = 1, 2, \ldots$, be such that

$$\sum_{n=1}^{\infty} \|X_n\|_q < +\infty.$$

Then:

1. The series $\sum_{n=1}^{\infty} X_n$ is absolutely *n-o*-convergent.

2. The rv $X = \sum_{n=1}^{\infty} X_n$ defined by the *n-o*-convergence of the series

belongs to $\mathscr{L}_q(\mathfrak{B}, \Sigma)$ and we have

$$\|X\|_q \leqslant \sum_{n=1}^{\infty} \|X_n\|_q.$$

3. The series $\sum_{n=1}^{\infty} X_n$ converges to X also in mean of order q.

6.5. We can prove that $\mathscr{L}_q(\mathfrak{B}, \Sigma)$ is a Banach space for every q with $1 \leqslant q < +\infty$. We shall prove first:

Theorem 6.9.

If $$X_n \in \mathscr{L}_q(\mathfrak{B}, \Sigma), \qquad n = 1, 2, \ldots,$$

is an N_q-fundamental sequence, then there exists a subsequence X_{n_k}, $k = 1, 2, \ldots$, which n-o-converges and N_q-converges to a rv $X \in \mathscr{L}_q(\mathfrak{B}, \Sigma)$.

Proof. Let ε be a positive number; then there exists a natural number n_0 such that $\|X_n - X_m\|_q < \varepsilon$ for all $n \geqslant n_0$ and $m \geqslant n_0$. We can now find by induction a strictly increasing sequence $n_1, n_2, \ldots$, in $\{1, 2, \ldots\}$ such that

$$\|X_{n_{k+1}} - X_{n_k}\|_q < \frac{1}{2^k}, \qquad k = 1, 2, \ldots.$$

Then

$$\sum_{k=1}^{\infty} \|X_{n_{k+1}} - X_{n_k}\|_q < +\infty.$$

By the exercises of Section 6.4, the series

$$\sum_{k=1}^{\infty} (X_{n_{k+1}} - X_{n_k})$$

N_q-converges to a rv $Y \in \mathscr{L}_q(\mathfrak{B}, \Sigma)$. Since

$$\sum_{i=1}^{k-1} (X_{n_{i+1}} - X_{n_i}) = X_{n_k} - X_{n_1},$$

$X_{n_k} - X_{n_1}$, $k = 1, 2, \ldots$, N_q-converges to Y; therefore the sequence X_{n_k}, $k = 1, 2, \ldots$, N_q-converges to $X = Y + X_{n_1}$.

From this theorem it follows easily:

Theorem 6.10.

The vector space $\mathscr{L}_q(\mathfrak{B}, \Sigma)$ is with respect to the norm $\| \ \|_q$ complete, i.e., a Banach space.

Proof. In fact, if

$$X_n \in \mathscr{L}_q(\mathfrak{B}, \Sigma), \qquad n = 1, 2, \ldots,$$

is N_q-fundamental, then there exists a subsequence X_{n_k}, $k = 1, 2, \ldots$, which *n-o-* and N_q-converges to a rv $X \in \mathscr{L}_q(\mathfrak{B}, \Sigma)$. Since the N_q-fundamental sequence X_n, $n = 1, 2, \ldots$, contains a subsequence converging to $X \in \mathscr{L}_q(\mathfrak{B}, \Sigma)$, it follows that X_n, $n = 1, 2, \ldots$, itself N_q-converges to $X \in \mathscr{L}_q(\mathfrak{B}, \Sigma)$.

The following theorems which we give as exercises are also true.

Theorem 6.11.

If $$X_n \in \mathscr{L}_q(\mathfrak{B}, \Sigma), \qquad n = 1, 2, \ldots,$$

N_q-converges to $X \in \mathscr{L}_q(\mathfrak{B}, \Sigma)$, then there exists a subsequence X_{n_k}, $k = 1, 2, \ldots$, n-o-converging to X.

Theorem 6.12.

If $X_n \in \mathscr{L}_q(\mathfrak{B}, \Sigma)$, $n = 1, 2, \ldots$, is an N_q-fundamental sequence n-o-converging to a rv $X \in \mathscr{V}^*(\mathfrak{B}, \Sigma)$, then $X \in \mathscr{L}_q(\mathfrak{B}, \Sigma)$ and the sequence X_n, $n = 1, 2, \ldots$, N_q-converges to X.

Theorem 6.13.

The vector space $\mathscr{S}(\mathfrak{B}, \Sigma)$ is n-o- and N_q-dense in $\mathscr{L}_q(\mathfrak{B}, \Sigma)$, i.e., for every $X \in \mathscr{L}_q(\mathfrak{B}, \Sigma)$ there exists a sequence $X_n \in \mathscr{S}(\mathfrak{B}, \Sigma)$ such that

$$X_n \xrightarrow{n\text{-}o} X \quad \text{and} \quad X_n \xrightarrow{Nq} X.$$

Theorem 6.14.

If
$$X_n \in \mathscr{L}_q(\mathfrak{B}, \Sigma), \qquad n = 1, 2, \ldots,$$
n-u-converges to
$$X \in \mathscr{V}^*(\mathfrak{B}, \Sigma),$$
then
$$X \in \mathscr{L}_q(\mathfrak{B}, \Sigma)$$
and
$$N_q\text{-}\lim X_n = X.$$

6.6. *Exercises.* (1) Let
$$X = \sum_{i \geqslant 1} \xi_i I_{a_i} \in \mathscr{E}(\mathfrak{B}, \Sigma)$$
and
$$Y^* = \sum_{j \geqslant 1} \eta_j^* I_{b_j} \in \mathscr{E}(\mathfrak{B}, \Sigma^*),$$

where Σ is a Banach space and Σ^* the conjugate space of Σ; then we define
$$\langle X, Y^* \rangle = \sum_{i, j} \langle \xi_i, \eta_j^* \rangle . I_{a_i \wedge b_j} \in \mathscr{E}(\mathfrak{B}, R).$$

Let now
$$X \in \mathscr{V}^*(\mathfrak{B}, \Sigma) \quad \text{and} \quad Y^* \in \mathscr{V}^*(\mathfrak{B}, \Sigma^*)$$

and let any sequence
$$X_n \in \mathscr{E}(\mathfrak{B}, \Sigma), \qquad \text{resp } Y_n^* \in \mathscr{E}(\mathfrak{B}, \Sigma^*), \qquad n = 1, 2, \ldots,$$
such that
$$X = n\text{-}o\text{-}\lim X_n, \qquad \text{resp } Y^* = n\text{-}o\text{-}\lim Y_n^*;$$
then there exists
$$o\text{-}\lim \langle X_n, Y_n^* \rangle \quad \text{in} \quad \mathscr{V}(\mathfrak{B}, R)$$
and is independent of the choice of the sequences X_n, Y_n^*, $n = 1, 2, \ldots$. Thus we may define
$$\langle X, Y^* \rangle = o\text{-}\lim \langle X_n, Y_n^* \rangle.$$

Prove: If q and s are real numbers with $1 < q < +\infty$, $1 < s < +\infty$ and
$$\frac{1}{q} + \frac{1}{s} = 1,$$
then for any
$$X \in \mathscr{L}_q(\mathfrak{B}, \Sigma) \quad \text{and} \quad Y^* \in \mathscr{L}_s(\mathfrak{B}, \Sigma^*)$$
we have
$$\langle X, Y^* \rangle \in \mathscr{L}_1(\mathfrak{B}, R)$$
and
$$|E(\langle X, Y^* \rangle)| \leqslant E(|\langle X, Y^* \rangle|) \leqslant \|X\|_q \, \|Y^*\|_s.$$

(2) Let Σ_1 and Σ_2 be two Banach spaces and $\upsilon : \Sigma_1 \to \Sigma_2$ a continuous linear mapping. If

$$X \in \mathscr{L}_q(\mathfrak{B}, \Sigma_1),$$

then
$$\upsilon \circ X \in \mathscr{L}_q(\mathfrak{B}, \Sigma_2)$$

and
$$\|\upsilon \circ X\|_q \leqslant \|\upsilon\| \, \|X\|_q.$$

VIII

COMPLEMENTS

1. THE RADON–NIKODYM THEOREM FOR THE BOCHNER INTEGRAL

1.1. In Chapter V, Section 4, Theorem 4.1, the so-called Radon–Nikodym theorem was formulated and proved in the case of real-valued rv's. This theorem may be formulated and proved in different directions in the case of rv's having values in a Banach space. Recently, a formulation of this theorem was given by M. A. Rieffel [1] with respect to the Bochner integral theory, which seems to be general and suitable for applications in probability theory. In the following, we shall state Rieffel's results without proofs and under the assumption that the measure space is a pr σ-algebra.

1.2. Let $(\mathfrak{B}, p)$ be a pr σ-algebra and Σ be a real Banach space. A function ψ defined on $\mathfrak{B}$ and having values in Σ is said to be a Σ-*valued measure* (or *signed measure*) on $\mathfrak{B}$ if and only if it is σ-additive, i.e.,

$$\psi\left(\bigvee_{n=1}^{\infty} a_n\right) = \sum_{n=1}^{\infty} \psi(a_n)$$

if $a_1, a_2, \ldots$ are pairwise disjoint.

Let $\mathcal{M}(\mathfrak{B}, \Sigma)$ be the set of all Σ-valued measures on $\mathfrak{B}$; then it is easy to prove that $\mathcal{M}(\mathfrak{B}, \Sigma)$ is a vector space. It may be proved that, for every $\psi \in \mathcal{M}(\mathfrak{B}, \Sigma)$, we have $\sup_{x \in \mathfrak{B}} \|\psi(x)\| < +\infty$, i.e., a Σ-valued measure on $\mathfrak{B}$ is always bounded.

We first notice that it is true for $\Sigma = R$, for a real-valued measure (signed measure) $\psi \in \mathcal{M}(\mathfrak{B}, R)$ can be represented as the difference of its upper and lower variations (*cf.* Chapter V, Section 3.3), i.e., $\psi = \psi^+ - \psi^-$, where ψ^+ and ψ^- are non-negative measures on $\mathfrak{B}$, hence bounded.

Now, for any $\psi \in \mathcal{M}(\mathfrak{B}, \Sigma)$, if Σ^* is the conjugate space of Σ, we have: $\xi^*(\psi(b))$, $b \in \mathfrak{B}$, is a real-valued measure on $\mathfrak{B}$, hence bounded, and this is true for every $\xi^* \in \Sigma^*$. This implies that $\psi(b)$, $b \in \mathfrak{B}$, is also bounded (*cf.* Dunford-Schwartz [1] Part I, p. 66, Theorem 20).

1.3. Let $\psi \in \mathcal{M}(\mathfrak{B}, \Sigma)$; then for every $b \in \mathfrak{B}$ the *total variation* of ψ on b, denoted by $v(\psi, b)$ is defined as

$$v(\psi, b) = \sup \sum_{i=1}^{n} \|\psi(b_i)\|$$

where the supremum is taken over all finite sequences $b_1, b_2, \ldots, b_n$ of pairwise disjoint elements in $\mathfrak{B}$ with $b_i \leqslant b$, $i = 1, 2, \ldots, n$.

It is easy to prove that:

the total variation

$$v(\psi, b) \equiv \bar{\psi}(b), \quad b \in \mathfrak{B},$$

considered as an extended-real-valued† function on $\mathfrak{B}$, is an extended non-negative measure on $\mathfrak{B}$; if

$$\bar{\psi}(b) = v(\psi, b) < +\infty,$$

then we say that ψ is with finite variation. A measure $\psi \in \mathcal{M}(\mathfrak{B}, \Sigma)$ is said to be *p-continuous* if and only if for every $\varepsilon > 0$ there exists a $\delta > 0$ such that $|\bar{\psi}(x)| < \varepsilon$, for all $x \in \mathfrak{B}$ with $p(x) < \delta$. Since $\bar{\psi}$ is bounded and σ-additive if ψ is with finite variation, in this case this definition is equivalent to the property $\psi(\varnothing) = 0$. Hence every $\psi \in \mathcal{M}(\mathfrak{B}, \Sigma)$ with finite variation is p-continuous.

1.4. We can now formulate the Radon–Nikodym theorem as follows:

Theorem. 1.1.

Let $(\mathfrak{B}, p)$ be a pr σ-algebra and Σ be a Banach space. Let $\psi \in \mathcal{M}(\mathfrak{B}, \Sigma)$; then there exists a rv $X \in \mathcal{L}_1(\mathfrak{B}, \Sigma)$ such that $\psi(b) = E(XI_b)$ for every

† i.e., having values in the interval $(-\infty, +\infty]$.

$b \in \mathfrak{B}$ *if and only if*

(1) ψ *is with finite variation, i.e., the total variation* $\bar{\psi}$ *of* ψ *is a finite measure on* $\mathfrak{B}$,

(2) *locally* ψ *somewhere has a compact average range, i.e., given* $b \neq \emptyset$ *there exists an element* $a \leqslant b$ *such that* $a \neq \emptyset$ *and*

$$A_a(\psi) \equiv \left\{ \frac{\psi(x)}{p(x)} : x \leqslant a,\ x \neq \emptyset \right\}$$

is relatively (norm) compact, or equivalently

(2*) *locally* ψ *somewhere has compact direction, i.e., given* $b \in \mathfrak{B}$, $b \neq \emptyset$, *there exists an* $a \leqslant b$ *and a compact subset* $K \subseteq \Sigma$ *not containing* θ, *such that* $a \neq \emptyset$ *and* $\psi(x)$ *is contained in the cone generated by* K *for all* $x \leqslant a$.

We notice that for any $X \in \mathscr{L}_1(\mathfrak{B}, \Sigma)$ the function $\psi_X(b) = E(XI_b)$, $b \in \mathfrak{B}$, is a Σ-valued measure, i.e., $\psi_X \in \mathscr{M}(\mathfrak{B}, \Sigma)$.

We notice that condition (2*) is satisfied for every real-valued signed measure $\psi \in \mathscr{M}(\mathfrak{B}, R)$, i.e., Rieffel's theorem contains the Theorem 4.1 of Chapter V.

2. CONDITIONAL PROBABILITY

2.1. Let $(\mathfrak{B}, p)$ be a pr σ-algebra. For any $a \in \mathfrak{B}$, $a \neq \emptyset$, the ratio $p(a \wedge b)/p(a)$ is called the *conditional probability* of $b \in \mathfrak{B}$ *given* a, or simply, *probability* of b *given* a and is denoted by

$$p_a(b) = \frac{p(a \wedge b)}{p(a)}, \quad b \in \mathfrak{B}.$$

Obviously we have

$$p(a \wedge b) = p(a)\, p_a(b).$$

By induction we obtain the multiplication rule:

If $a_i \in \mathfrak{B}$, $i = 1, 2, \ldots, n$, with $a_1 \wedge a_2 \wedge \ldots \wedge a_n \neq \emptyset$, then

$$p(a_1 \wedge a_2 \wedge \ldots \wedge a_n) = p(a_1)\, p_{a_1}(a_2) p_{a_1 \wedge a_2}(a_3) \ldots p_{a_1 \wedge a_2 \wedge \ldots \wedge a_{n-1}}(a_n).$$

Let $a \neq \emptyset$ be fixed; then $p_a(b)$ for all $b \in \mathfrak{B}$ is a function p_a on $\mathfrak{B}$, called the conditional probability given a on $\mathfrak{B}$. The function p_a is normed, non-negative and σ-additive on $\mathfrak{B}$, i.e., a σ-additive quasi-probability on $\mathfrak{B}$. Therefore $(\mathfrak{B}, p_a)$ is a quasi-probability σ-algebra.

2.2. Let

$$a\mathfrak{B} = \{x \in \mathfrak{B};\ \ x \leqslant a\}, \quad a \neq \emptyset;$$

then $p_a(x)$, for all $x \in a\mathfrak{B}$, is a probability on the Boolean σ-algebra $a\mathfrak{B}$, i.e., $(a\mathfrak{B}, p_a)$ is a pr σ-algebra. If $\mathfrak{N}_a$ is the σ-ideal of all $x \in \mathfrak{B}$ such that $p_a(x) = 0$, then the pr σ-algebra $(a\mathfrak{B}, p_a)$ is isometric to the quotient pr σ-algebra $(\mathfrak{B}/\mathfrak{N}_a, p_a)$. We notice that if the pr σ-algebra is homogeneous, then $(a\mathfrak{B}, p_a)$ is isometric to $(\mathfrak{B}, p)$ for every $a \neq \emptyset$, $a \in \mathfrak{B}$. The pr σ-algebra $(a\mathfrak{B}, p_a)$ is called the pr σ-algebra given a.

2.3. Let $\mathbf{a} = \{a_1, a_2, \ldots\}$ be an experiment in $(\mathfrak{B}, p)$; then for every $b \in \mathfrak{B}$ the erv

$$\sum_{i \geqslant 1} p_{a_i}(b)\, I_{a_i} \in \mathscr{E}(\mathfrak{B}, R)$$

is said to be the *conditional probability of b given the experiment* **a** and will be denoted by

$$P_{\mathbf{a}}(b) = \sum_{i \geqslant 1} p_{a_i}(b)\, I_{a_i}.$$

Let
$$\mathbf{a}_i = \{a_{i1}, a_{i2}, \ldots\}, \quad i = 1, 2, \ldots, k,$$

be experiments in $(\mathfrak{B}, p)$ and consider the experiments

$$\mathbf{a}_1 \wedge \mathbf{a}_2 \wedge \ldots \wedge \mathbf{a}_m, \quad m = 1, 2, \ldots, k.$$

Prove: the experiments $\mathbf{a}_1, \mathbf{a}_2, \ldots, \mathbf{a}_k$ are p-independent if and only if

$$P_{\mathbf{a}_1 \wedge \mathbf{a}_2 \wedge \ldots \wedge \mathbf{a}_{m-1}}(a_{mj}) = p(a_{mj})$$

for all $m = 2, 3, \ldots, k$ and all j.

The sequence of experiments

$$\mathbf{a}_i = \{a_{i1}, a_{i2}, \ldots\}$$

constitutes a *Markov's chain* if and only if

$$P_{\mathbf{a}_1 \wedge \mathbf{a}_2 \wedge \ldots \wedge \mathbf{a}_{n-1}}(a_{nq}) = P_{\mathbf{a}_{n-1}}(a_{nq})$$

for all $n = 2, 3, \ldots$ and $q = 1, 2, \ldots$.

Exercise. Let $\mathfrak{A} = \{a_1, a_2, \ldots\}$ be a countable subset of $\mathfrak{B}$, with $a_i \neq \emptyset$, $i = 1, 2, \ldots$; then the events of $\mathfrak{A}$ are p-independent if and only if, for every finite subset $\{i_1, i_2, \ldots, i_n\}$ of indices,

$$p_{a_{i_1} \wedge a_{i_2} \wedge \ldots \wedge a_{i_{n-1}}}(a_{i_n}) = p(a_{i_n}).$$

3. CONDITIONAL EXPECTATION

3.1. Let now Σ be a Banach space and

$$X = \sum_{i=1}^{n} \xi_i I_{a_i}$$

a rv in $\mathscr{S}(\mathfrak{B}, \Sigma)$; then, for any $a \neq \emptyset$, the equality

$$E_a(X) = \sum_{i=1}^{n} p_a(a_i)\, \xi_i = \frac{1}{p(a)} \cdot \sum_{i=1}^{n} p(a \wedge a_i)\, \xi_i$$

defines the so-called *expectation of X given a*. Let now $X \in \mathscr{E}(\mathfrak{B}, \Sigma)$ and

$$X = \sum_{j \geqslant 1} \xi_j I_{a_j}$$

be a representation of X by indicators; assume $E(X)$ exists in Σ; then the series

$$\sum_{j \geqslant 1} p_a(a_j)\, \xi_j = \frac{1}{p(a)} \cdot \sum_{j \geqslant 1} p(a \wedge a_j)\, \xi_j$$

converges and defines the conditional expectation of X given $a \neq \emptyset$.

Since

$$I_a X = \sum_{j \geqslant 1} \xi_i I_{a_i \wedge a},$$

we have

$$E_a(X) = \frac{1}{p(a)}\, E(I_a X),$$

i.e.,

$$p(a)\, E_a(X) = E(I_a X).$$

3.2. Let now $X \in \mathscr{L}_1(\mathfrak{B}, \Sigma)$; then $I_a X \in \mathscr{L}_1(\mathfrak{B}, \Sigma)$ and the equality

$$E_a(X) = \frac{1}{p(a)}\, E(I_a X)$$

defines the expectation of X given a.

In general, if for any $X \in \mathscr{V}^*(\mathfrak{B}, \Sigma)$ and $a \in \mathfrak{B}$, $a \neq \emptyset$, we have $I_a X \in \mathscr{L}_1(\mathfrak{B}, \Sigma)$, then we can define

$$E_a(X) = \frac{1}{p(a)}\, E(I_a X).$$

Consider $\mathscr{L}_1(a\mathfrak{B}, \Sigma)$ for any $a \in \mathfrak{B}$, $a \neq \emptyset$; then the set

$$\{X \in \mathscr{V}^*(\mathfrak{B}, \Sigma) : I_a X \in \mathscr{L}_1(\mathfrak{B}, \Sigma)\},$$

if we define $X = Y$ if and only if $I_a X = I_a Y$, is isomorphic to $\mathscr{L}_1(a\mathfrak{B}, \Sigma)$.

Let X be any rv in $\mathscr{L}_1(\mathfrak{B}, \Sigma)$, and $\mathbf{a} = \{a_1, a_2, \ldots\}$ be an experiment; then $I_{a_i} X \in \mathscr{L}_1(\mathfrak{B}, \Sigma)$, hence there exists $E_{a_i}(X)$ and

$$\sum_{i \geqslant 1} E_{a_i}(X) I_{a_i} = E_{\mathbf{a}}(X) \in \mathscr{E}(\mathfrak{B}, \Sigma);$$

this erv $E_{\mathbf{a}}(X)$ is said to be the *conditional expectation of X given the experiment* **a**. Obviously we have

$$E_{\mathbf{a}}(X) = \sum_{i \geqslant 1} \left(\frac{1}{p(a_i)} E(I_{a_i} X) \right) I_{a_i},$$

and for the conditional probability $P_{\mathbf{a}}(b)$ of $b \in \mathfrak{B}$ given the experiment **a** we have

$$P_{\mathbf{a}}(b) = E_{\mathbf{a}}(I_b).$$

3.3. Let now $\mathfrak{A}$ be the smallest Boolean σ-subalgebra of $\mathfrak{B}$ σ-generated by $\mathbf{a} = \{a_1, a_2, \ldots\}$; then for every $a \in \mathfrak{A}$, $a \neq \emptyset$, there exists a subset $\{a_{i_1}, a_{i_2}, \ldots\}$ of **a** such that $a = \bigvee_{j \geqslant 1} a_{i_j}$. Obviously, we have

$$E_a(X) = \frac{1}{p(a)} E(I_a X) = \frac{1}{p(a)} \sum_{j \geqslant 1} E(I_{a_{i_j}} X)$$

$$= \frac{1}{p(a)} \sum_{j \geqslant 1} p(a_{i_j}) E_{a_{i_j}}(X),$$

i.e.,
$$p(a) E_a(X) = E(I_a X) = \sum_{j \geqslant 1} p(a_{i_j}) E_{a_{i_j}}(X);$$

then we have

$$\sum_{j \geqslant 1} p(a_{i_j}) E_{a_{i_j}}(X) = E(I_a E_{\mathbf{a}}(X)),$$

hence
$$E(I_a X) = E(I_a E_{\mathbf{a}}(X)) \text{ for all } a \in \mathfrak{A}.$$

Let us now put $E_{\mathfrak{A}}(X)$ instead of $E_{\mathbf{a}}(X)$; then we have

$$E(I_a X) = E(I_a E_{\mathfrak{A}}(X));$$

i.e., given any rv $X \in \mathscr{L}_1(\mathfrak{B}, \Sigma)$ and a Boolean σ-subalgebra $\mathfrak{A}$ of $\mathfrak{B}$ σ-generated by the experiment $\mathbf{a} = \{a_1, a_2, \ldots\}$, there exists a rv

$$Y = E_{\mathfrak{A}}(X) \in \mathscr{L}_1(\mathfrak{A}, \Sigma)$$

such that

$$E(I_a X) = E(I_a Y) \text{ for all } a \in \mathfrak{A}.$$

We may also consider Y as the conditional expectation of $X \in \mathscr{L}_1(\mathfrak{B}, \Sigma)$ given the Boolean σ-subalgebra $\mathfrak{A}$ of $\mathfrak{B}$. This leads to the following definition.

3.4. Let $(\mathfrak{B}, p)$ be any pr σ-algebra and $(\mathfrak{A}, p)$ any pr σ-subalgebra of $(\mathfrak{B}, p)$. Then, given any $X \in \mathscr{L}_1(\mathfrak{B}, \Sigma)$, then any $Y \in \mathscr{L}_1(\mathfrak{A}, \Sigma)$ for which $E(I_a Y) = E(I_a X)$ for all $a \in \mathfrak{A}$ is said to be the *conditional expectation of X given* $\mathfrak{A}$.

Let us now put

$$\psi_X(b) = E(I_b X), \quad b \in \mathfrak{B},$$

and consider ψ_X restricted on $\mathfrak{A} \subseteq \mathfrak{B}$ as a measure on $\mathfrak{A}$ having values in Σ; then if this measure fulfills all the assumptions of the Radon–Nikodym theorem, the rv X possesses a conditional expectation of X given $\mathfrak{A}$ and it will be denoted by $E_{\mathfrak{A}} X$.

3.5. We can prove that for any real-valued rv $X \in \mathscr{L}_1(\mathfrak{B}, R)$ and any pr σ-subalgebra $\mathfrak{A}$ of $\mathfrak{B}$, the conditional expectation $E_{\mathfrak{A}} X$ always exists in $\mathscr{L}_1(\mathfrak{A}, R)$.

Proof. Obviously the stochastic space $\mathscr{V}(\mathfrak{A}, R)$ of all real-valued rv's over $(\mathfrak{A}, p)$ may be mapped isomorphically into the stochastic space $\mathscr{V}(\mathfrak{B}, R)$ of all real-valued rv's over $(\mathfrak{B}, p)$, so that $\mathscr{L}_1(\mathfrak{A}, R)$ is mapped into the subspace $\mathscr{L}_1(\mathfrak{B}, R)$ of $\mathscr{V}(\mathfrak{B}, R)$, and if X^* is the image of $X \in \mathscr{L}_1(\mathfrak{A}, R)$ by this mapping, then we have

$$E(I_a X^*) = E(I_a X) \text{ for all } a \in \mathfrak{A}.$$

Hence $\mathscr{L}_1(\mathfrak{A}, R)$ can be considered as a subspace of $\mathscr{L}_1(\mathfrak{B}, R)$. Let now $X \in \mathscr{L}_1(\mathfrak{B}, R)$; then, for every $a \in \mathfrak{A}$, we have $I_a X \in \mathscr{L}_1(\mathfrak{B}, R)$, hence $\psi_X(a) = E(I_a X)$ is defined for all $a \in \mathfrak{A}$ and is a real (signed) measure on $\mathfrak{A}$. Moreover, ψ_X is finite and absolutely continuous with respect to the probability p on $\mathfrak{A}$; hence, according to Theorem 4.1, Chapter V, there exists a rv $Y \in \mathscr{L}_1(\mathfrak{A}, R)$, such that

$$\psi_X(a) = E(I_a Y), \quad a \in \mathfrak{A}.$$

4. DISTRIBUTIONS OF RANDOM VARIABLES

4.1. Let $(\mathfrak{B}, p)$ be a pr σ-algebra, where the Boolean σ-algebra $\mathfrak{B}$ is without atoms. Let Σ be a Banach space and $\mathscr{V}^*(\mathfrak{B}, \Sigma)$, resp $\mathscr{E}(\mathfrak{B}, \Sigma)$ be the stochastic space of all Σ-valued rv's, resp of all Σ-valued erv's over $\mathfrak{B}$. Then, for every $X \in \mathscr{V}^*(\mathfrak{B}, \Sigma)$, there exists a sequence

$$X_n \in \mathscr{E}(\mathfrak{B}, \Sigma), \quad n = 1, 2, \ldots;$$

such that

$$n\text{-}o\text{-}\lim_{n \to \infty} X_n = X;$$

hence, according to Section 4.5, Chapter VII, $n\text{-}p\text{-}\lim X_n = X$.

Let now
$$X_n = \sum_{j \geqslant 1} \xi_{nj} I_{a_{nj}}$$
be a representation of X_n by indicators, for every $n = 1, 2, \ldots$. Let $\mathbf{B}_\Sigma$ be the Boolean σ-algebra of all Borel subsets of Σ. We shall say an element $\Psi \in \mathbf{B}_\Sigma$ possesses the property P_X if and only if

given $\delta > 0$ there exist an index n_0 and a real number $\varepsilon > 0$ such that for all $n \geqslant n_0$
$$p\Big(\bigvee_j a_{nj} : \xi_{nj} \in \Psi_\varepsilon\Big) < \delta$$
where
$$\Psi_\varepsilon = \{\xi \in \Sigma : \text{there exists an } \eta \in \overline{\Psi} \cap \overline{(\Psi^c)} \quad \text{with} \quad \|\xi - \eta\| < \epsilon\}.$$

It may be proved that the property P_X is independent of the choice of the sequence $X_n \in \mathscr{E}(\mathfrak{B}, \Sigma)$, $n = 1, 2, \ldots$, with n-o-$\lim X_n = X$ (*cf.* Georgiou [1]).

Let now $\mathbf{B}_X$ be defined as follows:
$$\mathbf{B}_X = \{\Psi \in \mathbf{B}_\Sigma : \Psi \text{ possesses the property } P_X\};$$
then $\mathbf{B}_X$ is a Boolean subalgebra of $\mathbf{B}_\Sigma$ and for every $\Psi \in \mathbf{B}_X$ there exists the
$$\lim_{n \to \infty} p\Big(\bigvee_j \{a_{nj} : \xi_{nj} \in \Psi\}\Big)$$
and is independent of the choice of the sequence
$$X_n \in \mathscr{E}(\mathfrak{B}, \Sigma), \quad n = 1, 2, \ldots.$$
Hence we may define
$$p_X(\Psi) = \lim_{n \to \infty} p\Big(\bigvee_j \{a_{nj} : \xi_{nj} \in \Psi\Big).$$
We notice that p_X is set-theoretically σ-additive and normed, i.e., $p_X(\Sigma) = 1$, on $\mathbf{B}_X$, i.e., $(\Sigma, \mathbf{B}_X, p_X)$ is a pr space, the so-called *distribution pr space* or *sample pr space* of X.

4.2. The pr space $(\Sigma, \mathbf{B}_X, p_X)$ can be extended to a Borel pr space $(\Sigma, \tilde{\mathbf{B}}_X, p_X)$, where $\tilde{\mathbf{B}}_X$ is the smallest Boolean σ-subalgebra of $\mathfrak{P}(\Sigma)$ containing $\mathbf{B}_X$. Moreover the pr space $(\Sigma, \tilde{\mathbf{B}}_X, p_X)$ can be extended to a Lebesgue pr space $(\Sigma, \mathbf{LB}_X, p_X)$, where $\mathbf{LB}_X$ is the smallest Boolean σ-subalgebra of $\mathfrak{P}(\Sigma)$ containing the Boolean σ-algebra $\tilde{\mathbf{B}}_X$ and satisfying the condition: $\Psi \in \mathbf{LB}_X$ with $p_X(\Psi) = 0$ implies $\Phi \in \mathbf{LB}_X$ for every $\Phi \subseteq \Psi$.

5. BOOLEAN HOMOMORPHISMS OF RV'S

5.1. In Section 1.1, Chapter VII, we noticed that a real-valued random variable can be characterized uniquely by a Boolean σ-homomorphism of $\mathbf{B}_R$ (the Boolean σ-algebra of all Borel subsets of the real line R) into $\mathfrak{B}$. Some analogous relation exists between a Σ-valued rv X and a Boolean σ-homomorphism of $\tilde{\mathbf{B}}_X$ into $\mathfrak{B}$. Namely, let $X \in \mathscr{V}^*(\mathfrak{B}, \Sigma)$ with

$$X = n\text{-}p\text{-}\lim_{n \to \infty} X_n, \quad X_n \in \mathscr{E}(\mathfrak{B}, \Sigma), \quad n = 1, 2, \ldots,$$

and

$$X_n = \sum_{j \geq 1} \xi_{nj} I_{a_{nj}}$$

be a representation of X_n by indicators, for $n = 1, 2, \ldots$; for every X_n we define the following function:

$$h_n^X : \mathbf{B}_X \ni \Psi \to h_n^X(\Psi) = \bigvee_j \{a_{nj} : \xi_{nj} \in \Psi\} \in \mathfrak{B};$$

then it is easy to prove that the sequence

$$h_n^X(\Psi) \in \mathfrak{B}, \quad n = 1, 2, \ldots,$$

converges in probability and defines a function

$$h^X(\Psi) = p\text{-}\lim h_n^X(\Psi), \text{ for all } \Psi \in \mathbf{B}_X,$$

independent of the choice of the sequence

$$X_n \in \mathscr{E}(\mathfrak{B}, \Sigma), \quad n = 1, 2, \ldots .$$

The function h^X is a Boolean σ-homomorphism of $\mathbf{B}_X$ into $\mathfrak{B}$ and can be extended to a Boolean σ-homomorphism $\tilde{h}^X$ of $\tilde{\mathbf{B}}_X$ into $\mathfrak{B}$ such that

$$p_X(\Psi) = p(\tilde{h}^X(\Psi)) \text{ for every } \Psi \in \tilde{\mathbf{B}}_X.$$

Moreover we have

$$X \neq Y \text{ implies } \tilde{h}^X \neq \tilde{h}^Y.$$

APPENDIX I

LATTICES

The prerequisites for my lectures are a knowledge of the elementary concepts of algebra, set theory and classical measure and integration theory. We shall mention here briefly the main concepts and theorems of lattice theory that are necessary for understanding the lectures. Most theorems are given without proof. The reader who is interested in a more detailed account of these theories may consult the following books: P. R. Halmos [3], Dwinger [1], Sikorski [5], Birkhoff [1], Carathéodory [6].

1. PARTIALLY ORDERED SETS

1.1. A non-empty set $\mathfrak{P}$ is said to be a *partially ordered set*, herein called a *po-set*, if and only if a binary relation $\leqslant$ is defined on $\mathfrak{P}$ satisfying the following conditions:

(1) $x \leqslant x$, for all $x \in \mathfrak{P}$ (reflexive).

(2) If $x \leqslant y$ and $y \leqslant x$, then $x = y$ (antisymmetric).

(3) If $x \leqslant y$ and $y \leqslant z$, then $x \leqslant z$ (transitive).

The dual (converse) relation is defined, as follows: $x \geqslant y$ if and only if $y \leqslant x$. It is also reflexive, antisymmetric and transitive.

A map h of a po-set $\mathfrak{P}$ into another po-set $\mathfrak{Q}$ is called *order preserving* if and only if the following condition is satisfied: If $x \leqslant y$ then $h(x) \leqslant h(y)$; the map h is an *isomorphic map* if it is one-one and if h and h^{-1} are order preserving.

1.2. A po-set $\mathfrak{C}$ is said to be a *chain* (also: a *totally* or *linearly ordered* set) if and only if the following condition holds:

(c) For every pair $(x, y) \in \mathfrak{C} \times \mathfrak{C}$ either $x \leqslant y$ or $y \leqslant x$.

Theorem 1.1.

Any non-empty subset of a po-set is itself a po-set under the same po-order relation.

2. LATTICES

2.1. Let $\mathfrak{L}$ be a po-set. An element $v \in \mathfrak{L}$ is called a *least upper bound* (lub) of the elements $x \in \mathfrak{L}$, $y \in \mathfrak{L}$ if and only if

(1) $x \leqslant v$, $y \leqslant v$;

(2) if $x \leqslant z$ and $y \leqslant z$ for an element $z \in \mathfrak{L}$, then $v \leqslant z$.

The concept of *greatest lower bound* (*glb*) is defined dually.

A *lattice* is a po-set $\mathfrak{L}$ in which every two elements have a lub and a glb. Instead of lub and glb we shall also use respectively the terms *supremum* or *join* and *infimum* or *meet*. The join and the meet of two elements x, y will be denoted by $x \vee y$ and $x \wedge y$, respectively.

Example. The set $\mathfrak{P}(E)$ of all subsets of a non-empty set E. Here join and meet are the usual set-theoretic union and intersection respectively.

Every finite sequence $x_i \in \mathfrak{L}$, $i = 1, 2, \ldots, n$, has a join and meet in $\mathfrak{L}$ denoted by $\bigvee_{i=1}^{n} x_i$ and $\bigwedge_{i=1}^{n} x_i$ respectively.

In a lattice, the operations join and meet satisfy the following relations:

commutative : $x \vee y = y \vee x, \quad x \wedge y = y \wedge x$

associative : $x \vee (y \vee z) = (x \vee y) \vee z, \quad x \wedge (y \wedge z) = (x \wedge y) \wedge z$

idempotent : $x \vee x = x, \quad x \wedge x = x$

absorption laws : $x \vee (x \wedge y) = x, \quad x \wedge (x \vee y) = x.$

Moreover, $x \leqslant y$ if and only if $x \wedge y = x$, equivalently $x \vee y = y$.

Exercise. Prove that every algebra $\mathfrak{L}$ with two binary operations $\vee$ and $\wedge$, satisfying the above relations is a lattice under the partial ordering defined by: $x \leqslant y$ if and only if $x \wedge y = x$. Show that $x \vee y$ and $x \wedge y$ are lub and glb respectively.

A *sublattice* of a lattice $\mathfrak{L}$ is a non-empty subset $\mathfrak{M} \subseteq \mathfrak{L}$, which is closed under the lattice operations $\vee$ and $\wedge$. A map h of a lattice $\mathfrak{L}_1$ into or onto another lattice $\mathfrak{L}_2$ is called a *homomorphic map* if and only if it preserves the lattice operations, that is:

$$h(x \vee y) = h(x) \vee h(y)$$

$$h(x \wedge y) = h(x) \wedge h(y).$$

The order relation is obviously also preserved by a lattice homomorphic map.

If the homomorphic map h is one-one, then we say h is an *isomorphic map* of $\mathfrak{L}_1$ into or onto $\mathfrak{L}_2$.

2.2. Now let $\mathfrak{L}$ be a lattice and a_i, $i \in I$, be a family of elements of $\mathfrak{L}$. If the lub and the glb of this family exist in $\mathfrak{L}$ we write:

$$(\mathfrak{L}) \bigvee_{i \in I} a_i \text{ or } \sup_{i \in I}^{\mathfrak{L}} a_i \quad \text{and} \quad (\mathfrak{L}) \bigwedge_{i \in I} a_i \text{ or } \inf_{i \in I}^{\mathfrak{L}} a_i,$$

respectively. A lattice $\mathfrak{L}$ is called $\aleph$-*complete*,† where $\aleph$ is a cardinal number $\geqslant \aleph_0$, if every family a_i, $i \in I$, with $0 < |I| \leqslant \aleph$ ($|I| \equiv$ cardinal number of I) has a supremum and an infimum in $\mathfrak{L}$. A lattice $\mathfrak{L}$ is called *complete* if it is $\aleph$-complete for every cardinal $\aleph$.

Exercise 1. Prove: If the lattices $\mathfrak{L}_1$ and $\mathfrak{L}_2$ are isomorphic as po-sets, then they are isomorphic as lattices.

2.3. An element $e \in \mathfrak{L}$ ($\emptyset \in \mathfrak{L}$) is said to be the *largest* or *unit* (*smallest* or *zero*) element of $\mathfrak{L}$ if and only if $x \leqslant e$ ($\emptyset \leqslant x$) for every $x \in \mathfrak{L}$.

Exercise 2. Prove: A complete lattice has a unit and a zero element.

Theorem 2.1.

Suppose that $\mathfrak{P}$ is a po-set with unit e and suppose $\mathfrak{P}$ is complete with respect to the operation $\wedge$, i.e., every family $a_i \in \mathfrak{P}$, $i \in I$, has an infimum in $\mathfrak{P}$. Then $\mathfrak{P}$ is also complete with respect to the operation $\vee$, i.e., it is a complete lattice.

2.4. *Distributive lattices.* A lattice $\mathfrak{L}$ is called *distributive* if the following condition holds:

(d) $x \wedge (y \vee z) = (x \wedge y) \vee (x \wedge z)$.

† instead of complete, the term *saturated* is also used.

We remark that (d) is equivalent to its dual:

(d′) $x \vee (y \wedge z) = (x \vee y) \wedge (x \vee z)$.

Complete lattices may satisfy various infinite distributive properties, for example:

(α) $y \wedge \bigvee_{i \in I} a_i = \bigvee_{i \in I} (y \wedge a_i)$ whenever $y \in \mathfrak{L}$, $a_i \in \mathfrak{L}$ $(i \in I)$.

(β) $y \vee \bigwedge_{i \in I} a_i = \bigwedge_{i \in I} (y \vee a_i)$ whenever $y \in \mathfrak{L}$, $a_i \in \mathfrak{L}$ $(i \in I)$.

(γ) $\bigvee_{i \in I} \bigwedge_{j \in J} a_{ij} = \bigwedge_{\phi \in \mathbf{F}} \bigvee_{i \in I} a_{i,\phi(i)}$ whenever $a_{ij} \in \mathfrak{L}$, $(i, j) \in I \times J$.

(δ) $\bigwedge_{i \in I} \bigvee_{j \in J} a_{ij} = \bigvee_{\phi \in \mathbf{F}} \bigwedge_{i \in I} a_{i,\phi(i)}$ whenever $a_{ij} \in \mathfrak{L}$, $(i, j) \in I \times J$.

In (γ) and (δ), $\mathbf{F}$ is the set of all mappings of I into J.

(α) and (β) are not, in general, equivalent, but each of them implies (d) and (d′). A complete lattice satisfying one of the relations (γ) and (δ) satisfies the other one and is said to be a *completely distributive* lattice.

2.5. A lattice $\mathfrak{L}$ with zero element $\emptyset$ and unit element e is said to be *complemented* if for every $x \in \mathfrak{L}$, there exists an element $x^* \in \mathfrak{L}$ such that $x \vee x^* = e$ and $x \wedge x^* = \emptyset$. x^* is called a *complement* of x.

Theorem 2.2.

If $\mathfrak{L}$ is a distributive lattice with a unit and a zero element then the complementation is unique, i.e., there exists at most one complement for every element.

2.6. *Orthocomplemented and orthomodular lattices.* A lattice $\mathfrak{L}$ with unit and zero elements is orthocomplemented if there exists a mapping $a \to a^\perp$ of $\mathfrak{L}$ onto itself such that $a^{\perp\perp} = a$, $a^\perp \vee a = e$, $a^\perp \wedge a = \emptyset$ and if $a \leqslant b$ then $a^\perp \geqslant b^\perp$. The element $a^\perp$ is called the *orthocomplement* of a. An *orthomodular* lattice is an orthocomplemented lattice which satisfies the condition:

(m) If $x \leqslant z$ then $x \vee (y \wedge z) = (x \vee y) \wedge z$.

Any lattice which satisfies the condition (m) is said to be a *modular* lattice. Hence an orthomodular lattice is orthocomplemented and modular.

3. BOOLEAN ALGEBRAS

3.1. A lattice-theoretic definition of a Boolean algebra is the following:

A *Boolean algebra* $\mathfrak{B}$ is a lattice with unit e and zero $\varnothing$, which is distributive and complemented. Now it follows from the preceding sections that a Boolean algebra can also be defined as an algebra with two binary operations, $\vee$, $\wedge$ and with one unitary operation (formation of complements, i.e., a map $a \to a^c$ of $\mathfrak{B}$ into itself) satisfying the following properties:

$$x \vee y = y \vee x \qquad x \wedge y = y \wedge x$$

$$x \vee (y \vee z) = (x \vee y) \vee z \qquad x \wedge (y \wedge z) = (x \wedge y) \wedge z$$

$$x \vee x = x \qquad x \wedge x = x$$

$$x \vee (x \wedge y) = x \qquad x \wedge (x \vee y) = x$$

$$x \wedge (y \vee z) = (x \wedge y) \vee (x \wedge z) \qquad x \vee (y \wedge z) = (x \vee y) \wedge (x \vee z)$$

$$x \vee \varnothing = x \qquad x \wedge e = x$$

$$x \wedge \varnothing = \varnothing \qquad x \vee e = e$$

$$x \vee x^c = e \qquad x \wedge x^c = \varnothing.$$

In $\mathfrak{B}$ we can introduce a partial ordering as follows: $a \leqslant b$ if and only if $a \vee b = b$, or equivalently $a \wedge b = a$. The complementation is unique in $\mathfrak{B}$ and to every $a \leqslant b$ there exists a relative complement, i.e., an element x such that $b = a \vee x$ and $a \wedge x = \varnothing$; we call this relative complement "the difference $b - a$". For arbitrary a and b in $\mathfrak{B}$ we can now define $a - b = a - a \wedge b$.

3.2. The following relations hold in $\mathfrak{B}$:

(1) $x \vee y = x \vee (x^c \wedge y)$

(2) $x \wedge y = x \wedge (x^c \vee y)$

(3) $x \leqslant y$ if and only if $x \wedge y^c = \varnothing$, equivalently $x^c \vee y = e$

(4) If $x \leqslant y$ then $x^c \geqslant y^c$

(5) $x^{cc} = x$.

In a distributive lattice $\mathfrak{R}$ with zero $\varnothing$, in which for every $a \leqslant b$ there exists a complement of a relative to b, i.e., an element $x \in \mathfrak{R}$ such that $a \vee x = b$ and $a \wedge x = \varnothing$, the relative complement x is also uniquely determined and denoted by $b - a$; furthermore, to every two elements x and y of $\mathfrak{R}$, there corresponds uniquely a difference $x - y = (x - x \wedge y)$.

We define a third important binary operation, the so-called *addition modulo* 2, or *symmetric difference*, as follows:

$$x+y = (x^c \wedge y) \vee (x \wedge y^c).$$

Then $\mathfrak{B}$, relative to $+$ as addition and $\wedge$ as multiplication, is an algebraic commutative ring with unit e and zero $\emptyset$. That is, denoting $x \wedge y$ by xy for convenience, we have the following properties:

$$x+y = y+x \qquad xy = yx$$

$$x+(y+z) = (x+y)+z \qquad x(yz) = (xy)z$$

$$x+\emptyset = x \qquad xe = x$$

$$x(y+z) = xy+xz.$$

Moreover $x+x = \emptyset$, i.e., algebraic difference and addition are identical. Therefore, $a+b = x$ implies $a = x+b$, $b = x+a$ and $a+b+x = \emptyset$. The algebraic multiplication is identical to the meet, hence idempotent. We have $x^c = e+x$.

We notice that the addition $+$ can be defined also in a distributive and relatively complemented, with zero element, lattice $\mathfrak{R}$, namely:

$$x+y = (x-y) \vee (y-x) = (x-x \wedge y) \vee (y-x \wedge y);$$

then $\mathfrak{R}$ is with respect to the operations $+$ as addition and $\wedge$ as multiplication an algebraic commutative ring perhaps without unit e (neutral element of the multiplication). Such an algebraic ring is called also a *Boolean ring*; it is also an idempotent ring, but not necessarily with unit.

3.3. Conversely, we can define a Boolean algebra as an idempotent algebraic ring with unit e, i.e., as an algebra with two operations " + " and " . ", satisfying the properties:

A. Addition

(1) $a+b = b+a$

(2) $a+(b+c) = (a+b)+c$

(3) For every pair a, b of elements in $\mathfrak{B}$, there exists at least one element $x \in \mathfrak{B}$ satisfying $a+x = b$.

Thus $\mathfrak{B}$, relative to the operation $+$, is a group and hence there exists a unique element $\emptyset \in \mathfrak{B}$ such that $a+\emptyset = a$ for every $a \in \mathfrak{B}$, i.e., there exists a zero element $\emptyset$ in $\mathfrak{B}$.

B. Multiplication

(1) $a(bc) = (ab)\,c$

(2) $(a+b)\,c = ac+bc$

(3) $c(a+b) = ca+cb$

(4) There exists an element $e \in \mathfrak{B}$ such that $ea = ae$ for every $a \in \mathfrak{B}$

(5) $aa = a^2 = a$.

We can now prove the commutativity of multiplication:

(6) $ab = ba$

together with

(7) $a\varnothing = \varnothing$

(8) $a+a = \varnothing$

and the fact that

(9) The equation $a+x = b$ has a solution x which is identical with $b+a$, i.e., algebraic difference and addition are identical.

3.4. We can introduce in $\mathfrak{B}$ a partial ordering relation by defining: $a \leqslant b$ if and only if $ab = a$, and prove that $\varnothing \leqslant x \leqslant e$, for every $a \in \mathfrak{B}$. Multiplication is identical with meet (infimum), i.e., $a \wedge b = ab$, and the join $\vee$ is related to the two operations + and . by the identity:

$$a \vee b = a+b+ab.$$

We also have

$$(a \vee b) \wedge x = (a \wedge x) \vee (b \wedge x),$$

i.e., $\mathfrak{B}$ is a distributive lattice. Further, for every $a \in \mathfrak{B}$ there exists a complement, given by $a^c = e+a$, i.e., $\mathfrak{B}$ as a po-set is a Boolean algebra. For proofs consult the books of Carathéodory ([6], Chapter I) and Dwinger [1] mentioned at the beginning of this Appendix.

Prove that: In any idempotent algebraic ring $\mathfrak{R}$ without unit e, the lattice structure can also be introduced and it can be proved that it is, with respect to this lattice structure, a relatively complemented and distributive lattice with zero.

3.5. We can prove that Boolean algebras always satisfy the distributive laws (α) and (β) of Section 2.4 (as far as the joins and meets involved exist). More precisely we have the following:

Theorem 3.1.

Let $\mathfrak{B}$ be a Boolean algebra and let $a_i \in \mathfrak{B}$, $i \in I$, be any family of elements of $\mathfrak{B}$ such that $(\mathfrak{B}) \bigvee_{i \in I} a_i$ exists. Then $(\mathfrak{B}) \bigvee_{i \in I} (x \wedge a_i)$ exists for every $x \in \mathfrak{B}$ and we have:

$$x \wedge \bigvee_{i \in I} a_i = \bigvee_{i \in I} (x \wedge a_i).$$

The dual theorem also holds.

Theorem 3.2.

Let $\mathfrak{B}$ be an $\aleph$-complete Boolean algebra; then we have:

$$\left(\bigvee_{i \in I} x_i\right) \wedge \left(\bigvee_{j \in J} y_j\right) = \bigvee_{i,\, j} (x_i \wedge y_j)$$

and dually, for every I and J with $|I| \leqslant \aleph$, $|J| \leqslant \aleph$.

Theorem 3.3.

Let $\mathfrak{B}$ be a Boolean algebra with $|\mathfrak{B}| \geqslant \aleph_0$; then there exists an infinite set $\mathfrak{S}$ of pairwise disjoint elements of $\mathfrak{B}$.

We shall call an $\aleph_0$-complete Boolean algebra a *Boolean σ-algebra.*

Theorem 3.4.

If a Boolean algebra $\mathfrak{B}$ is $\aleph$-complete relative to meet, then $\mathfrak{B}$ is $\aleph$-complete relative to join, and conversely.

Theorem 3.5.

Let $\mathfrak{B}$ be a Boolean σ-algebra with $|\mathfrak{B}| \geqslant \aleph_0$; then $|\mathfrak{B}| > \aleph_0$, i.e., the set $\mathfrak{B}$ is uncountable.

We say that a Boolean algebra satisfies the *countable chain condition* if every set of pairwise disjoint elements of $\mathfrak{B}$ is countable.

Theorem 3.6.

Let $\mathfrak{B}$ be a Boolean σ-algebra satisfying the countable chain condition; then $\mathfrak{B}$ is complete.

We have proved this theorem for the Boolean algebra of a pr σ-algebra in Section 2.2, Chapter 2.

3.6. We shall call an element $a \neq \emptyset$ of a Boolean algebra $\mathfrak{B}$ an *atom* in $\mathfrak{B}$ if and only if the following condition holds:

(α) If $x \leqslant a$, then either $x = a$ or $x = \emptyset$.

A Boolean algebra $\mathfrak{B}$ is said to be *atomic* if and only if for every $b \in \mathfrak{B}$ with $b \neq \emptyset$ there exists an atom a such that $a \leqslant b$. The Boolean algebra $\mathfrak{P}(E)$ of all the subsets of a set E is atomic; all one-point subsets $\{x\} \subseteq E$, $x \in E$, are atoms. A Boolean algebra without atoms is said to be *atomless*.

One can prove that a Boolean algebra $\mathfrak{B}$ is atomic if and only if the unit element e is representable as the join of all atoms of $\mathfrak{B}$.

3.7. Let $\mathfrak{B}$ be a Boolean $\aleph$-algebra, i.e., a Boolean algebra which, considered as a lattice, is $\aleph$-complete. A subset $\mathfrak{A}$ of $\mathfrak{B}$ is said to be a *Boolean* $\aleph^*$ *sub-algebra* of $\mathfrak{B}$, where $2 \leqslant \aleph^* \leqslant \aleph$, if and only if:

(I) $\mathfrak{A}$ is a Boolean subalgebra of $\mathfrak{B}$, (i.e., (1) $e \in \mathfrak{A}$, $\emptyset \in \mathfrak{A}$; (2) if $a \in \mathfrak{A}$ and $b \in \mathfrak{A}$, then $a \vee b$ and $a \wedge b \in \mathfrak{A}$; (3) if $a \in \mathfrak{A}$ then $a^c \in \mathfrak{A}$), and

(II) For every family $a_i \in \mathfrak{A}$, $i \in I$, with $|I| \leqslant \aleph^*$, $(\mathfrak{B}) \bigwedge_{i \in I} a_i \in \mathfrak{A}$.

Obviously we then have

$$(\mathfrak{A}) \bigwedge_{i \in I} a_i = (\mathfrak{B}) \bigwedge_{i \in I} a_i \in \mathfrak{A}.$$

Let $\mathfrak{A}$ be a Boolean $\aleph^*$-subalgebra of a Boolean $\aleph$-algebra $\mathfrak{B}$, where $\aleph^* \leqslant \aleph$, and let $\aleph^{**}$ be another cardinal number. We shall say that $\mathfrak{A}$ is an $\aleph^{**}$*-regular* (or $\aleph^{**}$*-invariant*) Boolean $\aleph^*$-subalgebra of the Boolean $\aleph$-algebra $\mathfrak{B}$ if and only if the following condition holds:

If $(\mathfrak{A}) \bigwedge_{i \in I} a_i$ exists with $a_i \in \mathfrak{A}$, $i \in I$, and $|I| \leqslant \aleph^{**}$, then $(\mathfrak{B}) \bigwedge_{i \in I} a_i$ exists and we have

$$(\mathfrak{B}) \bigwedge_{i \in I} a_i = (\mathfrak{A}) \bigwedge_{i \in I} a_i.$$

If $\aleph^{**} \leqslant \aleph^* \leqslant \aleph$, then every Boolean $\aleph^*$-subalgebra of a Boolean $\aleph$-algebra is $\aleph^{**}$-regular. In particular, a Boolean σ-subalgebra of a Boolean σ-algebra is always σ-regular.

A Boolean subalgebra of a Boolean $\aleph$-algebra is always finitely regular, but not always $\aleph^*$-regular if $\aleph^* \geqslant \aleph_0$.

Example. Let $\mathfrak{P}(\Omega)$ be the Boolean algebra of all subsets of

$$\Omega \equiv \{\xi \in R : 0 \leqslant \xi < 1\}$$

and $\mathfrak{A}$ the Boolean subalgebra of $\mathfrak{P}(\Omega)$ generated by the system of all half-open subintervals $[\alpha, \beta) \subseteq [0, 1) = \Omega$. $\mathfrak{A}$ is not a σ-regular Boolean subalgebra of $\mathfrak{P}(\Omega)$, because

$$(\mathfrak{A}) \bigwedge_{\nu=1}^{\infty} \left[0, \frac{1}{\nu}\right) = \varnothing,$$

but

$$(\mathfrak{P}(\Omega)) \bigwedge_{\nu=1}^{\infty} \left[0, \frac{1}{\nu}\right) = \{0\} \neq \varnothing,$$

that is the one-point subset $\{0\}$ of $[0, 1) = \Omega$.

3.8. Let $\mathfrak{B}$ be a Boolean $\aleph$-algebra with $\aleph \geqslant 2$ and $\mathfrak{K}$ a non-empty subset of $\mathfrak{B}$; then there always exists a smallest Boolean $\aleph^*$-subalgebra of $\mathfrak{B}$ containing $\mathfrak{K}$, for every $\aleph^*$ with $2 \leqslant \aleph^* \leqslant \aleph$; this Boolean $\aleph^*$-subalgebra is uniquely determined and will be denoted by $b_{\aleph^*}(\mathfrak{K}) \subseteq \mathfrak{B}$.

If we have $\mathfrak{B} = b_{\aleph^*}(\mathfrak{K})$, then we shall say that the subset $\mathfrak{K}$ is a $\aleph^*$-*generating basis* of $\mathfrak{B}$ or $\mathfrak{B}$ is $\aleph^*$-*generated by* $\mathfrak{K}$. If $\aleph^{**} > \aleph^*$, then $b_{\aleph^*}(\mathfrak{K})$ is not always an $\aleph^{**}$-regular Boolean $\aleph^*$-subalgebra of $\mathfrak{B}$. For example, the smallest Boolean subalgebra $b(\mathfrak{K})$ of a Boolean σ-algebra containing a subset $\mathfrak{K}$ of $\mathfrak{B}$ is not always σ-regular (*cf.* Example, Section 3.7).

4. HOMOMORPHISMS AND IDEALS OF A BOOLEAN ALGEBRA

4.1. A map $h : \mathfrak{B} \ni x \Rightarrow h(x) \in \mathfrak{B}^*$ of a Boolean algebra $\mathfrak{B}$ into a Boolean algebra $\mathfrak{B}^*$ is a *Boolean homomorphic map*, if it preserves the three Boolean operations, i.e.,

$$h(x \vee y) = h(x) \vee h(y)$$

$$h(x \wedge y) = h(x) \wedge h(y)$$

$$[h(x)]^c = h(x^c).$$

Theorem 4.1.

A map h of a Boolean algebra $\mathfrak{B}$ into a Boolean algebra $\mathfrak{B}^*$ is homomorphic if and only if it preserves the two operations of addition mod 2 *and multiplication* $. = \wedge$, *i.e.*,

$$h(x+y) = h(x)+h(y)$$

$$h(xy) = h(x)\,h(y)$$

and, moreover, maps the unit of $\mathfrak{B}$ *onto the unit of* $\mathfrak{B}^*$, *i.e.* $h(e) = e^*$.

Obviously the image of the zero of $\mathfrak{B}$ in $\mathfrak{B}^*$, under a homomorphic map, is the zero of $\mathfrak{B}^*$. We remark that a Boolean homomorphism is a special type of lattice homomorphism. It is easy to prove that a Boolean homomorphic map of $\mathfrak{B}$ into $\mathfrak{B}^*$ can be defined as a lattice homomorphic map, i.e., a map preserving the lattice operations $\vee$ and $\wedge$, which maps the zero element and the unit element of $\mathfrak{B}$ onto the zero and unit element of $\mathfrak{B}^*$ respectively. The image $h(\mathfrak{B})$ of $\mathfrak{B}$ in $\mathfrak{B}^*$ under a homomorphic map h of $\mathfrak{B}$ into $\mathfrak{B}^*$ is a Boolean subalgebra of $\mathfrak{B}^*$.

A map of $\mathfrak{B}$ into $\mathfrak{B}^*$ is an *isomorphic* map, if and only if it is a one-one homomorphic map. Let $\mathfrak{A}$ and $\mathfrak{B}$ be Boolean algebras. If there exists an isomorphic map of $\mathfrak{A}$ into the Boolean algebra $\mathfrak{B}$, then we shall say that the Boolean algebra $\mathfrak{A}$ can be embedded isomorphically into $\mathfrak{B}$. The Boolean subalgebra $h(\mathfrak{A})$ is then isomorphic to $\mathfrak{A}$. If $\mathfrak{B}$ is a Boolean σ-algebra and $h(\mathfrak{A})$ a σ-generating basis of $\mathfrak{B}$, i.e., $b_\sigma[h(\mathfrak{A})] = \mathfrak{B}$, then we shall define $\mathfrak{B}$ to be a σ-extension of $\mathfrak{B}$. If $h(\mathfrak{A})$ is a σ-regular Boolean subalgebra of $\mathfrak{B}$, then we shall say $\mathfrak{B}$ is a *σ-regular σ-extension* of $\mathfrak{A}$.

Every Boolean algebra $\mathfrak{A}$ has a σ-regular σ-extension.

4.2. A subset $\mathfrak{J}$ of a Boolean algebra $\mathfrak{B}$ is said to be an *ideal* in $\mathfrak{B}$ if and only if the following conditions are satisfied:

(1) If $x \in \mathfrak{J}$ and $y \in \mathfrak{J}$, then $x \wedge y \in \mathfrak{J}$.

(2) If $x \in \mathfrak{J}$ and $y \in \mathfrak{B}$ with $y \leqslant x$, then $y \in \mathfrak{J}$.

The following two conditions are equivalent to (1) and (2) together:

(1*) If $x \in \mathfrak{J}$ and $y \in \mathfrak{J}$, then $x+y \in \mathfrak{J}$.

(2*) If $x \in \mathfrak{J}$ and $y \in \mathfrak{B}$, then $xy \in \mathfrak{J}$.

An ideal $\mathfrak{J}$ in $\mathfrak{B}$ is said to be *proper* if $\mathfrak{J} \neq \mathfrak{B}$.

Let $\mathfrak{X}$ be a subset of $\mathfrak{B}$. Then there exists a smallest ideal, $\mathfrak{J}(\mathfrak{X})$, in $\mathfrak{B}$ containing $\mathfrak{X}$, the ideal generated by $\mathfrak{X}$.

A *principal ideal* is an ideal generated by one element $a \in \mathfrak{B}$ and is denoted by $a\mathfrak{B}$. Obviously we have $a\mathfrak{B} = \{x \in \mathfrak{B} : x \leqslant a\}$. The principal ideal $a\mathfrak{B}$ with $a \neq \emptyset$ is itself a Boolean algebra with a as unit. The principal ideal $\emptyset\mathfrak{B}$ is called the zero ideal and the principal ideal $e\mathfrak{B}$ is identical with $\mathfrak{B}$.

4.3. Let $h: \mathfrak{B} \ni x \Rightarrow h(x) \in \mathfrak{B}^*$ be a homomorphic map of $\mathfrak{B}$ into $\mathfrak{B}^*$. The *kernel* $\mathfrak{H}$ of h is the set of all elements of $\mathfrak{B}$, which are mapped onto the zero element of $\mathfrak{B}^*$, i.e., $\mathfrak{H} = \{x \in \mathfrak{B} : h(x) = \emptyset^*\}$. It is easy to verify that the kernel $\mathfrak{H}$ of h is an ideal in $\mathfrak{B}$.

Suppose now that $\mathfrak{I}$ is an ideal in a Boolean algebra $\mathfrak{B}$. We define a binary relation in $\mathfrak{B}$ as follows: $x \equiv y$ if and only if $x+y \in \mathfrak{I}$. The relation " $\equiv$ " is obviously an equivalence relation. Moreover, it is easy to prove that it is a *congruence relation*, i.e., if $x \equiv y$ and $a \equiv b$, then $x+a \equiv y+b$, $xa \equiv yb$ and $x^c \equiv y^c$. The set $\mathfrak{B}/\mathfrak{I}$ of all equivalence classes of $\mathfrak{B}$ modulo $\mathfrak{I}$ is made into a Boolean algebra if we define:

$$x/\mathfrak{I} + y/\mathfrak{I} = x+y/\mathfrak{I}$$
$$x/\mathfrak{I} \cdot y/\mathfrak{I} = xy/\mathfrak{I}$$
$$[x/\mathfrak{I}]^c = x^c/\mathfrak{I}$$

where $x/\mathfrak{I}$ denotes the equivalence class of $\mathfrak{B}$ mod $\mathfrak{I}$ containing the element $x \in \mathfrak{B}$.

There exists a natural (canonical) map h of $\mathfrak{B}$ onto $\mathfrak{B}/\mathfrak{I}$ defined by

$$h: \mathfrak{B} \ni x \Rightarrow x/\mathfrak{I} \in \mathfrak{B}/\mathfrak{I}$$

and this map is obviously a homomorphic map of $\mathfrak{B}$ onto $\mathfrak{B}/\mathfrak{I}$; the kernel $\mathfrak{H}$ of h is $\mathfrak{I}$. We shall call the Boolean algebra $\mathfrak{B}/\mathfrak{I}$ the *quotient* (or *factor*) algebra of $\mathfrak{B}$ modulo $\mathfrak{I}$. Thus

Theorem 4.2.

(i) *Every ideal $\mathfrak{I}$ of a Boolean algebra $\mathfrak{B}$ defines a Boolean quotient algebra $\mathfrak{B}^* = \mathfrak{B}/\mathfrak{I}$ and a homomorphic map h of $\mathfrak{B}$ onto $\mathfrak{B}^*$ such that the kernel of h is $\mathfrak{I}$, and*

(ii) *Conversely, every homomorphic map h of a Boolean algebra $\mathfrak{B}$ onto a Boolean algebra $\mathfrak{A}$ defines an ideal $\mathfrak{I}$ of $\mathfrak{B}$, namely the kernel of h; the quotient algebra $\mathfrak{B}/\mathfrak{I}$ is then isomorphic to $\mathfrak{A}$.*

4.4. Let $\mathfrak{B}$ be a Boolean σ-algebra. Moreover, let $\mathfrak{I}$ be an ideal in $\mathfrak{B}$. We shall say that $\mathfrak{I}$ is a *σ-ideal* if and only if $x_i \in \mathfrak{I}$, $i = 1, 2, \ldots$, implies ($\mathfrak{B}$) $\bigvee_{i=1}^{\infty} x_i \in \mathfrak{I}$. One can easily prove: If $\mathfrak{B}$ is a Boolean σ-algebra and $\mathfrak{I}$ a σ-ideal in $\mathfrak{B}$, then the quotient Boolean algebra $\mathfrak{B}/\mathfrak{I}$ is also a σ-algebra.

We shall say that a homomorphic map h of a Boolean algebra $\mathfrak{B}$ onto a Boolean algebra $\mathfrak{B}^*$ is a *σ-homomorphism* if and only if the following condition holds:

$$\text{If} \quad (\mathfrak{B}) \bigwedge_{i=1}^{\infty} a_i = \emptyset, \quad \text{then} \quad (\mathfrak{B}^*) \bigwedge_{i=1}^{\infty} h(a_i) = \emptyset^*.$$

Obviously we have then:

$$\text{If} \quad (\mathfrak{B}) \bigwedge_{i=1}^{\infty} a_i = a \in \mathfrak{B}, \quad \text{then} \quad (\mathfrak{B}^*) \bigwedge_{i=1}^{\infty} h(a_i) = h(a) \in \mathfrak{B}^*.$$

and dually:

$$\text{If} \quad (\mathfrak{B}) \bigvee_{i=1}^{\infty} a_i = b \in \mathfrak{B}, \quad \text{then} \quad (\mathfrak{B}^*) \bigvee_{i=1}^{\infty} h(a_i) = h(b) \in \mathfrak{B}^*.$$

If $\mathfrak{B}^*$ is a Boolean subalgebra of a Boolean algebra $\tilde{\mathfrak{B}}$, then

$$(\mathfrak{B}) \bigwedge_{i=1}^{\infty} a_i = a \in \mathfrak{B}$$

does not always imply

$$(\tilde{\mathfrak{B}}) \bigwedge_{i=1}^{\infty} h(a_i) = h(a);$$

the implication is valid if and only if $\mathfrak{B}^*$ is a σ-regular Boolean subalgebra of $\tilde{\mathfrak{B}}$.

4.5. An ideal $\mathfrak{J}$ of a Boolean algebra $\mathfrak{B}$ is said to be a *prime ideal* if and only if the following conditions hold:

(1) $e \notin \mathfrak{J}$, i.e., e is not in $\mathfrak{J}$,

(2) For every $x \in \mathfrak{J}$, either $x \in \mathfrak{J}$ or $x^c \in \mathfrak{J}$.

According to (1), if $x \in \mathfrak{J}$, then $x^c \notin \mathfrak{J}$.

The following two conditions are equivalent to the conditions (1) and (2):

(1*) $\mathfrak{J} \neq \mathfrak{B}$

(2*) If $x \wedge y \in \mathfrak{J}$, then either x or y (or both) belongs to $\mathfrak{J}$.

The concepts of filter and ultrafilter are defined as the duals of the concepts of ideal and prime ideal respectively, i.e., a subset $\mathfrak{F}$ of a Boolean algebra $\mathfrak{B}$ is a *filter* in $\mathfrak{B}$ if and only if

(I) $x \in \mathfrak{F}$ and $y \in \mathfrak{F}$, implies $x \wedge y \in \mathfrak{F}$,

(II) $x \in \mathfrak{F}$ and $y \in \mathfrak{B}$ with $y \geqslant x$ implies $y \in \mathfrak{F}$.

The subset $\mathfrak{F}$ is an *ultrafilter* in $\mathfrak{B}$ if and only if $\mathfrak{F}$ is a filter $\mathfrak{F} \neq \mathfrak{B}$ and $x \vee y \in \mathfrak{F}$ implies $x \in \mathfrak{F}$ or $y \in \mathfrak{F}$. The following theorem holds:

Theorem 4.3.

An ideal $\mathfrak{J}$ in a Boolean algebra $\mathfrak{B}$ is a prime ideal if and only if the quotient algebra $\mathfrak{B}/\mathfrak{J}$ is isomorphic to the Boolean algebra $\mathfrak{U} = \{\emptyset, e\}$, i.e., to the Boolean algebra with two elements $\emptyset$ and e.

4.6. Let $\mathfrak{B}$ be a Boolean algebra and Ω the set of all prime ideals of $\mathfrak{B}$. We can prove that Ω is not empty and that, for every $x \neq \emptyset$, there exists a prime ideal $\mathfrak{J} \in \Omega$ such that $x^c \in \mathfrak{J}$. We define now a map h of $\mathfrak{B}$ into $\mathfrak{P}(\Omega)$ as follows:

$$\mathfrak{B} \ni x \Rightarrow h(x) = X \equiv \{\mathfrak{J} \in \Omega, \ x^c \in \mathfrak{J}\}. \tag{1}$$

Then h is an isomorphic map of $\mathfrak{B}$ into $\mathfrak{P}(\Omega)$. The image $h(\mathfrak{B})$ of the Boolean algebra $\mathfrak{B}$ in the Boolean algebra $\mathfrak{P}(\Omega)$ is a Boolean subalgebra of $\mathfrak{P}(\Omega)$, i.e., a field $h(\mathfrak{B}) = \mathfrak{F}$ of subsets $X \subseteq \Omega$ isomorphic to $\mathfrak{B}$. Hence the following theorem holds:

Theorem 4.4. (Stone).

Every Boolean algebra $\mathfrak{B}$ is isomorphic to a field $\mathfrak{F}$ of subsets of a set Ω with $\Omega \in \mathfrak{F}$. The field $\mathfrak{F} = h(\mathfrak{B})$ defined by (1) is said to be the Stone field *or the* Stone representation *of $\mathfrak{B}$ by a field.*

There exist other fields of subsets of a set E, which are isomorphic to a Boolean algebra $\mathfrak{B}$. However, the Stone representation is important, because if we take the system $\mathfrak{K} = h(\mathfrak{B})$ as a topological basis of open sets, then Ω, as a topological space, is a compact Hausdorff space which is totally disconnected.

Loomis proved the following representation theorem for Boolean σ-algebras:

Theorem 4.5. (Loomis).

Let $\mathfrak{B}$ be a Boolean σ-algebra. Then there exists a σ-field $\mathfrak{K}$ of subsets of a set E (i.e., a Boolean σ-subalgebra of the Boolean algebra $\mathfrak{P}(E)$ of all subsets of E) and a σ-ideal $\mathfrak{N}$ of subsets of E such that $\mathfrak{B}$ is isomorphic to the quotient Boolean σ-algebra $\mathfrak{K}/\mathfrak{N}$.

5. ORDER CONVERGENCE

Let $\mathfrak{B}$ be an $\aleph_0$-complete lattice or a Boolean σ-algebra. Let $x_\nu \in \mathfrak{B}$, $\nu = 1, 2, \ldots$, be a sequence. Then

$$\tilde{x} = \bigwedge_{\rho=1}^{\infty} \bigvee_{\nu \geqslant \rho} x_\nu \quad \text{and} \quad \underset{\sim}{x} = \bigvee_{\rho=1}^{\infty} \bigwedge_{\nu \geqslant \rho} x_\nu$$

exist and are called the o-lim sup $x_\nu = \tilde{x}$ and the o-lim inf $x_\nu = \underset{\sim}{x}$ respectively of the sequence $x_\nu \in \mathfrak{B}$, $\nu = 1, 2, \ldots$.

If we have $\tilde{x} = \underset{\sim}{x}$, i.e.,

$$\bigwedge_{\rho=1}^{\infty} \bigvee_{\nu \geqslant \rho} x_\nu = \bigvee_{\rho=1}^{\infty} \bigwedge_{\nu \geqslant \rho} x_\nu = x,$$

then we shall say that the sequence x_ν, $\nu = 1, 2, \ldots$ is *order convergent* to x and we shall denote it by:

$$(\mathfrak{B})\ o\text{-lim}\, x_\nu = x \quad \text{or} \quad x_\nu \underset{\mathfrak{B}}{\overset{o}{\rightarrow}} x.$$

If $x_\nu\uparrow$, i.e , $x_\nu \leqslant x_{\nu+1}$, $\nu = 1, 2, \ldots$, then

$$o\text{-lim}\, x_\nu = x = \bigvee_{\nu=1}^{\infty} x_\nu$$

exists. We also write $x_\nu \underset{\mathfrak{B}}{\overset{o}{\uparrow}} x$. Dually if $x_\nu\downarrow$, i.e., $x_\nu \geqslant x_{\nu+1}$, $\nu = 1, 2, \ldots$,

$$o\text{-lim}\, x_\nu = x = \bigwedge_{\nu=1}^{\infty} x_\nu$$

exists; we also write $x_\nu \underset{\mathfrak{B}}{\overset{o}{\downarrow}} x$.

Order convergence can be defined in any lattice $\mathfrak{B}$ without the assumption that $\mathfrak{B}$ is $\aleph_0$-complete as follows:

Let $x_\nu \in \mathfrak{B}$, $\nu = 1, 2, \ldots$; then we say

$$\mathfrak{B}\text{-}o\text{-lim}\, x_\nu = x \quad \text{or} \quad x_\nu \underset{\mathfrak{B}}{\overset{o}{\rightarrow}} x$$

for an element $x \in \mathfrak{B}$, if and only if there is an increasing sequence $a_\nu \in \mathfrak{B}$, $\nu = 1, 2, \ldots$, and a decreasing sequence $b_\nu \in \mathfrak{B}$, $\nu = 1, 2, \ldots$, such that

$$a_\nu \leqslant x_\nu \leqslant b_\nu, \quad \nu = 1, 2, \ldots,$$

and

$$(\mathfrak{B}) \bigvee_{\nu=1}^{\infty} a_\nu = x = (\mathfrak{B}) \bigwedge_{\nu=1}^{\infty} b_\nu.$$

It is easy to prove that, if $\mathfrak{B}$ is $\aleph_0$-complete, then this definition is equivalent to that given previously.

6. CLOSURES

6.1. Let $\mathfrak{B}$ be a Boolean σ-algebra and $\mathfrak{S}$ be a non-empty subset of $\mathfrak{B}$. We assume that $\emptyset$ and e belong to $\mathfrak{S}$. Then we shall denote by $\mathfrak{S}^\wedge$, $\mathfrak{S}^\vee$, $\mathfrak{S}^+$ the so-called $\wedge$-closure, $\vee$-closure. $+$-closure of $\mathfrak{S}$ in $\mathfrak{B}$, respectively, i.e., the smallest subset of $\mathfrak{B}$ which is closed for the operation $\wedge$, $\vee$, $+$, respectively, and contains $\mathfrak{S}$. We can construct $\mathfrak{S}^\wedge$, $\mathfrak{S}^\vee$, $\mathfrak{S}^+$ if we adjoin to $\mathfrak{S}$ all those elements $x \in \mathfrak{B}$ that can be written in the form $x = s_1 \wedge s_2 \wedge \ldots \wedge s_n$, $x = s_1 \vee s_2 \vee \ldots \vee s_n$, $s = s_1 + s_2 + \ldots + s_n$ respectively, where $s_\nu \in \mathfrak{S}$, $\nu = 1, 2, \ldots, n$. Likewise, we define the countable $\wedge$-closure (or briefly the δ-closure) $\mathfrak{S}^\delta$ and the countable $\wedge$-closure (or briefly σ-closure) $\mathfrak{S}^\sigma$ of $\mathfrak{S}$ in $\mathfrak{B}$ as the smallest subset of $\mathfrak{B}$ which is closed for meets and joins respectively of a countable number of elements and contains $\mathfrak{S}$. We can construct $\mathfrak{S}^\delta$, $\mathfrak{S}^\sigma$ if we adjoin to $\mathfrak{S}$ all those elements $x \in \mathfrak{B}$ that can be written in the form $x = s_1 \wedge s_2 \wedge \ldots$, $x = s_1 \vee s_2 \vee \ldots$ respectively, where $x_\nu \in \mathfrak{S}$, $\nu = 1, 2, \ldots$.

The so-called order-convergence closure (briefly o-closure) $\mathfrak{S}^0$ of $\mathfrak{S}$ is defined as the smallest subset of $\mathfrak{S}$ containing $\mathfrak{S}$ and satisfying the property: $x_\nu \in \mathfrak{S}$, $\nu = 1, 2, \ldots$, and $o\text{-lim}\, x_\nu = x$ imply $x \in \mathfrak{S}$.

6.2. It is easy to prove that the smallest Boolean subalgebra $b(\mathfrak{S})$ of $\mathfrak{B}$ containing $\mathfrak{S}$ is equal to $(\mathfrak{S}^\wedge)^+ = (\mathfrak{S}^\vee)^+$. The smallest Boolean σ-subalgebra $b_\sigma(\mathfrak{S})$ of $\mathfrak{B}$ containing $\mathfrak{S}$ can be generated by transfinite induction as follows: we set

$$\mathfrak{A}_0 = b(\mathfrak{S}) = \mathfrak{S}^{\wedge\,+} = \mathfrak{S}^{\vee\,+};$$

then $\mathfrak{A}$ is a Boolean subalgebra of $\mathfrak{B}$. We then set

$$\mathfrak{A}_1 = \mathfrak{A}_0{}^0,\ \mathfrak{A}_2 = \mathfrak{A}_1{}^0,\ \ldots,\ \mathfrak{A}_\xi = \mathfrak{A}_{\xi-1}^0$$

if ξ is an isolated ordinal number,

$$\mathfrak{A}_\xi = \bigcup_{\eta<\xi} \mathfrak{A}_\eta$$

if ξ is a limit ordinal number $\left(\bigcup_{\eta<\xi} \mathfrak{A}_\eta\right.$ is the set-theoretical union of all $\left.\mathfrak{A}_\eta \text{ with } 0 \leqslant \eta < \xi\right)$. Obviously we have:

$$\mathfrak{A}_0 \subseteq \mathfrak{A}_1 \subseteq \ldots \subseteq \mathfrak{A}_\xi \subseteq \ldots \subseteq \mathfrak{A}_{\omega_1},\quad 0 \leqslant \xi < \omega_1$$

where ω_1 is the first uncountable ordinal number. Every $\mathfrak{A}_\xi$ is a Boolean subalgebra of $\mathfrak{B}$ containing $\mathfrak{S}$ and $\mathfrak{A}_{\omega_1} = b_\sigma(\mathfrak{S})$ is the smallest Boolean σ-subalgebra of $\mathfrak{B}$ containing $\mathfrak{S}$. $b_\sigma(\mathfrak{S})$ can be generated also as follows: we set $\mathfrak{A} = b(\mathfrak{S}) = \mathfrak{S}^{\wedge +} = \mathfrak{S}^{\vee +}$, and then we set

$$\mathfrak{A}_\xi = \Big(\bigcup_{\eta<\xi} \mathfrak{A}_\eta\Big)^{\sigma \wedge +}$$

for every ξ with $0 < \xi \leqslant \omega_1$. Every $\mathfrak{A}_\xi$, $0 \leqslant \xi \leqslant \omega_1$ is a Boolean subalgebra of $\mathfrak{B}$ containing $\mathfrak{S}$ and moreover $\mathfrak{A}_{\omega_1}$ is the smallest Boolean σ-subalgebra of $\mathfrak{B}$ containing $\mathfrak{S}$, i.e., we have $\mathfrak{A}_{\omega_1} = b_\sigma(\mathfrak{S})$.

Let $\mathfrak{B}$ be a Boolean algebra and $\mathfrak{S}$ a subset of $\mathfrak{B}$ such that $b(\mathfrak{S}) = \mathfrak{B}$; then $\mathfrak{S}$ is said to be a *basis* of $\mathfrak{B}$. If $\mathfrak{B}$ is a Boolean σ-algebra and $\mathfrak{S}$ a subset of $\mathfrak{B}$ such that $b_\sigma(\mathfrak{S}) = \mathfrak{B}$, then $\mathfrak{S}$ is said to be a *σ-basis* of $\mathfrak{B}$. We shall say also that $\mathfrak{S}$ generates $\mathfrak{B}$ in the first case and $\mathfrak{S}$ σ-generates $\mathfrak{B}$ in the second case.

APPENDIX II

LATTICE GROUPS, VECTOR LATTICES

We shall also mention briefly here the main concepts and theorems of the theory of lattice groups and vector lattices. The reader who is interested in a more detailed account of these theories may consult the following books: Birkhoff [1], Carathéodory [6], Fuchs [1], Peressini [1], Cristescu [1], and the paper by Kantorovic-Vulikh-Pinsker [1].

1. LATTICE GROUPS

1.1. We shall consider additive (commutative) groups. The identity of the group will be called zero element and will be denoted by θ. An additive group Σ is said to be a *lattice group* (briefly *l-group*) if and only if a partial ordering relation $\geqslant$ is defined in Σ, which satisfies the following axioms:

(1) If $x \geqslant y$, then $x+z \geqslant y+z$ for every $z \in \Sigma$.

(2) $x \in \Sigma$ and $y \in \Sigma$ imply that there exist

$$\sup(x, y) = x \vee y \in \Sigma$$

and

$$\inf(x, y) = x \wedge y \in \Sigma.$$

Let us now define: $x > y$ if and only if $x \geqslant y$ but $x \neq y$; then we have:

(3) $x > \theta$ and $y > \theta$ imply $x+y > \theta$.

(4) $x > y$ if and only if $x-y > \theta$.

The dual relations $\leqslant$ and $<$ can also be defined and satisfy the dual to (1) to (4) properties. Moreover, we have:

(5) $x < y$ if and only if $-x > -y$.

The following concepts play an important role in the theory of l-groups:

$$x^+ \equiv x \vee \theta$$

$$x^- \equiv (-x) \vee \theta$$

$$|x| \equiv x^+ + x^-,$$

which are called the *positive part*, the *negative part*, and the *absolute value* (*modulus*) of x, respectively. Obviously, we have $x^+ \geqslant \theta$, $x^- = (-x)^+ \geqslant \theta$ and $x = x^+ - x^-$. The absoute value satisfies the following conditions:

(6) $|x| = x \vee (-x) = x^+ \vee x^-$.

(7) $|x| \geqslant \theta$, and $|x| = \theta$ if and only if $x = \theta$.

(8) $|x+y| \leqslant |x| + |y|$.

(9) $|nx| \leqslant |n|\,|x|$, where n any integer.

The following relations are valid:

(10) $x+y = (x \vee y) + (x \wedge y)$.

(11) $(x \vee y)+z = (x+z) \vee (y+z)$ and dually.

(12) $-(x \vee y) = (-x) \wedge (-y)$ and dually.

(13) $|x \vee a - y \vee a| \leqslant |x-y|$, $|x \wedge a - y \wedge a| \leqslant |x-y|$.

(14) $|x \vee y| \leqslant |x| \vee |y|$.

(15) $|x-y| \leqslant a$ if and only if $y-a \leqslant x \leqslant y+a$.

(16) $a \leqslant x \leqslant b$ and $a \leqslant y \leqslant b$ imply $|x-y| \leqslant b-a$.

1.2. The elements x and y are said to be *orthogonal* (*disjoint*) denoted by $x \perp y$, if and only if $|x| \wedge |y| = \theta$. We have:

(17) If $x \perp z$ and $y \perp z$, then $x \vee y \perp z$, $x \wedge y \perp z$ and $x+y \perp z$.

(18) If $x \perp y$, then $mx \perp ny$, where m and n are integers.

(19) We always have $x^+ \perp x^-$.

(20) $x \perp y$ if and only if $|x| \vee |y| = |x| + |y|$.

(21) An l-group is, as a lattice, distributive, i.e.,

$$x \vee (y \wedge z) = (x \vee y) \wedge (x \vee z) \text{ and dually.}$$

Moreover, if $x_i \in \Sigma$, $i \in I$, and there exist $\bigvee_{i \in I} x_i$ and $\bigwedge_{i \in I} x_i$ in Σ, then there also exist $\bigvee_{i \in I} (x_i + a)$, $\bigwedge_{i \in I} (x_i + a)$, $\bigvee_{i \in I} (x_i \wedge a)$, $\bigvee_{i \in I} (x_i \vee a)$, $\bigvee_{i \in I} x_i^+$, $\bigvee_{i \in I} x_i^-$, $\bigwedge_{i \in I} x_i^+$, $\bigwedge_{i \in I} x_i^-$, $\bigwedge_{i \in I} (-x_i)$, and $\bigvee_{i \in I} (-x_i)$ in Σ, and we have:

$$\bigvee_{i \in I} (x_i + a) = \Big(\bigvee_{i \in I} x_i\Big) + a, \quad \bigwedge_{i \in I} (x_i + a) = \Big(\bigwedge_{i \in I} x_i\Big) + a$$

$$\Big(\bigvee_{i \in I} x_i\Big) \wedge a = \bigvee_{i \in I} (x_i \wedge a), \quad \Big(\bigwedge_{i \in I} x_i\Big) \vee a = \bigwedge_{i \in I} (x_i \vee a)$$

$$\bigvee_{i \in I} x_i^+ = \Big(\bigvee_{i \in I} x_i\Big)^+, \quad \bigwedge_{i \in I} x_i^- = \Big(\bigvee_{i \in I} x_i\Big)^-, \quad \bigwedge_{i \in I} x_i = -\bigvee_{i \in I} (-x_i).$$

1.3. (I) An l-group Σ is said to be *archimedean* if and only if the following condition is valid:

(α) $x \geqslant \theta$ and $ny \leqslant x$, for every $n = 1, 2, \ldots$, imply $y \leqslant \theta$, or equivalent $x \geqslant \theta$ and $ny \leqslant x$ for every $n = \pm 1, \pm 2, \ldots$ imply $y = \theta$.

(II) An l-group Σ is said to be *saturated*, resp σ-saturated†, if and only if for every subset $T \subseteq \Sigma$, resp for every countable subset $T \subseteq \Sigma$, which is upper bounded in Σ there exists the least upper bound $\bigvee_{t \in T} t$ in Σ.

It is easy to see that this condition implies its dual condition. A saturated l-group is, obviously, σ-saturated.

(III) An l-group Σ^* is an *l-subgroup* of an l-group Σ or the l-group Σ is an *extension* of the l-group Σ^* if and only if Σ^* is a subgroup and at the same time a sublattice of Σ.

(IV) Let Σ^* be an l-subgroup of an l-group Σ. We say Σ^* is a *regular* (*invariant*), resp *σ-regular* (*σ-invariant*)‡, l-subgroup of Σ if and only if the following condition is valid:

If for any decreasing directed family (net) $x_i \in \Sigma^*$, $i \in I$, resp decreasing sequence $x_n \in \Sigma^*$, $n = 1, 2, \ldots$, we have

$$\bigwedge_{i \in I} x_i = \theta, \text{ resp } \bigwedge_{n=1}^{\infty} x_n = \theta, \text{ in } \Sigma^*$$

---

† Instead of "saturated," "complete with respect to the lattice operations" is also used.

‡ Instead of " regular " or " invariant ", " correct " is also used.

then we have also

$$\bigwedge_{i \in I} x_i = \theta, \text{ resp } \bigwedge_{n=1}^{\infty} x_n = \theta, \text{ in } \Sigma;$$

the l-group Σ as an extension of the l-group Σ^* is then a regular resp σ-regular extension.

We notice that every l-subgroup of a σ-saturated l-group is always archimedean. Conversely: every archimedean l-group can always be extended regularly to a saturated l-group.

(V) An l-subgroup Σ^* of the l-group Σ is said to be *normal* in Σ if and only if $x \in \Sigma$, $x^* \in \Sigma^*$, and $|x| \leqslant |x^*|$ imply $x \in \Sigma^*$. A normal l-subgroup in Σ is also called an *l-ideal* in Σ.

(VI) An l-subgroup Σ^* of the lattice group Σ is said to be a *component* or a *band* in Σ if and only if Σ^* is regular and normal in Σ.

1.4. *Units.* Very useful is the concept of a unit in an l-group Σ. An element $u \in \Sigma$ is said to be a *strong unit* if and only if for every $x \in \Sigma$ there exists a positive integer n such that $nu > x$. A positive element $e > \theta$ is said to be a *weak unit* if and only if e is orthogonal only to the zero element, i.e., $e \wedge |x| = \theta$ implies $x = \theta$ for any $x \in \Sigma$. It is easy to prove that any strong unit is a weak unit. In fact, let u be a strong unit; then, if $u \wedge |x| = \theta$, for some $x \in \Sigma$, there exists an integer $n_x > 0$ such that $n_x u > |x|$, i.e., $(n_x u) \wedge |x| = |x|$; but $u \wedge |x| = \theta$, implies $(n_x u) \wedge |x| = \theta$; hence $|x| = \theta$, i.e., $x = \theta$.

2. VECTOR LATTICES OR LINEAR LATTICE SPACES

2.1. A *vector lattice* is a real vector (linear) space U which is an l-group with respect to the operation of addition and which, furthermore, satisfies the following condition:

(v) If $x \geqslant y$ and $\lambda \in R$, $\lambda \geqslant 0$, then $\lambda x \geqslant \lambda y$.

It is easy to see that the following properties are valid in a vector lattice:

(1) If $x > \theta$ and $\lambda \in R$, $\lambda > 0$, then $\lambda x > \theta$.

(2) If $\lambda \in R$, $\lambda \geqslant 0$, then $\lambda(x \vee y) = \lambda x \vee \lambda y$ and dually.

(3) If $\lambda \leqslant 0$, $\lambda \in R$, then $\lambda(x \vee y) = \lambda x \wedge \lambda y$ and dually.

(4) $|\lambda x| = |\lambda|\,|x|$, $\lambda \in R$, $x \in U$.

(5) $0x = \theta$, and if $\lambda \neq 0$ with $\lambda x = \theta$, then $x = \theta$.

(6) If there exists $\bigvee_{i \in I} x_i$ in U and $\lambda \geqslant 0$, $\lambda \in R$, then there exists $\bigvee_{i \in I} \lambda x_i$ and we have

$$\lambda \bigvee_{i \in I} x_i = \bigvee_{i \in I} \lambda x_i$$

and dually, if $\lambda \leqslant 0$, then there exists $\bigwedge_{i \in I} \lambda x_i$ in U and we have

$$\lambda \bigvee_{i \in I} x_i = \bigwedge_{i \in I} \lambda x_i.$$

We notice that all concepts and results in the theory of l-groups are carried over to the theory of vector lattices. It may happen that a certain l-subgroup U of the vector lattice U is itself a vector lattice; in this case we say that U is a *vector sublattice* of U. We may also speak about *regular vector sublattices* and also about *normal* or *component vector sublattices* in U. A component (i.e., regular and normal) vector sublattice B of a vector lattice U is said to be a *band* in U.

BIBLIOGRAPHICAL NOTES

The literature indicated below has been used in the preparation of this monograph. It must not be considered as a complete bibliography. Further references can be found in the literature indicated.

CHAPTER I

Probability algebras as Boolean algebras with an additive, non-negative probability (normed measure) are considered by Birkhoff [1], p. 197, Halmos [1], Kappos [2, 4, 7, 8], Kolmogoroff [2], Segal [1], and others. Measure theory in Boolean algebras is studied by Carathéodory [1, 2, 3, 6]. For further literature on this subject, see also Sikorski [5], Section 42. Horn and Tarski [1] have considered Boolean algebras with a strictly positive measure and have studied structure problems on these algebras. Probability algebras with the assumption that the probability is strictly positive are systematically studied by Kappos [8]. For Section 8.4 and the question of the existence of Boolean algebras which cannot be endowed with a probability, *cf.* also Maharam [2] and Kelley [1]. We notice that for every Boolean algebra A there exists a linearly ordered algebraic field F (in general non-archimedean), such that A can be endowed with a strictly positive, finitely additive F-valued measure; compare O. Nikodym [3] and Luxemburg [1, 2].

CHAPTER II

A metric distance defined with the help of a measure was first used by O. Nikodym [1]. Kappos [4, 6, 8] and Kolmogoroff [2] have used this metric to extend probability algebras to probability σ-algebras in which the probability is σ-additive. For the remarks of Section 2, see Horn and Tarski [1]. For Sections 3 and 4, see Carathéodory [2], Horn and Tarski [1], and Halmos and v. Neumann [1].

CHAPTER III

Cartesian products of Boolean algebras have been studied by Sikorski [2, 3, 4], Kappos [1, 4, 8], and Ridder [1]. Sikorski and Kappos have studied Cartesian products of probability algebras and their relation to the concept of algebraic or probability independence.

Section 3: A general classification of measure algebras, hence also of probability algebras, has been first obtained by Dorothy Maharam [1, 4]; see also Kappos [8] and Zink [1].

CHAPTER IV

Kolmogoroff [1], p. 6, has defined a finite partition of the sure event as a trial (Versuch) and has used this concept to define the simple (with a finite number of values) random variables. Carathéodory [1, 3, 4, 6] used later finite partitions of elements (somata) of a Boolean algebra to define simple place functions (Ortsfunktionen) over a Boolean algebra. To define all place functions over a Boolean algebra he has used other processes [1, 2, 3]. Analogous processes have been used by Wecken [1], Olmsted [1], Gomes [1], Sikorski [1], Ridder [1, 2], Bishof [1], and Nikodym [2]. Carathéodory has noticed in [4] that elementary random variables can be used to define by Dedekind cuts all place functions; see also Kappos [3]. Kappos [3, 11] has used countable partitions in measure algebras (probability σ-algebras) to define the set of all elementary random variables and has extended it with the help of uniform convergence to obtain the set of all random variables; see also Goffman [1].

In the present chapter we apply a process, which is well-known in the completion theory of abelian lattice-groups (see Papangelou [1, 2], Banaschewski [1], and Everett [1]), to complete the set of all elementary random variables and obtain the set of all random variables by using the order relation and order convergence. This process can be applied to define generalised random variables having values in abelian groups or Banach lattices; see Kappos [13, 14] and, moreover, the theory stated in Chapter VII of the present book.

CHAPTERS V AND VI

To define the expectation and moments of random variables we have used an integration theory, the details of which are stated in Kappos [1, II] and which was first used to define the integral in the classical measure theorey by O. Nikodym [1].

For Chapter V, Sections 3 and 4, and Chapter VI, see also Halmos [2], Loève [1], Carathéodory [6], Krickeberg [1], and Neveu [1]. For Chapter V, Section 3, we notice that O. Onicescu and his school in Bucharest have used as fundamental concept that of a signed measure (fonction-somme) to define random variables in Boolean algebras; see Onicescu [1, 2] and Cristescu [2].

CHAPTERS VII AND VIII

In the text of these chapters and particularly in Chapter VII, Section 1, the main literature on generalized random variables has been given. A Lebesgue integration theory for functions having values in vector spaces has been first studied by Bochner [1] in 1933. For recent literature, compare Dunford–Schwarz [1], Chapter III, Section 15, Hille–Phillips [1], Chapter III, Section 1, Dinculeanu [1], and Bourbaki, N.: " Eléments de Mathématique, Intégration," Livre VI, Ch. 6, Intégration vectorielle, Hermann, Paris 1959. Compare also two mimeographed editions of lectures given by Laurent Schwartz on Radon measures on arbitrary topological spaces, 1964–65, in Institut Henri Poincaré and later in Maryland University, U.S.A..

Applications of this theory to the definition of random variables having values in Banach spaces and to the study of probability problems related to this theory first began systematically by Fréchet in 1944 and are continued by his school in Paris. Compare Fréchet, M.: "Abstrakte Zufallselemente. Bericht über die Tagung Wahrscheinlichkeitsrechnung und math. Statistik in Berlin " vom 19 bis 22. Okt. 1954, 23–28, Veb deutscher Verlag der Wissenschaften, Berlin 1956. Compare also the literature given in this paper, Hanš [1], and Metivier [1, 2]. For random variables over Boolean σ-algebras having values in Banach spaces and the problems related to this theory see Georgiou [1, 2, 3].

Remark: For Appendices I and II the literature is given in the text. Compare also the new edition of Birkhoff's book [1] and a translation of a book by Vulikh [1].

BIBLIOGRAPHY

Banaschewski, B.

1. Über die Vervollständigung geordneter Gruppen. *Math. Nachrichten*, **16,** 51–71 (1957).

Birkhoff, G.

1. " Lattice Theory." Amer. Math. Soc. Colloquium Publications, New York, second edition 1948; also third (new) edition 1967.

Bishof, A

1. Beiträge zur Carathédoryschen Algebraisierung des Integralbegriffs. *Schr. Math. Inst. u. Inst. angew. Math. Univ. Berlin* **5,** 237–262 (1941).

Bochner, S.

1. Integration von Funktionen, deren Werte die Elemente eines Vektorraumes sind. *Fund. Math.* **20,** 262–276 (1933).

Carathédory, C.

1. Entwurf für eine Algebraisierung des Integralbegriffs. *S.-B. bayer. Akad. Wiss.* 1938, pp. 27–69.
2. Die Homomorphien von Somen und die Multiplikation von Inhaltsfunktionen. *Ann. Scuola Norm. Sup. Pisa* (2), **8,** 105–130 (1939).
3. Über die Differentiation von Massfunktionen. *Math. Z.* **46,** 181–189 (1940).
4. Bemerkungen zum Riesz-Fischerschen Satz und zur Ergodentheorie. *Abh. Math. Sem. Hansischen Universität* **14,** 351–389 (1941).
5. Bemerkungen zum Ergodensatz von G. Birkhoff. *S.-B. Math.-Nat. Kl. bayer. Akad. Wiss.* 1944, pp. 189–208.
   (All the above papers can be found in Carathéodory's: " Gesammelte mathematische Schriften ", Band IV, C. H. Beck, München, 1956.)
6. " Mass und Integral und ihre Algebraisierung." Birkhäuser, 1956. Also, translated into English: "Algebraic Theory of Measure and Integration ", Chelsea, 1963.

Cristescu, R.

1. Spatii liniare ordonate. *Edit. Acad. R.P.R.*, 1959.
2. Sur les fonctions-sommes de O. Onicescu, *Bull. Sci. Math.* **89,** 49–63 (1965).

Dinculeanu, N.

1. " Vector Measures." Pergamon Press, 1967.

Dubins, L. E.

1. Generalized random variables. *Trans. Am. Math. Soc.* **84,** 273–309 (1957).

Dunford, N. and Schwartz, J. T.

1. " Linear operators," Part I. Interscience, New York, 1958.

Dwinger, Ph.

1. " Introduction to Boolean algebras." Physica-Verlag, Würzburg, 1961.

Everett, C. J.

1. Sequence completion of lattice modules. *Duke Math. J.* **11,** 109–119 (1944).

Everett, C. J. and Ulam, S.

1. On ordered groups. *Trans. Amer. Math. Soc.* **57,** 208–216 (1945).

Fuchs, L.

1. " Teilweise geordnete algebraische Strukturen." Vandenhoeck und Ruprecht, Göttingen, 1966.

Georgiou, P.

1. Zufällige Variable mit Werten in Banach–Räumen und ihre Warhscheinlichkeitsverteilungen. (Greek, German abstract, doctoral dissertation at Athens University.) *Bull. Soc. Math. de Grèce, N.S.* **4,** 1–68 (1963).
2. Vektorwertige Zufallsvariable und ihre Wahrscheinlichkeitsverteilung. *Archiv Math.* **18,** 70–77 (1966).

3. Vektorwertige Masse und Zufallsvariablen auf Booleschen Algebren und der Satz von Radon–Nikodym. *Symposium on Probability Methods in Analysis, Lecture Notes in Mathematics* 31. Springer-Verlag, 1967.

Goffman, C.

1. Remarks on lattice ordered groups and vector lattices. I. Carathéodory functions. *Trans. Amer. Math. Soc.* **88,** 107–120 (1958).

Gomes, A. P.

1. " Naçâo de functional em espaços sem pontos," Porto, 1946.

Halmos, P. R.

1. The foundations of probability. *Am. Math. Monthly* **51,** 497–510 (1944).
2. " Measure Theory." Van Nostrand, New York, 1950.
3. " Lectures on Boolean algebras." Van Nostrand, Toronto–New York–London, 1963.

Halmos, P. R. and Neumann, J. v.

1. Operator methods in classical mechanics II. *Ann. Math.* **43,** 332–350 (1942).

Hanš, O.

1. Generalized random variables. *Trans. First Prague Conference on Inform. Theory*, 61–103, Prague 1956.

Hille, E. and Phillips, R. S.

1. " Functional analysis and semigroups." Revised Ed., Amer. Math. Soc. Colloquium Publications, Providence, R.I., 1957.

Horn, A. and Tarski, A.

1. Measures in Boolean algebras. *Trans. Am. Math. Soc.* **64,** 467–497 (1948).

Kantorovič, L. V., Vulikh, B. Z. and Pinsker, A. G.

1. Partially ordered groups and partially ordered linear spaces. *Amer. Math. Soc. Translations, Series* 2, **27,** 51–124.

Kappos, D. A.

1. Die Cartesischen Produkte und die Multiplikation von Massfunktionen in Booleschen Algebren. I. *Math. Ann.* **120,** 43–74 (1947); II. *Math. Ann.* **121,** 223–333 (1949).
2. Zur mathematischen Begründung der Wahrscheinlichkeitstheorie. *S.-B. bayer. Akad. Wiss., Math.-Nat. Kl., München* 1948, pp. 309–320.
3. Ein Beitrag zur Carathéodoryschen Definition der Ortsfunktion in Booleschen Algebren. *Math. Z.* **51,** 616–634 (1949).
4. Über die Unabhängigkeit in der Wahrscheinlichkeitstheorie. *S.-B. bayer. Akad. Wiss., Math.-Nat. Kl., München* 1950, pp. 157–185.
5. Über äquimessbare (verteilungsgleiche) Funktionen. Ebenda 1951, pp. 113–128.
6. Erweiterung von Massverbänden. *J. reine angew. Math.* **191,** 97–109 (1953).
7. Die Totaladditivität der Wahrscheinlichkeit. *Bull. de la Soc. Math. de Grèce* **28,** 63–80 (1953).
8. " Strukturtheorie der Wahrscheinlichkeitsfelder und -räume." Springer Verlag, 1960.
9. Verteilungsgleiche Zufallsvariablen. *Bull. Soc. Math. de Grèce, N.S.*, **2,** 104–113 (1961).
10. Remarks on the representation of probability fields and of spaces of random variables. *Colloquium on Combinatorial Methods in Probability Theory*, pp. 84–89, Aarhus Universitet, Danmark, August 1–10, 1962.
11. Strukturtheorie der Räume von Zufallsvariablen. *Trans. of the Third Prague Conference on Information Theory*, June 5 to 13, 1962, Prague 1964, 359–375.
12. " Probability Algebras and Stochastic Spaces." Mimeographed text of lectures given at the Cath. Univers. of America, Washington D.C., 1963–64.
13. Remarks on the structure of spaces of generalized random variables. *Bull. Soc. Math. de Grèce, N.S.* **6,** 129–142 (1965).

14. Random variables having values in ordered abelian lattice-groups or vector lattices. Report at the international congress of Mathematics, Moskow 1966 (unpublished).

Kelley, J. L.

1. Measures in Boolean algebras. *Pacific J. Math.* **9,** 1165–1177 (1959).

Kolmogoroff, A. N.

1. " Grundbegriffe der Wahrscheinlichkeitsrechnung." Springer-Verlag, Berlin 1933.
2. Algèbres dè Boole métriques complètes. *Dotatek. Rocznika polk. Towarz mat.* **22,** 21–30 (1951).

Krickeberg, K.

1. " Wahrscheinlichkeitstheorie." Teubner Verl., Stuttgart, 1963.

Loève, M.

1. " Probability Theory." Third Ed., van Nostrand, New York 1963.

Luxemburg, W. A. J.

1. " Non-standard analysis." California Institute of Technology, Pasadena, 1962.
2. Two applications of the method of construction by ultrapowers to analysis. *Bull. Am. math. Soc.* **68,** 416–419 (1962).
3. A remark on Sikorski's extension theorem for homomorphisms in the theory of Boolean algebras. *Fund. Math.* **55,** 239–247 (1964).

MacNeille, H. M.

1. Partially ordered sets. *Trans. Am. math. Soc.* **42,** 416–460 (1937).

Maharam, D.

1. On homogeneous measure algebras. *Proc. nat. Acad. Sci.* (*Wash.*) **28,** 108–111 (1942).
2. An algebraic characterization of measure algebras. *Ann. Math.* **48,** 154–167 (1947).
3. The representation of abstract measure functions. *Trans. Am. math. Soc.* **65,** 279–330 (1949).
4. Decompositions of measure algebras and spaces. *Trans. Am. math. Soc.* **69,** 142–160 (1950).

Metivier, M.

1. Convergence de martingales à valeurs vectorielles. *Bull. Soc. Math. de Grèce, N.S.* **5,** 57–74 (1964).
2. Martingales à valeurs vectorielles. Application à la dérivation. Symposium on Probability Methods in Analysis. Lecture Notes 31, pp. 239–255. Springer-Verlag 1967.

Nedoma, J.

1. Note on generalized random variables. *Trans. of the First Prague Conference* 1956 *on Inform. Theory, Stat. Dec. Functions, Random Processes*, pp. 139–141.

Neveu, J.

1. " Bases mathematiques du calcul des probabilités." Masson Èd., Paris, 1964.

Nikodym, O.

1. Sur une généralisation des intégrals de M. J. Radon. *Fund. Math.* **15,** 131–179 (1930).
2. Sur les êtres fonctionnoïdes. *Compt. rend.* **226,** 375–377, 458–460, 541–543 (1948).
3. Sur le mesure non-archimédienne effective sur une tribu de Boole arbitraire. *Compt. rend.* **251,** 2113–2115 (1960).

Olmsted, J. M. H.

1. Lebesgue theory on a Boolean algebra. *Trans. Am. math. Soc.* **51,** 164–193 (1942).

Onicescu, O.

1. " Seminar de functii Suma." R. P. Rom (Ed.). Bucaresti, 1963.
2. Funzioni-somma e loro applicazioni alla teoria dei processi stocastici. *Rendiconti di Mathematica* **26,** 1–77 (1967).

Papadopoulou, S.

1. Remarks on the completion of commutative lattice-groups with respect to order convergence. *Bull. Soc. Math. de Grèce, N.S.,* **9,** 138–142 (1968).

Papangelou, F.

1. Concepts of algebraic convergence and completion of abelian lattice-groups and Boolean algebras. (In Greek, doctoral dissertation at Athens University, Greece 1963.) *Bull. Soc. Math. de Grèce, N.S.* **3,** 26–114 (1962).
2. Order convergence and topological completion of commutative lattice-groups. *Math. Annalen* **155,** 81–107 (1964).

Peressini, A. L.

1. " Ordered Topological Vector Spaces." Harper and Row, New York, 1967.

Ridder, J.

1. Mass-und Integraltheorie in Strukturen. *Acta Math.* **73,** 131–171 (1941).
2. Zur Mass- und Integrationstheorie in Strukturen. *Ind. Math.* **8,** 64–81 (1946).

Rieffel, M. A.

1. The Radon–Nikodym theorem for the Bochner integral. *Trans. Am. math. Soc.* **131,** 466–487 (1968).

Segal, I. E.

1. Abstract probability spaces and a theorem of Kolmogoroff. *Am. J. Math.* **76,** 721–732 (1954).

Sikorski, R.

1. The integral in a Boolean algebra. *Coll. Math.* **2,** 20–26 (1949).
2. Independent fields and Cartesian products. *Studia Math.* **11,** 171–184 (1950).
3. Cartesian products of Boolean algebras. *Fund. Math.* **37,** 25–54 (1950).
4. On measures in Cartesian products of Boolean algebras. *Coll. Math.* **2,** 124–129 (1951).
5. " Boolean algebras." Second edition, Springer-Verlag, 1964.

Vulikh, B. Z.

1. " Introduction to the theory of partially ordered spaces." (Translation from the Russian), Groningen, 1967.

Wecken, F.

1. Abstrakte Integrale und fastperiodische Funktionen. *Math. Z.* **45,** 377–404 (1939).

Zink, R. E.

1. On the structure of measure spaces. *Acta Math.* **107,** 53–71 (1962).

LIST OF SYMBOLS

Symbols are defined on the pages listed.

INDEX

A

B

C